Stochastic Models for Learning

ROBERT R. BUSH
Assistant Professor of Social Relations

FREDERICK MOSTELLER
Professor of Mathematical Statistics

DEPARTMENT OF SOCIAL RELATIONS
HARVARD UNIVERSITY

Martino Publishing
Mansfield Centre, CT
2012

Martino Publishing
P.O. Box 373,
Mansfield Centre, CT 06250 USA

www.martinopublishing.com

ISBN 978-1-61427-319-6

Cover design by T. Matarazzo

Printed in the United States of America On 100% Acid-Free Paper

Stochastic
Models
for Learning

ROBERT R. BUSH
Assistant Professor of Social Relations

FREDERICK MOSTELLER
Professor of Mathematical Statistics

DEPARTMENT OF SOCIAL RELATIONS
HARVARD UNIVERSITY

New York · John Wiley & Sons, Inc.
London · Chapman & Hall, Limited

To

SAMUEL A. STOUFFER

Preface

MATHEMATICS IS OFTEN REGARDED AS A SCIENCE, BUT TO DO SO
is an error. Science is concerned with the empirical world, whereas
mathematics is the study of relationships among empirically undefined
quantities. The bridge between pure mathematics and empirical science
is the identification of mathematical constructs with observables; when
such identifications are made we shall say that we have a mathematical
model of a situation. This book is concerned with such a model.
Nothing new is offered to the mathematician in his professional capac-
ity, because no new mathematical methods have been introduced; the
kinds of mathematics employed are well known and rather well developed. The model raises some new problems that have interested
mathematicians, and they have given us the benefit of their results.
Usually their proofs require a degree of mathematical sophistication
beyond the level we have set for this volume. In addition, our model
brings new problems of statistical estimation, but for the most part we
have used slight variations on well-known methods to solve these problems. In the future, mathematical statisticians may discover new and
better ways of solving some of the problems we describe.

To many experimental psychologists a book such as this may appear
unnecessary because they may feel that answers to the important questions of psychology are to be found in the laboratory or in the field and
in the collection of more and better data, rather than in mathematical
formulas. We have no quarrel with such views; rather, this book is
for readers who feel that mathematical analysis may contribute to the
development of theory in one of the many fields of psychology—learning. We have no intention of trying to convert psychologists who feel
that human behavior cannot be mathematically described. A possible
mathematical framework for analyzing data from a variety of experiments on animal and human learning is presented. We have tried to
give more than a collection of "local" models, each designed to analyze
one special type of data. The system we describe in rather general
terms is applied to a number of particular experimental problems, but
we make no pretense at completeness or finality and shall feel much
rewarded if we provide a start on a good approach.

We are aware of numerous earlier attempts to construct mathematical models for learning. The history of such attempts would require an extensive monograph, and so we have not attempted to review such work in this volume, though some of it can be shown to be closely related to our model, notably the work of Hull, Gulliksen and Wolfle, Rashevsky, and Thurstone.

We wish to say a few words about the level of mathematics used in this book. For the most part we have tried to write our book for the experimental psychologist with only a limited background in mathematics. We assume that our reader has taken, at one time or another, an elementary course in differential and integral calculus and a course in applied statistics. Actually, we make relatively little use of calculus, but we assume a degree of facility with mathematical manipulations characteristic of a person who has completed a course in the subject. We will make use of certain kinds of mathematics such as matrix algebra and set theory ordinarily studied in more advanced courses, but we will present as complete an exposition of these topics as we feel is necessary for the analysis given. These expositions occur in the text when the needs arise. Because we are writing mainly for readers with limited mathematical preparation, we include many more steps in the various mathematical developments than we would otherwise. Our attempt to include so many equations is to aid the reader rather than to hinder him. Frequently we repeat an equation rather than force the reader to look back fifty pages.

Perhaps a word of advice about reading mathematical material will not be out of order for the psychologist whose mathematics is rusty but who decides he wants to follow certain derivations in detail. Start with a fast reading of the sections concerned to get the general orientation. Then take paper and pencil and work through each derivation step by step. A good understanding of a mathematical development often goes with a feeling that one has invented it oneself, and that the original authors were somewhat opaque.

This book is divided into two main parts. In Part I we present the general model, describe some of its mathematical properties, and consider a number of special cases; in Part II we apply the model to a number of specific experimental problems in learning and devote considerable attention to the statistical problems of estimating model parameters and measuring goodness of fit. Not all the mathematical machinery developed in Part I is used in the applications of Part II. Our goal in Part I extends beyond the derivation of formulas to be used in analyzing data; we have attempted to subject the general model to careful mathematical analysis in order to study its properties. We feel that this study is important for making decisions about where and how to apply the model.

Various methods may be used in reading this book, depending on the goal of the reader and his particular background in mathematics. We have starred many sections which are mainly mathematical and unnecessary for the main development. An experimental psychologist who is reading this book only to make a general evaluation can profitably omit the starred sections or, in fact, may omit or merely scan most of Part I. We do consider Chapters 1 and 3 of Part I absolutely essential to an understanding of Part II, however. In Part II we present applications of the mathematical framework to certain experimental problems, and there the psychologist will probably find material of most interest to him. We recommend that he read Chapter 9 on statistical estimation before reading later chapters, but Chapters 10, 11, 12, 13, and 14 are essentially independent of one another.

The psychologist who plans to apply the mathematical analysis to data of his own may find it necessary to read much more of the book. We do not consider the starred sections necessary nor do we think that Chapters 2, 4, and 6 are absolutely essential for this purpose. But beyond this we have few omissions to recommend.

Those mathematicians who wish to examine the contents of this book may be mainly interested in Part I. The major mathematical problems arise in Chapters 4 and 6. Statisticians, on the other hand, may find more interest in Part II, since we consider a number of new problems in estimation. Our handling of these problems follows rather conventional lines, but the statements of the problems and the difficulties encountered may attract the statistician.

The work we describe in this book began in the summer of 1949 with an attempt by one of us (F. M.) to analyze some data on the reports of hospital patients on the analgesic effects of certain drugs. It was postulated that a learning effect was superimposed upon the biological effects of the drugs. These data were ultimately analyzed with a quite different model, but the original model suggested to us the possibility of an "operator" model for handling data from learning experiments. As we became more and more aware of the large bulk of empirical information on learning which was available, the model was reformulated and modified many times. Independent of our initial attempts, William K. Estes at Indiana University developed a model of conditioning based upon the principles of association theory. We have been much influenced by Estes' work and have adopted many of his ideas from time to time. Also independent of our early work, George A. Miller when at Harvard University and at the Institute for Advanced Study investigated some of the sequential properties of behavior. This work too has influenced ours. Perhaps the period of greatest progress in our research occurred during the summer of 1951, when we had the privilege of working for two months with Cletus J. Burke, William K. Estes, George A. Miller, David Zeaman, William J. McGill, Katherine S. Harris, and Jane E. Beggs. This two-month seminar, entitled Mathematical Models for Behavior Theory, was made possible by the Social Science Research Council's program of Inter-University Summer Research Seminars supported by a grant from the John and Mary R. Markle Foundation. Tufts College, acting as host to the group, kindly made space and facilities available. Further progress was made, particularly on mathematical problems, during the summer of 1952, when we participated in a University of Michigan conference at Santa Monica, California. That conference was part of a program sponsored by the Ford Foundation, and it was conducted in cooperation with the RAND Corporation in Santa Monica. Persons especially helpful to us at that conference were Richard Bellman, Clyde H. Coombs, Robert L. Davis, William K. Estes, Merrill M. Flood, Theodore E. Harris, Samuel Karlin, Tjalling C. Koopmans, Roy Radner, Harold N. Shapiro, Gerald L. Thompson, and Robert M. Thrall. Dr. Flood has had continued interest in these problems and has developed several models related to ours.

From the beginning, the work which has resulted in this book was facilitated by support from the Laboratory of Social Relations at Harvard University. During the first two years, one of us (R. R. B.) was a recipient of a Fellowship in the Natural and Social Sciences awarded jointly by the National Research Council and the Social Science Research Council. During the several years of research we have had many useful suggestions and criticisms from our colleagues at Harvard and elsewhere. During the early stages of this work, William O. Jenkins, now at the University of Tennessee, spent many hours reading and criticizing our work, making us aware of numerous experimental problems and, more generally, giving us an orientation towards experimental problems of interest to psychologists. We express our sincerest thanks to Dr. Jenkins for this guidance. Suggestions on mathematical and statistical questions were made from time to time by John W. Tukey and Allan Birnbaum. Joseph Weizenbaum assisted us in the early stages of the work on Chapter 13. We are also grateful to Lotte Bailyn, David G. Hays, Solomon Weinstock, and Thurlow R.

Wilson for suggestions about revising the manuscript to make it more readable. George A. Miller critically read large portions of the semifinal draft and made numerous suggestions for the improvement of content and exposition. Doris Entwisle and Cleo Youtz prepared early drafts of the manuscript, carried out numerous computations, and made frequent helpful criticisms. The semifinal and final copies of the manuscript were prepared by Vernon L. Schonert.

In every sense, this book and the research which preceded it were joint efforts of the two authors. Neither of us alone would have had the patience to carry the work to completion. Furthermore, the frequent demolishment of each other's ideas has kept the book to a reasonable size.

ROBERT R. BUSH
FREDERICK MOSTELLER

Harvard University
February, 1955

Contents

Introduction

In the construction of mathematical models there are usually three main steps or levels: (1) the mathematical system, (2) the identifications or coordinating definitions, and (3) the specific applications.

First there must be a mathematical theory or system. Empirical phenomena, experimental quantities and variables, are not properly discussed within the mathematical system. The elements of the system are not operationally defined, for they are abstract concepts which acquire meaning only through their relationships with one another. Nevertheless, the empirical phenomena for which one hopes to build a model may suggest an appropriate mathematical system to use as a framework. We do not adhere to a strictly formal position, but rather we label the elements of our mathematical system so as to suggest possible *identifications* between them and observable quantities. Our mathematical system is concerned with classes of *responses* and *events*, for example. Since we believe that behavior is a statistical phenomenon, from a macroscopic point of view at least, we attempt to describe response tendencies by sets of probability variables. We introduce certain types of mathematical *operators* to correspond to the events which alter response tendencies during learning. The main outline of the mathematical system is given in Chapter 1.

The second step in the development of a mathematical model is to state the general correspondence between elements in the mathematical system and empirical phenomena. In the last paragraph we have practically done that for our model. We will spell it out a little further. To make the mathematical system into a general mathematical model for learning, we say that the *responses* in the system correspond to responses of organisms, that the *events* in the system correspond to the events in the real world, and that the response tendencies of organisms correspond to the sets of probability variables in the system. Once such identifications are set up we have a general mathematical model for learning. Some would call it a general theory of learning, but we try to avoid the word "theory" because to psychologists our model will seem very different from the more classical learning theories. In several places in this book we suggest correspondences between our model and psychological theories.

1

The third step is to make specific applications of the general model to actual situations. These applications might be regarded as specific models. The general model then generates many specific models. For example, the general gravitational model has specific applications to celestial bodies, to bodies on inclined planes, and to bodies falling near the earth. These are specific models that flow from a more general model. Similarly a general model of heat flow could be applied to describe temperature distributions in long rods, in thin sheets, in solid blocks, or under deserts. These specific models may or may not approximate empirical facts. Similarly, the general model developed in this book can be applied, for example, to describe behavior in rote learning, in avoidance training experiments, and in choice situations with risk.

Only after the identifications are made in specific situations is it relatively easy to demonstrate that a specific model is inadequate in the sense that observed results do not agree with the forecasts. Close agreement does not prove that the model is correct but suggests that it may be useful; poor agreement indicates that the specific model, including the identifications, is inappropriate. When similar identifications are made in many different situations and good results are achieved, we begin to feel that the method of identification is reasonable and that the mathematical system was wisely chosen. Poor agreement between fact and prediction does not demonstrate that the mathematical system is unsatisfactory, however, for the same mathematics with new identifications may give quite satisfactory results. It would not surprise us if at a later date someone were to claim that we had made poor choices of identifications in our analyses of experimental data, and then proceed to re-analyze the data with essentially the same mathematics but new identifications. Indeed, it is next to impossible to prove that a mathematical system cannot lead to useful models, for only after repeated failures with different sets of identifications would we have any evidence that a particular mathematical system was unsuitable. One mathematical system can lead to many models, and only the models can be judged as adequate or inadequate.

The preceding paragraphs try to distinguish among a mathematical system, a general model, and a specific model. The reader who is still confused about these distinctions should not be disturbed. Time spent arguing over such arbitrary categorizations contributes little to progress in science; such arguing is best done on convivial rather than professional occasions.

We wish to make a point or two about the relation of our model to mathematical statistics and probability. For the most part the data obtained from learning experiments have been analyzed in the past by standard statistical techniques. In many cases a graphical presentation is sufficient to demonstrate a point being made by the investigator, whereas in other cases something like a conventional t-test is used to compare two groups of subjects. These techniques have been useful and will continue to be. But statisticians have not developed special techniques for the special problems that arise from learning data. Nothing comparable to the statistical methodologies developed for sampling inspection, bioassay, or epidemiology is available to the experimental psychologist who wants to study learning and memory.

Data on animal and human learning present peculiar problems to the statistician; since irreversible changes take place while the data are being collected, repeated sampling is seldom possible. Organisms that can be considered "identical" at the start of an experiment do not remain completely "identical" because each has a different history during the course of the experiment. Observations such as these often throw doubts on the routine application of standard statistical procedures. More important, they suggest that, if methods specifically designed for handling these data were available, considerable gains in efficiency and meaningfulness would obtain. The specific models presented in this book are so designed. The mathematical developments in the first eight chapters are for the most part topics in the field of probability, or stochastic processes. We use the word "stochastic" to put some emphasis on the temporal nature of the probability problems we consider. Later, when estimation procedures are developed, we are working in the field of mathematical statistics. Finally, when we identify the mathematical components with experimental components we are describing specific mathematical models for learning.

In general terms we now define what we mean by learning and state our fundamental view of the learning process. We consider any systematic change in behavior to be learning whether or not the change is adaptive, desirable for certain purposes, or in accordance with any other such criteria. We consider learning to be "complete" when certain kinds of stability—not necessarily stereotypy—obtain. After we have described our model we can make these notions more explicit, but we wish to stress at the outset the generality with which we use the term learning. We do not take a position of strict determinism with respect to behavior and its prediction. We tend to believe that behavior is intrinsically probabilistic, although such an assumption is not a necessary part of our model. Whether behavior is statistical by its very nature or whether it appears to be so because of uncontrolled or uncontrollable conditions does not really matter to us. In either case we would hold that a probability model is appropriate for describing a variety of experimental results presently available.

In order to illustrate the type of approach we follow in this book, we now present two simple examples of the way we later handle data from a learning experiment.

FIRST EXAMPLE. For the first illustration we consider reward training of rats in the simplest type of T-maze experiment. (In Chapter 13 this problem is treated extensively.) We have an elevated T-maze with boxes at the two ends of the T. On each experimental trial a hungry rat is placed at the starting position (base of the T) and is allowed to run down the maze to the choice point, where it then turns right or left. On every trial a pellet of food is placed in the box on the right side of the maze; food is never placed in the box on the left side. Retracing is not allowed, and so the rat may not obtain food on some trials. A rat runs through many trials and eventually learns to turn right at the choice point. Dozens of aspects of the rat's behavior can be observed and measured, but we are concerned here with only one: whether it turned right or left on each trial. The data we consider then for a single rat are a sequence of right and left turns, and our problem is to analyze this sequence. We may be lucky enough to have such data for a large group of rats which we are willing to assume are identical.

A model for these data can now be constructed by making several abstractions. On each trial the rat must go either right or left, and so we say there is a probability p_n that the rat goes *right* on trial n, where n has the values $1, 2, 3, \cdots$. For this exposition, we assume that none of the rats has an initial position preference, and so we let $p_1 = 0.5$. We know that later in the sequence the rat is more likely to go right than left, and, in fact, after a sufficiently great number of trials, the rat is almost certain to go right. Hence we want p_n to be very near 1.00 when n is some large number (say 300).

The question now arises: What causes p_n to increase from 0.5 for $n = 1$ to 1.00 for n very large? Reasonably enough, we say that the increase is a consequence of what happens to the rat in the maze. Consider the first trial. The rat either goes right and finds food, or it goes left and does not find food. If the rat goes right, the reward should increase the probability of its going right on the second trial, that is, p_2 should be greater than 0.5. On the other hand, if the rat goes left, the absence of reward should decrease the probability of its going left again, and so in this case also the probability p_2 of going right on trial 2 should be greater than 0.5.

For this introductory exposition we assume that the increase in probability on the first trial is some fraction, say one-tenth, of the maximum possible increase. The maximum possible increase is $1.0 - 0.5$ or 0.5, and so we take $p_2 = 0.5 + 0.1(1.0 - 0.5) = 0.55$. We assume here that the effect of a reward on the right side is the same as the effect of a non-reward on the left side. This is a very special assumption, which is made here only for simplicity and is not made in most of the book for problems of this sort. These assumptions are extended to all trials; on any trial n we add to the probability p_n an amount which is one-tenth of $1 - p_n$ to obtain p_{n+1}. Thus, $p_3 = 0.55 + (0.1)(1 - 0.55) = 0.595$, and $p_4 = 0.595 + 0.1(1 - 0.595) = 0.6355$, and so on. We can compute the theoretical probabilities on as many trials as we like. In Table 1 we

TABLE 1

Probabilities p_n for the first 20 trials for T-maze example.

Trial number n	Probability p_n	Trial number n	Probability p_n
1	0.5000	11	0.8257
2	0.5500	12	0.8431
3	0.5950	13	0.8588
4	0.6355	14	0.8729
5	0.6720	15	0.8856
6	0.7048	16	0.8971
7	0.7343	17	0.9073
8	0.7609	18	0.9166
9	0.7848	19	0.9250
10	0.8063	20	0.9325

present the first twenty probabilities. This highly simplified model is now complete, and the problem that remains is to test the model.

The obvious way to test the model is to collect data on a large number of nearly identical rats, say 100. The proportion of rats that turn right on any trial n is an *estimate* of the probability p_n. For the first trial, $p_1 = 0.5$, and so running 100 rats on this trial is analogous to flipping 100 true coins; the proportion of heads obtained is an estimate of the true probability of getting a head. This proportion is likely to be quite near 0.5. Again on the second trial we observe the proportion of rats that go right, and this proportion gives us an estimate of p_2. We continue in this way and obtain estimates of p_n for n from 1 to whatever total number of trials we have had the patience to run in the experiment. Having obtained these experimental estimates of the probabilities p_n we can compare them with the values of p_n computed from the model above. All that remains is to measure the goodness-of-fit of the model to the data.

If we were to analyze actual data from the T-maze experiment described above, we would likely find that the model was a poor predictor, unless we were very lucky. How then do we patch up the model? One obviously arbitrary assumption is that we always add *one-tenth* of $1 - p_n$ to p_n to get p_{n+1}. This quantity 0.1 describes the rapidity of learning, and perhaps it is too small for the rats used in the experiment. We might try 0.2 or 0.3 and recompute the theoretical probabilities. More generally we could try many such values until we got good agreement with the data, or at least the best possible agreement. Fortunately this trial-and-error procedure is not necessary. We do some algebra instead. We let this fraction be a, and we compute p_1, p_2, $\cdots$ in terms of a. The problem is to choose the value of a that yields the best fit to the data. In the language of mathematical statistics, we need to *estimate* the *parameter a* from the data.

We might inquire into further possible generalizations of the model just described. Even the one generalization already introduced—letting a be determined from the data—may not be sufficient to account for the data. We could estimate the probability p_1 for the first trial from the data instead of assuming it to be 0.5, and this procedure might lead to better agreement between model and data because real animals often have position preferences. Beyond this, we could give up the assumption that reward on the right side causes exactly the same increase in probability of turning right as does non-reward on the left side. The value of the parameter a might be different for these two events. But it is just at this point that the model becomes much more complicated. For example, the probability p_2 depends on what the rat actually did on trial 1, and p_3 depends on what the rat did on trials 1 and 2. Different rats will produce different sequences of right and left turns, usually at least, and so different rats will have different probabilities on trial 3, for example. For a large population of rats we shall have a *distribution* of probabilities, not just a single probability p_n, on each trial. In most of this book we are concerned with such distributions, their properties from trial to trial, and with methods of estimating parameters of these distributions from experimental data.

SECOND EXAMPLE. We now present a second example of an experimental problem in learning and indicate how we construct a model and analyze the data. The experiment is one reported by Solomon and Wynne on the avoidance training of dogs. In Chapter 11 we analyze this experiment in more detail after we have developed our mathematical system and the general model. In this introduction we describe only the specific model for this problem and indicate one way the data can be analyzed.

The Solomon-Wynne experiment used an intense electric shock from which a dog could escape by jumping over a barrier. A conditioned stimulus preceded the onset of shock by 10 seconds. The dogs learned to avoid the electric shock by responding to this conditioned stimulus. The data which concern us here are simply whether each dog on each trial avoided or received shock. Solomon and Wynne observed a sequence of shocks and avoidances for each dog. These sequences eventually contained all avoidances, that is, the dogs learned to avoid with certainty.

Many statistics of the sequential data may be computed, and several of these are given by Solomon and Wynne. Examples are the mean number of trials before the first avoidance (mean for 30 dogs), mean number of trials before the second avoidance, mean number of shocks received during all acquisition trials, etc. Such statistics may be computed directly from the data without the use of a model. Moreover, when certain experimental conditions are varied, such as the intensity of the shock or the time between the conditioned stimulus and the onset of shock, these statistics may be obtained for each such condition. The only problem that remains is to interpret these statistics, that is, to infer what they mean in conditioning and learning terms. The following model provides one way to interpret the data.

We assume that on each trial n ($n = 1, 2, 3, \cdots$) there is a true probability q_n that a particular dog will receive shock, that is, that it will not jump soon enough to avoid shock. Since the dogs in fact do learn to avoid, the probabilities q_n must decrease as n gets large. What causes the q_n to decrease? We assume that the changes in the q_n are a consequence of the animals' previous experience. In particular, if a dog avoids the shock on trial n, then the probability that it will be shocked on the next trial is some constant α_1 times q_n. We require that α_1 be between zero and unity so that $\alpha_1 q_n$ is less than q_n but remains positive. Similarly we assume that if the dog receives a shock on trial n, then q_n is multiplied by some other constant α_2 to give the probability of shock on the next trial. Again α_2 must be between zero and unity. Thus, on each trial we multiply the probability of shock on that trial by either α_1 or α_2 to obtain the probability of shock on the next trial.

We make one further assumption which is strongly supported by the data of Solomon and Wynne. We assume that the probability q_1 of shock on the first trial is unity, that is, that each dog is certain to be shocked on the first trial. By our previous assumption, then, the probability q_2 of shock on the second trial is $\alpha_2 q_1$ or just α_2. Now on this second trial a dog may either avoid or be shocked. If the dog avoids on the second trial, the probability q_3 that it will be shocked on the third trial is $\alpha_1 q_2$ or $\alpha_1 \alpha_2$; if the dog is shocked on the second trial, the probability q_3 that it will be shocked on the third is $\alpha_2 q_2$ or α_2^2. Proceeding in this way, we may compute the probability of shock q_n on any trial n if we know how many previous shocks or avoidances the dog had. If we denote the number of previous shocks by j and the number of previous avoidances by k, we have $q_n = \alpha_2{}^j \alpha_1{}^k$. Previous to trial n there will have been $n - 1$ trials and so $j + k = n - 1$. With this information we can compute various statistics of the data in terms of the two parameters α_1 and α_2. For example, we could compute the expected fraction of dogs that are shocked on each trial. This fraction would tend to zero as the number of trials becomes large because both α_1 and α_2 are assumed to be less than one. This conclusion is required by the data; indeed, we have constructed the model so that after many trials avoidance would almost always occur.

The main problem which faces us is how to *estimate* α_1 and α_2 from the data in order to get a close fit between the data and the model. In Chapter 11 we discuss several ways of estimating these two parameters, but here we describe only one way. This procedure is not very efficient, but it is adequate for the present illustration. Consider first a statistic mentioned earlier, the mean number of trials before the first avoidance. We denote this statistic by F_1. This number F_1 depends only on our parameter α_2 and not upon α_1. We see this as follows. Prior to the first avoidance, a dog is shocked on every trial, and so its probability of shock on each of these trials is α_2 to some power j, where j is the number of previous shocks. The parameter α_1 is not involved until after the first avoidance. The statistic F_1, then, depends only on α_2. In Chapter 8 this function is derived, but here we provide only a table based upon that derivation. In Table 2 we show some computed values of F_1 for different values of α_2. (Table A at the end of the book is more extensive.) We enter Table 2 with the statistic F_1 which can be computed directly from the data, and

TABLE 2

Mean number of trials, F_1, before the first avoidance for different values of α_2.

α_2	F_1	α_2	F_1
0.81	2.81	0.91	4.13
0.82	2.88	0.92	4.39
0.83	2.97	0.93	4.70
0.84	3.07	0.94	5.08
0.85	3.17	0.95	5.57
0.86	3.29	0.96	6.23
0.87	3.42	0.97	7.21
0.88	3.56	0.98	8.84
0.89	3.73	0.99	12.52
0.90	3.91	1.00	infinity

so obtain an estimate of α_2. In the Solomon-Wynne experiment, the mean number of shocks before the first avoidance was $F_1 = 4.50$. From Table 2, therefore, we see that α_2 is approximately 0.92.

Now consider another statistic, T_2, defined as the mean total number of shocks. This statistic T_2 will be a function of both α_1 and α_2 in our model, because all the data are involved in obtaining T_2. Again we defer the derivation of the function until Chapter 8. However in Table 3 we give computed values of T_2 for different values of α_1 and α_2. (Table B is more extensive.)

Because T_2 is obtained directly from the experimental data, and since α_2 has already been estimated, Table 3 can be used to estimate α_1. From the

TABLE 3

Mean total number of shocks, T_2, for different values of α_1 and α_2.

α_1	α_2									
	0.90	0.91	0.92	0.93	0.94	0.95	0.96	0.97	0.98	0.99
0.75	6.11	6.39	6.70	7.07	7.51	8.05	8.73	9.64	10.98	13.41
0.76	6.25	6.54	6.87	7.25	7.70	8.26	8.96	9.90	11.30	13.82
0.77	6.41	6.70	7.04	7.44	7.90	8.48	9.21	10.18	11.63	14.25
0.78	6.57	6.88	7.23	7.63	8.12	8.72	9.47	10.49	11.99	14.72
0.79	6.75	7.06	7.42	7.85	8.35	8.97	9.75	10.81	12.38	15.22
0.80	6.93	7.26	7.64	8.08	8.60	9.24	10.06	11.16	12.79	15.77
0.81	7.13	7.47	7.86	8.32	8.87	9.54	10.39	11.54	13.24	16.36
0.82	7.35	7.70	8.11	8.59	9.16	9.85	10.74	11.95	13.73	17.00
0.83	7.58	7.95	8.38	8.87	9.47	10.20	11.13	12.39	14.27	17.71
0.84	7.83	8.22	8.66	9.19	9.81	10.57	11.55	12.88	14.85	18.48
0.85	8.11	8.51	8.98	9.53	10.18	10.99	12.02	13.41	15.50	19.34
0.86	8.41	8.84	9.33	9.90	10.59	11.44	12.53	14.00	16.22	20.30
0.87	8.75	9.19	9.71	10.32	11.05	11.94	13.10	14.66	17.02	21.38
0.88	9.12	9.59	10.14	10.78	11.55	12.51	13.73	15.40	17.92	22.59
0.89	9.53	10.03	10.62	11.30	12.12	13.14	14.45	16.24	18.94	23.98
0.90	10.00	10.54	11.16	11.89	12.77	13.86	15.27	17.20	20.12	25.58

Solomon-Wynne data the mean total number of shocks is $T_2 = 7.80$; we have just obtained the value $\alpha_2 = 0.92$. Hence we look down the column in Table 3 corresponding to $\alpha_2 = 0.92$ until we find the number closest to 7.80. This appears in the row for $\alpha_1 = 0.81$, and so this is our estimate of the parameter α_1.

The estimation procedures just described are by no means the best possible, as they have quite large sampling errors. However, we have made our point: the parameters α_1 and α_2 can be estimated from the data, in this case from the statistics F_1 and T_2. The question is whether α_1 and α_2 are more meaningful than F_1 and T_2. Our answer is in two parts. The first part has to do with the purpose of the model; in terms of two parameters, α_1 and α_2, the model predicts properties of the entire sequence of behavior and these predictions are testable. Numerous other statistics of the data can be computed from the obtained estimates of α_1 and α_2 and the model. In this sense, α_1 and α_2 are more useful than F_1 and T_2.

The second part of our answer to the question about the meaningfulness of α_1 and α_2 has to do with the meaning of these parameters when they were first introduced in the model. Consider α_1; we asserted that if avoidance occurs on trial n, the probability of shock on trial $n + 1$ is $q_{n+1} = \alpha_1 q_n$. The parameter α_1, therefore, is a measure of the "ineffectiveness" of an avoidance trial in reducing the probability of shock. The nearer α_1 is to unity, the less effective, and the nearer α_1 is to zero the more effective is one avoidance trial. Similarly, α_2 is a measure of the ineffectiveness of a shock trial in reducing the probability of shock.

We estimated that $\alpha_1 = 0.81$ and $\alpha_2 = 0.92$ for the Solomon-Wynne data. Hence, in that experiment, an avoidance trial has a greater effect than a shock trial. We can compare the effectiveness of shock and avoidance trials by finding out how many shock trials have an effect equivalent to one avoidance trial. We just need to know how many times to multiply 0.92 by itself to get 0.81. It turns out that $(0.92)^{2.5}$ is approximately 0.81, and so we infer that an avoidance trial has the same effect as 2.5 shock trials. Without a model, an inference such as this could not be made readily. The parameters α_1 and α_2 may change, of course, when experimental conditions are varied. In Chapter 11 we discuss estimates of these parameters for further data obtained by Solomon and his colleagues.

The foregoing two examples illustrate the general strategy we follow in applying our mathematical system to particular situations. In order to be as intelligible as possible to a non-mathematical reader, we here suppressed the mathematics and emphasized the problems of interpreting empirical data. The time has now come to make the mathematical structure explicit; we hope these two examples will provide guidance through mathematical developments that are sometimes unavoidably difficult and circuitous.

PART I

THE MATHEMATICAL
SYSTEM AND THE
GENERAL MODEL

CHAPTER 1

The Basic Model

1.1 INTRODUCTORY COMMENTS

This chapter presents the basic structure of the mathematical system used throughout this book. Descriptions of this mathematical system have already appeared in the literature [1, 2]. As each concept is introduced it is discussed both intuitively and formally. First we give intuitive arguments for the necessity or desirability of each concept for describing learning data and suggest possible identifications between the mathematical constructs and empirical quantities. As pointed out in the Introduction, specific identifications are not postulated until Part II, where the mathematical system is applied to various experiments. In addition to an intuitive introduction to each concept this chapter also contains a more formal description, a description that divorces the mathematical concept from experimental problems. The formal description may indicate the possible generality of the mathematical system and thus may suggest to some readers quite different applications from those discussed in Part II.

1.2 RESPONSE CLASSES AND PROBABILITIES

In order to describe behavioral changes, we must distinguish among various kinds of responses. For even the simplest type of learning, such as bar-pressing by rats, we need two classes of responses, those which terminate in a relay closure indicating a bar press and those which do not. Bar-pressing and not-bar-pressing are taken to be mutually exclusive and exhaustive classes of responses. It is not necessary to identify all responses with overt motor activity—doing "nothing" is a response, and in the example just cited falls into the class of not-bar-pressing. Although most of our analysis will deal with just two response categories, we shall define in general r mutually exclusive and exhaustive classes of responses. An example of a situation in which more than two response classes are necessary is an experiment in which a human subject is asked to choose on each trial among several words or abstract symbols.

Some readers may object at the outset to our taking the response classes as mutually exclusive; they may feel that this is a serious limitation on our framework. For example, a rat may press a bar and flick its tail simultaneously or a person may withdraw his hand from an electrified contact and say "ouch" at the same instant. Such considerations are inappropriate at this point, we feel, because we have not given the response classes explicit operational definitions, or, more precisely, we have not postulated identifications. We do take the position, however, that mutually exclusive classes of behavior can always be defined in any experimental problem. For instance, we might define one class as bar-pressing with a tail flick and another class as bar-pressing without a tail flick, and these two classes would be mutually exclusive. Or, perhaps more usefully, we could define a class as bar-pressing with or without tail flicks, that is, the union of the two classes just mentioned. In this example, the choice of definition would hinge upon what were being studied experimentally, bar presses, tail flicks, or both. In other words, mutually exclusive classes can always be found, but the serious question is whether or not such classes are useful.

The requirement that the r response classes be exhaustive causes no difficulties, for a residual category—"everything else"—can always be introduced. For example, in a runway experiment we may be focusing attention on a rat's response of leaving a starting box and we may record only the time at which this response occurred. Prior to an occurrence of such a response the rat does many other things which we classify together as not-leaving-starting-box.

In our mathematical system we represent the r response classes by a set of alternatives $A_1, A_2, \cdots, A_r$. The nature of these alternatives is not specified in the mathematical system, for they play a role analogous to that of points and lines in geometry. Later, when we apply our system to experimental problems we identify the alternatives A_j with certain classes of behavior. At times we call them response classes in anticipation of the applications.

As an index of behavior we have chosen a set of probabilities, p_j, where $j = 1, 2, \cdots, r$, one for each alternative or class of responses. In Part II we show how those probabilities can be related to experimental measures of behavior. We define p_j as the probability that alternative A_j will be chosen, for example, that a member of the jth class of responses will be performed on a trial. A trial is defined as an opportunity for choosing among the r alternatives. A trial so defined will correspond to an experimental trial in many problems, but in other problems in which time is an important variable a trial will be a short interval of time. More is said about these time problems in Chapter 14.

Since the r alternatives are exhaustive and mutually exclusive, some alternative must occur on each trial. That is to say, we have

THE PROBABILITY INVARIANCE RULE:

$$(1.1) \qquad p_1 + p_2 + \cdots + p_r = \sum_{j=1}^{r} p_j = 1, \qquad 0 \leq p_j \leq 1.$$

If the alternatives were not exhaustive the probabilities could add up to less than unity, and, if they were not mutually exclusive, they could add up to more than unity. If they were neither mutually exclusive nor exhaustive, the total could be any positive number. The probability invariance rule states that probability cannot be created or destroyed; the total probability is always the same on every trial. It can, however, be moved about from one alternative to another. Such a flow of probability from one class of responses to another is the basis for our description of learning. In fact, the most that a model derived from our mathematical system could predict is the set of r probabilities on every trial. Complete learning, as we use the expression, does not correspond necessarily to stereotyped behavior where one response always occurs. The final stable probability of a response might be 0.8, for example, and so that response would occur on the average 80 percent of the time. In other words, the completion of learning leads only to certain types of statistical stability. As an example of the behavior we might expect after complete learning, consider an experiment in which a person may (1) press a green button, (2) press a red button, or (3) press a white button. Let the probabilities of these responses stabilize at 0.6, 0.3, and 0.1, respectively. During an observation period of 100 trials after stability is reached, we would expect to find then that the person had pressed the green button about 60 times, the red button about 30 times, and the white about 10.

The probability, p_j, that a jth class of responses will occur on a particular trial cannot be directly measured. We conceive that every organism possesses a "true" probability p_j at the start of each trial.* As far as the mathematical system is concerned, the physical basis for this probability is irrelevant. A variety of physiological models might provide it. Any sort of fictitious device that aids us to think about probabilities is as acceptable as any other. For example, we might imagine that the organism has a small disc that it can rotate. The disc is part shaded and part white, as shown in Fig. 1.1; the area of the shaded sector is proportional to p_j and that of the white sector to $1 - p_j$. At the beginning of each trial, the organism whirls the disc, and, if a fixed marker points to the shaded part when the disc comes to rest, the jth response is made, and if it points

---

* Several discourses on probability theory are available [3, 4].

to white the jth response is not made. In other words, we assume that all sectors of the same area have the same chance of being pointed to after one whirl of the disc. This principle may be extended to as many response classes as we choose. For example, if we have four classes having probabilities 0.5, 0.3, 0.1, and 0.1, we have a disc as shown in Fig. 1.2. As before, a decision is made by rotating the disc and observing which portion of the disc appears opposite the marker when the disc comes to rest.

We could also think of the organism as possessing an urn of black and white balls with the proportion of black balls equal to p_j. The organism

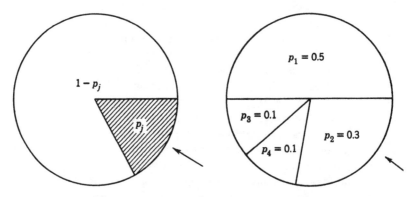

Fig. 1.1. Illustration of a disc which may be rotated for determining whether or not the jth response is to be made. The shaded area is proportional to p_j, and the unshaded area to $1 - p_j$. The fixed marker is indicated by the arrow.

Fig. 1.2. A "decision-making disc" for the example of four response classes and probabilities of 0.5, 0.3, 0.1, and 0.1.

draws from the urn to decide whether or not to make the jth response. Another possible analogy is that the organism uses a random number table for making decisions. The above heuristic devices have little to do, of course, either with the real world or with the mathematical system. However, we postulate that organisms behave *as if* they possessed such probability mechanisms.

We are able to obtain *estimates* of the probabilities in certain special cases. If p_j is constant, we may estimate p_j from the proportion of trials in which an animal makes responses of class A_j. We are, in effect, estimating p_j by sampling from a population of responses of a single animal. On the other hand, if we had a large number of animals, we could take as an estimate of the value of p_j the proportion of animals which made the jth response when the trial occurs. However, all these animals would have to be stochastically identical—by this we mean that each animal would necessarily have the same true probability p_j at the time of measurement.

1.3 FACTORS WHICH CHANGE THE PROBABILITIES

Our next task is to specify what factors in the process change the set of probabilities. We begin with a very general view: whenever certain *events* occur, the probabilities are altered in a determined way. The nature of these events depends upon the experimental situation being considered, but in the mathematical system we need only an abstract set of t events, $E_1, E_2, \cdots, E_t$.

For purposes of exposition we may take a more restricted view. We assume that every time a response occurs it has an outcome. This outcome may be a reward given by an experimenter, it may be a change in the external environment, or it may be merely proprioceptive stimulation. We denote the possible outcomes by $O_1, O_2, \cdots, O_s$. Whatever the outcome, we assume that it has some effect (including zero effect) on the r probabilities associated with the r response classes. It is our opinion that this assumption is not inconsistent with most current learning theories—reinforcement theorists emphasize the environmental events of reward and punishment, whereas the association theorists are concerned mainly with changes in external and internal stimulation. From our point of view these are all outcomes of trials.

We assume furthermore that a particular outcome following a given response changes the set of r probabilities in a unique way which is independent of earlier events in the process. In other words, if we are given a set of probabilities p_j on trial n, the new set on trial $n + 1$ is completely determined by the response occurring on trial n and its outcome. (Earlier events will, of course, determine the p_j on trial n.) This assumption we shall refer to as the *independence-of-path assumption*. Note that this assumption means that we are not concerned with the route or path used to achieve the value of p_j. The theory is ahistorical in the sense that the effects of past events can influence the future of the organism only if they have influenced its present state and are somehow embodied in the description of the present state. Just as the velocity of a freely falling body contains a "memory" of how far the body has fallen, the p_j contain a "memory" of the events that produced them.

We have now arrived at the following position. A learning experiment consists of a sequence of trials, on each of which one and only one response occurs. Each response occurrence has an outcome which alters the probabilities of the various responses. We now need some mathematical machinery to describe the effects of various outcomes on the set of r probabilities; this is the subject of the next section.

1.4 THE CONCEPT OF AN OPERATOR

All mathematical operations on a function f can be defined in terms of operators (see Davis [5]). A familiar operator is the differential operator $\dfrac{d}{dx}$; the quantity $\dfrac{df}{dx}$ is broken up into an operator $\dfrac{d}{dx}$ and an operand f. For a function f the operator $\dfrac{d}{dx}$ defines a new quantity $\dfrac{df}{dx}$. Whereas $f(x)$ denotes the value of the function at x, $\dfrac{df}{dx}$ represents the rate of change of the value of the function. Other familiar examples of operators are *log* and *sin*. Two special operators are the identity operator which leaves every operand unchanged, and the null operator which causes every operand to become zero. In general, an operator O when applied to an operand x defines a new quantity Ox. Hence O represents a transformation on all values of x. (We use the letter O because it is the first letter of the word Operator.)

Even simpler operations of addition and multiplication may be defined in terms of operators, and we introduce these operators here. The operator A when applied to a variable x will indicate the addition of a constant a (the symbol A is used to denote addition):

DEFINITION OF A:

(1.2) $$Ax = x + a.$$

The notation Ax does *not* mean A multiplied by x.

The meaning of the operator A may become clearer if we perform operation A on operand x more than once. The notation A^2x will be used to denote the application of A to a quantity Ax. We then have the definition

$$A^2x = A(Ax).$$

We note from the definition of A that an application of A to *any* operand x is identical to adding a quantity a to that operand. Hence, we have

$$A(Ax) = (Ax) + a.$$

But we know from the definition of A the value of Ax in terms of x, and so we obtain

$$A^2x = A(Ax) = (Ax) + a = (x + a) + a = x + 2a.$$

We may readily generalize the above to n applications of A to x and obtain

(1.3) $$A^nx = AAA \cdots Ax = x + na.$$

Similarly, an operator M when applied to x will serve to indicate that x is to be multiplied by a constant m.

DEFINITION OF M:

(1.4) $$Mx = mx.$$

Note that the left side of this equation does *not* mean "M times x" (it does mean "M operating on x"), but the right side does mean "m times x." When we apply M to x twice we obtain

$$M^2x = M(Mx) = m(Mx) = m(mx) = m^2x,$$

and when we apply M to x a total of n times, there results

(1.5) $$M^nx = m^nx.$$

The two operators, A and M, defined above, will be of considerable interest in our mathematical development, and so we shall here indicate what happens when both A and M are applied to an operand x. First, when we apply M to x and then apply A to Mx we get

$$A(Mx) = (Mx) + a = mx + a.$$

Second, when we apply A to x and then apply M to (Ax) we have

$$M(Ax) = m(Ax) = m(x + a) = mx + ma.$$

We see at once that the order of application of M and A is important. Because the operation AMx yields a different result from the operation MAx, the operators A and M are said to be non-commutative. If the order of application of a pair of operators can be reversed without affecting the result, the pair of operators is said to commute. The difference

$$AMx - MAx = (mx + a) - (mx + ma) = a(1 - m)$$

is called the commutator of A and M.

We may denote the operation AMx above by a new operator L applied to an operand x. We see that this leads to the most general linear function of x. A linear function of x is, by definition, a constant plus the product of x and another constant. Throughout our discussion we shall be concerned mainly with operators which represent a linear transformation on a probability variable p. From now on the general variable x we shall be working with is probability. These operators have the form

DEFINITION OF L:

(1.6) $$Lp = a + mp, \qquad 0 \leq p \leq 1.$$

The variable p is defined only for numbers from zero to unity, inclusive, for there are no negative probabilities nor any probabilities greater than unity. The quantity Lp will also be considered a probability and so it also is defined only for numbers from zero to unity, inclusive, a requirement that will place restrictions on the constants a and m. Our basic learning operators will be like L in the last equation; p may be the probability of a certain response on trial n, and Lp will be its probability on trial $n + 1$. The operator L, then, describes the effects of the outcome of whatever occurred on trial n on the probability p.

The question arises as to why we define L to represent a *linear* transformation on the probability rather than some other transformation. The answer is easy: linearity is assumed in order to simplify the mathematical analysis. Many functions that are used in applied work can be approximated by polynomials. The higher the degree of the polynomial, the better the approximation usually is. In fact many functions such as logarithms, sines, and exponentials can be represented by infinite series. From this point of view the linearity assumption we use represents the fitting of the first two terms of a series. To put it in symbols, if we regard an operation Op as expressible by a series,

$$Op = a_0 + a_1 p + a_2 p^2 + \cdots,$$

we have taken Lp to be an approximation to the true function Op.

The use of linear transformations in the learning model to be presented may seem to many readers as a very strong restriction. It is. However, it will be seen presently that even with these easy transformations, the mathematics becomes complicated and leads to many unsolved problems in stochastic processes. There is little hope indeed of solving these problems with non-linear transformations. Of course the ultimate justification of the linearity assumption depends upon the agreement between model and experiment, but, if it is any comfort to the reader, we hasten to point out that the whole of quantum physics is based upon an assumption of linear operators. (Although linear operators are basic in the mathematical machinery which follows, linearity is not essential to the general approach.)

One further remark is appropriate at this point. In the last section, we mentioned the independence-of-path assumption, according to which the probability on trial $n + 1$ depended upon the probability on trial n and not upon how it got there, that is, not upon earlier values of probability. We now see the desirability of that assumption, for without it, our operation Lp would not be a function of p alone as given in its definition (equation 1.6), it would also be a function of previous probabilities.

It might be mentioned that operators having the form of A might be used to construct a learning model. Such a model would, on the whole, be easy to use. If the operator were applied directly to probabilities, there would be one rather obvious objection. Suppose $a = 0.1$ and the initial probability of making the response is $p = 0.1$. When the event occurs that makes A applicable, the probability of the response changes to $p + a = 0.1 + 0.1 = 0.2$. Suppose after further learning the probability has reached 0.9. Again the event that makes A applicable occurs, and the probability is changed to 1.0. The example shows that the gain in probability due to the event is 0.1, no matter what the original probability was. This result is contrary to most experience which suggests that it is more difficult to close the gap between $p = 0.9$ and 1.0 than to close the gap between $p = 0.1$ and 0.2. Such a difficulty might be got around by operating on some variable related to p instead of p itself. For example, A might operate on x, where x is $1/(1 - p)$. Then when $p = 0.0$, $x = 1$. With $a = 0.1$, one operation of A leads to $Ax = 1.1$, which when solved for the new value of p gives 0.091. When $p = 0.5$, $x = 2$, and one operation of A gives $Ax = 2.1$, which yields a p of 0.524. This formulation then would lead to smaller increments in p the larger the value of p. Remarks analogous to these could be made about the operator M. We shall not pursue the matter further here. We actually use M as a special case.

1.5 MATRIX OPERATORS

A mathematically convenient type of operator is a matrix, for a matrix can operate on a whole set of variables at the same time. Moreover, matrices can be used to represent all linear transformations on a set of variables. Matrix operators will be used in much of the analysis which follows since we are assuming linearity and because we are interested in the changes which occur in our set of r probabilities, $p_j(j = 1, 2, \cdots, r)$. Because the matrices we shall use are event operators there will be a different matrix for each event. When an event occurs, its matrix will be applied to the whole set of probabilities for the r alternatives, and thus change all probabilities simultaneously. For the benefit of those readers unfamiliar with the use of matrices we shall describe some fundamental principles of matrix algebra.*

Simple arithmetic and algebra deal with single numbers and their combinations. We know the rules of addition and multiplication of these numbers so well that we seldom realize that those rules are arbitrary. Now a matrix is simply a rectangular array of numbers, called elements of

---

* For a treatment of matrix algebra, see Thurstone [6].

the matrix, as shown below:

$$\begin{bmatrix} u_{11} & u_{12} & \cdots & u_{1r} \\ u_{21} & u_{22} & \cdots & u_{2r} \\ \cdot & \cdot & & \cdot \\ \cdot & \cdot & & \cdot \\ \cdot & \cdot & & \cdot \\ u_{s1} & u_{s2} & \cdots & u_{sr} \end{bmatrix}.$$

Each element has two subscripts: the first indicates the row of the element, and the second tells us the column of the element. Nothing is implied about combining these elements. Just because the elements are thrown together in a matrix does not mean that they necessarily have anything to do with one another. They may or they may not. Whether they do or not comes from considerations apart from the matrix. If the matrix has s rows and r columns, it has $s \times r$ elements. The entire matrix plays the role in matrix algebra that is played by a single quantity x in ordinary algebra. Matrices acquire meaning as soon as we specify how two matrices are combined to give a third matrix. Thus, we need the rules of addition and multiplication for two matrices. We shall illustrate these rules for three rows ($s = 3$) and three columns ($r = 3$).

Consider two matrices $\mathbf{U}$ and $\mathbf{V}$ given by (a letter in bold-face type will denote a matrix)

$$\mathbf{U} = \begin{bmatrix} u_{11} & u_{12} & u_{13} \\ u_{21} & u_{22} & u_{23} \\ u_{31} & u_{32} & u_{33} \end{bmatrix},$$

$$\mathbf{V} = \begin{bmatrix} v_{11} & v_{12} & v_{13} \\ v_{21} & v_{22} & v_{23} \\ v_{31} & v_{32} & v_{33} \end{bmatrix}.$$

By definition the *sum* of $\mathbf{U}$ and $\mathbf{V}$ is

$$\mathbf{U} + \mathbf{V} = \begin{bmatrix} u_{11} + v_{11} & u_{12} + v_{12} & u_{13} + v_{13} \\ u_{21} + v_{21} & u_{22} + v_{22} & u_{23} + v_{23} \\ u_{31} + v_{31} & u_{32} + v_{32} & u_{33} + v_{33} \end{bmatrix}.$$

For example, the elements may have the values given by

$$\mathbf{U} = \begin{bmatrix} 2 & 1 & 3 \\ 0 & 1 & 2 \\ 4 & 0 & 6 \end{bmatrix}, \quad \mathbf{V} = \begin{bmatrix} 0 & 3 & 1 \\ 7 & 0 & 0 \\ 1 & 5 & 1 \end{bmatrix},$$

in which case their sum is

$$\mathbf{U} + \mathbf{V} = \begin{bmatrix} 2 & 4 & 4 \\ 7 & 1 & 2 \\ 5 & 5 & 7 \end{bmatrix}.$$

The rule of multiplication is a bit more complicated. Multiplication of two matrices $\mathbf{A}$ and $\mathbf{B}$ yields different results, depending on the order; usually $\mathbf{AB}$ does not equal $\mathbf{BA}$. If $\mathbf{AB} = \mathbf{BA}$, then $\mathbf{A}$ and $\mathbf{B}$ are said to commute. Moreover, the product $\mathbf{AB}$ is defined only when the number of columns in $\mathbf{A}$ is equal to the number of rows in $\mathbf{B}$. Referring to the 3×3 matrices $\mathbf{U}$ and $\mathbf{V}$ given above, we obtain the element in the ith row and the jth column of the product matrix $\mathbf{UV}$ by multiplying the ith row of $\mathbf{U}$ times the jth column of $\mathbf{V}$ in the following special way:

$$u_{i1}v_{1j} + u_{i2}v_{2j} + u_{i3}v_{3j} = \sum_{k=1}^{3} u_{ik}v_{kj}.$$

Therefore we have

$$\mathbf{UV} = \begin{bmatrix} (u_{11}v_{11} + u_{12}v_{21} + u_{13}v_{31}) & (u_{11}v_{12} + u_{12}v_{22} + u_{13}v_{32}) & (u_{11}v_{13} + u_{12}v_{23} + u_{13}v_{33}) \\ (u_{21}v_{11} + u_{22}v_{21} + u_{23}v_{31}) & (u_{21}v_{12} + u_{22}v_{22} + u_{23}v_{32}) & (u_{21}v_{13} + u_{22}v_{23} + u_{23}v_{33}) \\ (u_{31}v_{11} + u_{32}v_{21} + u_{33}v_{31}) & (u_{31}v_{12} + u_{32}v_{22} + u_{33}v_{32}) & (u_{31}v_{13} + u_{32}v_{23} + u_{33}v_{33}) \end{bmatrix}.$$

For the numerical values given above for the elements of $\mathbf{U}$ and $\mathbf{V}$ we have

$$\mathbf{UV} = \begin{bmatrix} (2\times0+1\times7+3\times1) & (2\times3+1\times0+3\times5) & (2\times1+1\times0+3\times1) \\ (0\times0+1\times7+2\times1) & (0\times3+1\times0+2\times5) & (0\times1+1\times0+2\times1) \\ (4\times0+0\times7+6\times1) & (4\times3+0\times0+6\times5) & (4\times1+0\times0+6\times1) \end{bmatrix}$$

$$= \begin{bmatrix} 10 & 21 & 5 \\ 9 & 10 & 2 \\ 6 & 42 & 10 \end{bmatrix}.$$

The product $\mathbf{VU}$ follows the same rule and is for our numerical example

$$\mathbf{VU} = \begin{bmatrix} 4 & 3 & 12 \\ 14 & 7 & 21 \\ 6 & 6 & 19 \end{bmatrix}.$$

We may multiply a matrix $\mathbf{U}$ by a scalar quantity c; such an operation tells us to multiply each element of $\mathbf{U}$ by c. For example, consider the matrix $\mathbf{U}$ above and a constant $c = 3$. Then

$$c\mathbf{U} = \begin{bmatrix} 6 & 3 & 9 \\ 0 & 3 & 6 \\ 12 & 0 & 18 \end{bmatrix}.$$

A special matrix which will be of considerable interest later is the identity matrix $\mathbf{I}$ defined by (for $s = r = 3$):

$$\mathbf{I} = \begin{bmatrix} 1 & 0 & 0 \\ 0 & 1 & 0 \\ 0 & 0 & 1 \end{bmatrix}.$$

The reader can easily verify that when $\mathbf{I}$ is multiplied with any other matrix $\mathbf{W}$ (for which the operation is defined) the product is identical to $\mathbf{W}$.

Although the rules of matrix addition and multiplication given above were for three rows and three columns, these rules are readily generalized for s rows and r columns.

RULE I: THE ADDITION OF MATRICES. If $\mathbf{U}$ and $\mathbf{V}$ are two $s \times r$ matrices and their sum is the matrix $\mathbf{S}$, we have for the elements of $\mathbf{S}$

$$(1.7) \qquad s_{ij} = u_{ij} + v_{ij} \qquad \begin{aligned} i &= 1, 2, \cdots, s \\ j &= 1, 2, \cdots, r. \end{aligned}$$

RULE II: THE MULTIPLICATION OF MATRICES. If $\mathbf{U}$ is an $s \times r$ matrix and $\mathbf{V}$ is an $r \times t$ matrix, the product $\mathbf{M} = \mathbf{UV}$ is an $s \times t$ matrix and the elements are

$$(1.8) \qquad m_{ij} = \sum_{k=1}^{r} u_{ik} v_{kj} \qquad \begin{aligned} i &= 1, 2, \cdots, s \\ j &= 1, 2, \cdots, t. \end{aligned}$$

RULE III: SCALAR MULTIPLICATION. If an $s \times r$ matrix $\mathbf{U}$ is multiplied by a scalar c, every element in that matrix is to be multiplied by c. Let $\mathbf{W} = c\mathbf{U}$; then the elements of $\mathbf{W}$ are

$$(1.9) \qquad w_{ij} = c u_{ij} \qquad \begin{aligned} i &= 1, 2, \cdots, s \\ j &= 1, 2, \cdots, r. \end{aligned}$$

It will be recalled that the product $\mathbf{UV}$ was not necessarily identical with the product $\mathbf{VU}$. Thus multiplication is not *commutative* for matrices. On the other hand, $\mathbf{U} + \mathbf{V} = \mathbf{V} + \mathbf{U}$, because we are just adding up single elements. Thus addition is commutative for matrices. Another important arithmetic rule has to do with *associativity*. If we add 3 and 2 and then add 6, we get the same result as adding the sum of 2 and 6 to 3. Such an associative law holds for matrices: $(U + V) + W = U + (V + W)$. Furthermore, the associative rule holds for multiplication as well: $(UV)W = U(VW)$. The interpretation is that if we first multiply V by U and then multiply W by the product matrix, we get the same result as multiplying W first by V and then multiplying this product matrix by U. Some readers may wonder whether there is any arithmetical operation for which associativity does not hold. The answer is yes; subtraction is a familiar example:

$$3 - [5 - 4] \neq [3 - 5] - 4.$$

Matrices are not necessarily square. In fact, a matrix may have only one row, in which case it is called a *row vector*. Or a matrix may have but one column, in which case it is called a *column vector*. (If the reader is familiar with the mathematical term "vector," no further explanation is required; if he is not, it will suffice if he merely regards the word as a synonym for "matrix with one row or one column." Vectors, like other matrices, are indicated by letters in bold-face type.) Consider the matrix U given above and a column vector

$$\mathbf{x} = \begin{bmatrix} x_1 \\ x_2 \\ x_3 \end{bmatrix}.$$

The product $U\mathbf{x}$ is obtained by Rule II stated above and is

$$U\mathbf{x} = \begin{bmatrix} u_{11} & u_{12} & u_{13} \\ u_{21} & u_{22} & u_{23} \\ u_{31} & u_{32} & u_{33} \end{bmatrix} \begin{bmatrix} x_1 \\ x_2 \\ x_3 \end{bmatrix} = \begin{bmatrix} u_{11}x_1 + u_{12}x_2 + u_{13}x_3 \\ u_{21}x_1 + u_{22}x_2 + u_{23}x_3 \\ u_{31}x_1 + u_{32}x_2 + u_{33}x_3 \end{bmatrix}.$$

The product $U\mathbf{x}$, then, is also a column vector. For the numerical values of the elements of U given above and the vector

$$\mathbf{x} = \begin{bmatrix} 2 \\ 1 \\ 3 \end{bmatrix},$$

we have

$$U\mathbf{x} = \begin{bmatrix} 14 \\ 7 \\ 26 \end{bmatrix}.$$

Note that the product $\mathbf{xU}$ is not defined since the number of columns in $\mathbf{x}$ does not equal the number of rows in $\mathbf{U}$.

We represent our set of r probabilities, for the r response classes, by a column vector $\mathbf{p}$:

$$\mathbf{p} = \begin{bmatrix} p_1 \\ p_2 \\ \cdot \\ \cdot \\ \cdot \\ p_r \end{bmatrix}.$$

In order to transform this column vector we need a class of operators. Such an operator is an $r \times r$ matrix $\mathbf{T}$:

$$\mathbf{T} = \begin{bmatrix} u_{11} & u_{12} & \cdots & u_{1r} \\ u_{21} & u_{22} & \cdots & u_{2r} \\ \cdot & \cdot & \cdot & \cdot \\ \cdot & \cdot & \cdot & \cdot \\ \cdot & \cdot & \cdot & \cdot \\ u_{r1} & u_{r2} & \cdots & u_{rr} \end{bmatrix}.$$

When we apply this matrix operator $\mathbf{T}$ to the column vector $\mathbf{p}$ we obtain a new column vector $\mathbf{Tp}$:

$$\mathbf{Tp} = \begin{bmatrix} u_{11}p_1 + u_{12}p_2 + \cdots + u_{1r}p_r \\ u_{21}p_1 + u_{22}p_2 + \cdots + u_{2r}p_r \\ \cdot \\ \cdot \\ \cdot \\ u_{r1}p_1 + u_{r2}p_2 + \cdots + u_{rr}p_r \end{bmatrix}.$$

(As a mnemonic device, $\mathbf{Tp}$ may be read as "transformed $\mathbf{p}$.") Once we have used $\mathbf{T}$ to operate on $\mathbf{p}$ it is easy to give an interpretation of the elements u_{ij}. They tell us how to weight up the probabilities of the various classes to get the new probabilities of those classes, or, more simply, they provide the instructions for transferring probability from one class to another. In the next section we use operators like the matrix $\mathbf{T}$ to discuss the special case of two alternatives A_1 and A_2.

1.6 TWO ALTERNATIVES OR RESPONSE CLASSES

Before continuing our general development, we illustrate the use of matrix operators for the simplest case of two mutually exclusive and exhaustive classes of responses. Actually, this amounts to much more than an illustration, as most of the experimental problems discussed in Part II involve only two types of responses. We define a probability vector

$$(1.10) \qquad\qquad \mathbf{p} = \begin{bmatrix} p \\ q \end{bmatrix},$$

where p and q are the probabilities of occurrence of A_1 and A_2, respectively. Note the distinction between the vector $\mathbf{p}$ and the element p. Here, as usual, we have the probability invariance rule:

$$p + q = 1.$$

The most general square matrix operator for this case is of the form

$$(1.11) \qquad\qquad \mathbf{T} = \begin{bmatrix} u_{11} & u_{12} \\ u_{21} & u_{22} \end{bmatrix}.$$

When the operator $\mathbf{T}$ is applied to the vector $\mathbf{p}$ we obtain a new vector given by

$$(1.12) \qquad \mathbf{Tp} = \begin{bmatrix} u_{11} & u_{12} \\ u_{21} & u_{22} \end{bmatrix} \begin{bmatrix} p \\ q \end{bmatrix} = \begin{bmatrix} u_{11}p + u_{12}q \\ u_{21}p + u_{22}q \end{bmatrix}.$$

Before the application of $\mathbf{T}$, the probability vector is $\mathbf{p}$. After the application of $\mathbf{T}$ to $\mathbf{p}$ the probability vector is $\mathbf{Tp}$ given above. Since alternatives A_1 and A_2 have been defined to be mutually exclusive and exhaustive, and the probability of occurrence of A_1 is $u_{11}p + u_{12}q$ whereas the probability of occurrence of A_2 is $u_{21}p + u_{22}q$, we must have, then, according to the invariance rule,

$$(u_{11}p + u_{12}q) + (u_{21}p + u_{22}q) = 1,$$

or

$$(1.13) \qquad (u_{11} + u_{21})p + (u_{12} + u_{22})q = 1.$$

This equation must hold for all values of p and q consistent with the condition that p and q sum to unity, and so in particular for $p = 1$ and $q = 0$ we get the restriction

$$u_{11} + u_{21} = 1,$$

whereas for $q = 1$ and $p = 0$, we have

$$u_{12} + u_{22} = 1.$$

These last two equations assert that the column sums of the matrix T must each be unity. (Matrices with columns which separately sum to unity are called "stochastic matrices.") We now make a change in notation, for reasons that will become clear presently. We let

$$a = u_{12}$$
$$b = u_{21}.$$

These substitutions plus the fact that the columns must sum to unity allow us to write the 2×2 matrix operator T given above (equation 1.11) in the form

$$T = \begin{bmatrix} 1 - b & a \\ b & 1 - a \end{bmatrix}.$$

Now in view of our discussion in Section 1.3 on factors which change the probabilities, it follows that we need as many operators as we have classes of events. In general, we may have t events and so we will have one operator for each. All these operators will have the form of T in the last equation, but the elements of the matrices will differ. For occurrences of the ith event ($i = 1, 2, \cdots, t$) we apply the operator

$$(1.14) \qquad T_i = \begin{bmatrix} 1 - b_i & a_i \\ b_i & 1 - a_i \end{bmatrix}.$$

If we now apply one of these operators to the operand $\mathbf{p}$ defined in equation 1.10, we obtain

$$(1.15) \qquad T_i\mathbf{p} = \begin{bmatrix} (1 - b_i)p + a_i q \\ b_i p + (1 - a_i)q \end{bmatrix} \qquad (i = 1, 2, \cdots, t).$$

In the analysis in the following chapters we can save considerable space if we denote the first element of the foregoing vector as $Q_i p$ and the second element by $\tilde{Q}_i q$. Since the elements of the probability vector $T_i\mathbf{p}$ sum to unity, $\tilde{Q}_i q = 1 - Q_i p$. Hence we write*

$$(1.16) \qquad \begin{aligned} Q_i p &= (1 - b_i)p + a_i q \\ \tilde{Q}_i q &= b_i p + (1 - a_i)q \end{aligned} \qquad (i = 1, 2, \cdots, t).$$

---

* Strictly speaking, the operators Q_i and $\tilde{Q}_i$ are not linear operators even though they are derived from a linear operator T_i. Mathematicians and physicists define a linear operator R as one which satisfies the two relations

$$R(u + v) = Ru + Rv,$$
$$Rcu = cRu,$$

where u and v are arbitrary operands and c is a constant. (See, for example, Davis [5, p. 7].) Matrices are linear operators, but our operators Q_i and $\tilde{Q}_i$ do not satisfy these relations and so are not linear operators. Of course, $Q_i p$ and $\tilde{Q}_i q$ are linear *functions* of p and q, respectively.

The terms on the right side of these equations may be rearranged in various ways which are useful for different purposes. First, if we let

(1.17) $$\alpha_i = 1 - a_i - b_i,$$

and use the relation $q = 1 - p$, we can write the operation in the

SLOPE-INTERCEPT FORM:

(1.18) $$Q_i p = a_i + \alpha_i p \qquad (i = 1, 2, \cdots, t).$$

It may be noted that this expression is a linear function of p, analogous to the expression given earlier for the operator L, where $Lp = a + mp$. Second, we may rewrite the first equation of 1.16 in the

GAIN-LOSS FORM:

(1.19) $$Q_i p = p + a_i(1 - p) - b_i p \qquad (i = 1, 2, \cdots, t).$$

This form will be particularly convenient for the development in Chapter 2, but it may be helpful at this point for us to interpret the three terms on the right side of this equation. The first term is simply the probability of alternative A_1 before an operator is applied. Because p can only get as large as unity, the second term, $a_i(1 - p)$, represents a gain in probability which is proportional to the largest possible gain, $(1 - p)$. The third term, $-b_i p$, denotes a loss proportional to the largest possible loss, $-p$, because p can only become as small as zero. Thus $Q_i p$ is made of three terms, the original value of p, a gain proportional to the greatest possible gain in p, and a loss proportional to the greatest possible loss. When described in this way, the operators we are using may appear more familiar to many readers. Increments proportional to $(1 - p)$ lead to growth curves which for small values of a_i approximate an increasing exponential function, whereas decrements proportional to p lead to decay curves which for small values of b_i approximate a decreasing exponential function. Hull [7] assumed increments in "habit strength" to follow such a growth law.

Still another form of writing the function $Q_i p$ is useful. We replace a_i in the slope-intercept form by $1 - \alpha_i$ times a new constant λ_i:

$$a_i = (1 - \alpha_i)\lambda_i.$$

This equation is simply the definition of λ_i. Substituting this in $Q_i p$ gives the

FIXED-POINT FORM:

(1.20) $$Q_i p = \alpha_i p + (1 - \alpha_i)\lambda_i.$$

We now indicate our reason for introducing this form of writing. If it should happen that $p = \lambda_i$, we see that this equation (1.20) tells us that $Q_i p = \lambda_i$. Conversely, if we require that $Q_i p = p$, we have $p = \lambda_i$,

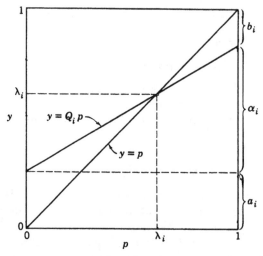

Fig. 1.3.　Geometrical interpretation of the parameters a_i, b_i, α_i, and λ_i.

provided only that $\alpha_i \neq 1$. Therefore λ_i is a *fixed point* of the operator Q_i—if p should ever equal λ_i, the operator Q_i would not change p. We shall be interested in the fixed points λ_i throughout much of this book.

In this fixed-point form, the coefficients α_i and $1 - \alpha_i$ can be regarded as weights summing to unity. When so regarded we see that $Q_i p$ is just a weighted average of p and λ_i. When α_i is between 0 and 1 this average will, of course, lie between p and λ_i.

Fig. 1.4.　Linear plot of the probability p, showing the fixed point λ_i.

A geometrical interpretation of the various parameters we have introduced may be seen in Fig. 1.3. We have plotted the function $Q_i p$ against p. The slope of this line is α_i and the intercept at $p = 0$ is a_i. The value of $Q_i p$ at $p = 1$ is $1 - b_i$. We also show the line $y = p$, the main diagonal of the unit square. The point at which the line $Q_i p$ intersects this line $y = p$ is the fixed point λ_i. Another way of visualizing the fixed point λ_i is shown in Fig. 1.4. Here we plot values of p along a line from 0 to 1. Some point on the line has the value λ_i, and another point the value p. If p is to the left of λ_i and we apply Q_i to p, the new value $Q_i p$ is to the right of p. When p is to the right of λ_i, $Q_i p$ is to the left of p. Hence the effect of the operator Q_i is to generate a new point in the direction of λ_i from p. If p is at λ_i, then $Q_i p$ is also at λ_i. This is the reason for calling λ_i the fixed point of the operator Q_i.

The complement of $Q_i p$ is $\tilde{Q}_i q = 1 - Q_i p$, and from equation 1.20 we have (remembering that $p = 1 - q$)

$$\tilde{Q}_i q = \alpha_i q + (1 - \alpha_i)(1 - \lambda_i).$$

The vector $T_i p$ is thus

$$T_i p = \begin{bmatrix} \alpha_i p + (1 - \alpha_i)\lambda_i \\ \alpha_i q + (1 - \alpha_i)(1 - \lambda_i) \end{bmatrix}.$$

This vector may be written as the sum of two vectors. We remove the first term from the expression for each element (Rule I) and get

$$T_i p = \begin{bmatrix} \alpha_i p \\ \alpha_i q \end{bmatrix} + \begin{bmatrix} (1 - \alpha_i)\lambda_i \\ (1 - \alpha_i)(1 - \lambda_i) \end{bmatrix}.$$

Next we remove the scalar multipliers according to Rule III and get

$$T_i p = \alpha_i \begin{bmatrix} p \\ q \end{bmatrix} + (1 - \alpha_i) \begin{bmatrix} \lambda_i \\ 1 - \lambda_i \end{bmatrix},$$

and so if we denote by $\boldsymbol{\lambda}_i$ the vector with components λ_i and $1 - \lambda_i$ we have the vector equation

(1.21) $T_i \mathbf{p} = \alpha_i \mathbf{p} + (1 - \alpha_i)\boldsymbol{\lambda}_i.$

EXAMPLE 1.   Show that

$$\lambda_i = \frac{a_i}{a_i + b_i},$$

when $a_i + b_i \neq 0$. It has been shown that λ_i is the value of p such that the operator $Q_i p = p$. When we write the gain-loss form

$$Q_i p = p + a_i(1 - p) - b_i p$$

we can ask that the right side be equal to p:

$$p + a_i(1 - p) - b_i p = p$$

$$a_i(1 - p) = b_i p$$

$$p = \frac{a_i}{a_i + b_i},$$

which is the desired result.

EXAMPLE 2. When $\alpha_i = 1$, all values of p are fixed points of the operator Q_i. When $\alpha_i = 1$ is substituted into the fixed-point form

$$Q_i p = \alpha_i p + (1 - \alpha_i)\lambda_i$$

we get

$$Q_i p = p.$$

EXAMPLE 3. Suppose $a_i = -1.4$, $b_i = -0.5$, find $Q_i p$ when $p = 0.6$.

$$Q_i p = p + a_i(1 - p) - b_i p.$$

Substituting into the right side we get $0.6 - 1.4(0.4) + 0.5(0.6) = 0.6 - 0.56 + 0.30 = 0.34$.

EXAMPLE 4. Do Example 3 with $p = 0.1$. We get $0.1 - 1.4(0.9) + 0.5(0.1) = -1.11$. Since probabilities cannot be negative, such a result is impossible. This example and the preceding one show that there are values of a_i and b_i that will give admissible results for some values of p, but not for others. In the next section we regard this situation as undesirable, and outlaw any values of a_i and b_i, and therefore any values of α_i and λ_i that do not give admissible results for every value of p when Q_i is applied.

EXAMPLE 5. Let $\lambda_i = 0.6$, $\alpha_i = 0.4$, compute $Q_i p$ for $p = 0.5$ and for $p = 0.7$. We use the fixed-point form

$$Q_i p = \alpha_i p + (1 - \alpha_i)\lambda_i.$$

When $p = 0.5$ we get $0.4(0.5) + 0.6(0.6) = 0.56$. When $p = 0.7$ we get $0.4(0.7) + 0.6(0.6) = 0.64$. As in the discussion of Fig. 1.4, the important point in this example is that $Q_i p$ is larger than p in the first case, but smaller than p in the second case, even though the same operator was applied in both cases. After we have restricted our parameters, it will turn out that application of the operator brings $Q_i p$ between p and λ_i when α_i is positive.

EXAMPLE 6. When $\lambda_i = 0$ the operator $Q_i p$ reduces to the same form as the operator M, where $m = \alpha_i$. We substitute $\lambda_i = 0$ in the fixed-point form

$$Q_i p = \alpha_i p + (1 - \alpha_i)\lambda_i$$

to get

$$Q_i p = \alpha_i p.$$

EXAMPLE 7. Interpret $\lambda_i = 1$, $\alpha_i = 0$. Then $Q_i p = 1$. This means that no matter what p was, the probability is changed to 1. This in turn means that the operator Q_i might correspond to insightful learning, or "one-trial learning," if Q_i is to be applied on the first trial. Similarly $\lambda_i = 0$ and $\alpha_i = 0$ correspond to "extinction" in one application of the operator.

EXAMPLE 8.   Let $a_i = 0.4$, $\alpha_i = -0.3$, $p = 0.5$, and find the new probabilities after applying Q_i once and twice.   Using the slope-intercept form

$$Q_i p = a_i + \alpha_i p,$$

we have after one application, $0.4 - 0.3(0.5) = 0.25$.   The second application gives $0.4 - (0.3)(0.25) = 0.325$.   The sequence of probabilities after 0, 1, 2 applications of Q_i is 0.5, 0.25, 0.325.   The important point to notice is that the p's first decrease, then increase.   This oscillatory feature is peculiar to operators with negative values of α_i.   In practical applications we have not used negative α's.

1.7 RESTRICTIONS ON THE PARAMETERS

In this section we discuss restrictions on the several sets of parameters we have introduced, restrictions which arise from the fact that probabilities must remain between zero and unity.   Readers not interested in the derivations of these restrictions may omit this section; the results are summarized in Figs. 1.5 and 1.6.   Similar discussions may be found in the literature [1, 2, 8, 9].

All values of probability must lie in the closed interval from zero to unity.   The term *closed interval* means that zero and unity are included as possible values, and such an interval is designated by square brackets: [0, 1].   This means that p and $Q_i p$ must not be smaller than zero or larger than unity.   This requirement places some restrictions on the allowed values of the parameters a_i and b_i in the gain-loss form $Q_i p = p + a_i(1 - p) - b_i p$.   Consider first the situation when $p = 0$. We see that we then have $Q_i p = a_i$.   Since $Q_i p$ must be in the closed interval [0, 1], we must have the restriction

(1.22) $0 \leq a_i \leq 1.$

Then consider what happens when $p = 1$; we see that this implies that $Q_i p = 1 - b_i$.   Again $Q_i p$ must be in the interval [0, 1] and so $(1 - b_i)$ must be in that interval.   From this we conclude

(1.23) $0 \leq b_i \leq 1.$

We have obtained the foregoing restrictions on a_i and b_i by considering special limiting values of p.   Therefore, these restrictions are necessary conditions, but we have yet to demonstrate that they are sufficient to keep $Q_i p$ in the closed interval [0, 1] for *all* values of p.   This sufficiency is easily demonstrated, however.   First, we note that for any allowed p, the expression $p + a_i(1 - p) - b_i p$ assumes its largest value when a_i is as large as possible and when b_i is as small as possible.   Thus, for given p in the closed interval [0, 1], the probability $Q_i p$ will be largest when

$a_i = 1$ and $b_i = 0$. Substitution of these values gives $Q_i p = 1$. We conclude then that the restrictions on a_i and b_i are sufficient to keep $Q_i p$ from getting larger than unity for all allowed values of p. Furthermore, we note that for a given allowed value of p, the probability will be as small as possible when $a_i = 0$ and $b_i = 1$. Substituting these values we obtain $Q_i p = 0$. Therefore, the restrictions on a_i and b_i are sufficient to keep $Q_i p$ from being negative for any allowed value of p.

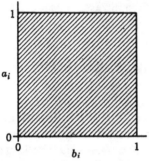

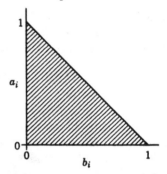

Fig. 1.5a. A plot showing restrictions on the parameters a_i and b_i. The square shaded area represents the admissible region.

Fig. 1.5b. A plot showing restrictions on the parameters a_i and b_i when the function $Q_i p$ is restricted to have positive slope, α_i. The triangular shaded area represents the admissible region.

To summarize, necessary and sufficient conditions for $0 \leq Q_i p \leq 1$, for all p, $0 \leq p \leq 1$, are $0 \leq a_i \leq 1$, $0 \leq b_i \leq 1$. These restrictions on a_i and b_i may be depicted geometrically if we plot values of a_i on an ordinate and values of b_i on an abscissa as shown in Fig. 1.5a.

The parameter α_i, defined earlier by $1 - a_i - b_i$ above, also has a restricted range of values. We see at once from that definition and from the above restrictions on a_i and b_i that we must have

$$-1 \leq \alpha_i \leq 1.$$

However, this condition, along with the restriction on a_i, is not sufficient to keep $Q_i p (= a_i + \alpha_i p)$ in the closed interval $[0, 1]$. For example, if $a_i = 0$ and $\alpha_i = -1$, then $Q_i p = -p$, and this is impossible. This suggests that a_i and α_i cannot be chosen independently, and this is indeed the case. We see that $Q_i p$ increases linearly with p provided that α_i is positive. In that case $Q_i p$ has a maximum value of $a_i + \alpha_i$ when $p = 1$. Hence for α_i positive

$$a_i + \alpha_i \leq 1$$

or

$$\alpha_i \leq 1 - a_i.$$

So we see that if α_i is positive it must be less than or equal to $1 - a_i$. Furthermore, when α_i is positive, $Q_i p$ has a minimum value of a_i when $p = 0$. Therefore we must have

$$0 \leq a_i.$$

Now when α_i is negative, $Q_i p$ has a maximum value of a_i when $p = 0$ and so

$$a_i \leq 1.$$

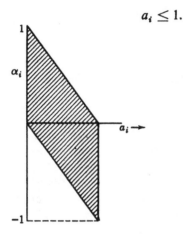

Fig. 1.6a.　Plot showing the possible values of α_i and a_i. The shaded area indicates the values consistent with the restrictions.
$$0 < a_i < 1,$$
$$-a_i < \alpha_i < 1 - a_i.$$

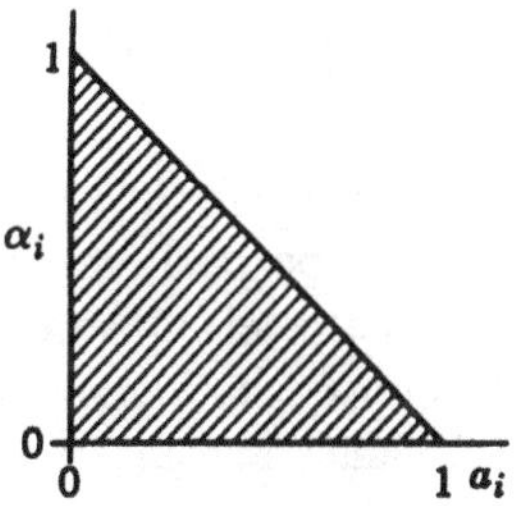

Fig. 1.6b.　Plot showing the possible values of α_i and a_i when α_i is restricted to positive values. The shaded area represents the values consistent with the restrictions
$$0 < a_i < 1,$$
$$0 < \alpha_i < 1 - a_i.$$

Finally, when α_i is negative, $Q_i p$ has a minimum value of $a_i + \alpha_i$ when $p = 1$ and so if α_i is negative we must have

$$0 \leq a_i + \alpha_i$$

or

$$-a_i \leq \alpha_i.$$

Hence, if α_i is negative it must be greater than $-a_i$. We may now combine the foregoing inequalities to obtain

$$0 \leq a_i \leq 1$$

and

(1.24) $$-a_i \leq \alpha_i \leq 1 - a_i.$$

The range of possible values of a_i and α_i is shown in Fig. 1.6a.

The two sets of conditions just derived, $0 \leq a_i \leq 1$, $0 \leq b_i \leq 1$, and $0 \leq a_i \leq 1$, $-a_i \leq \alpha_i \leq 1 - a_i$, will be assumed throughout the remainder of this book.

The conditions on the fixed points λ_i in the equation

$$Q_i p = \alpha_i p + (1 - \alpha_i)\lambda_i$$

follow rather easily from our earlier discussion. We must have

(1.25) $$0 \leq \lambda_i \leq 1,$$

because p must also be between 0 and 1.

Throughout most of this book we shall use one further restriction on the parameters, a restriction not imposed by the mathematics but rather by the fact that we are interested in describing learning phenomena. The condition is simply that α_i be non-negative:

(1.26) $$0 \leq \alpha_i.$$

This restriction means that the slope of the line $Q_i p$ versus p, shown in Fig. 1.3, must be positive or zero. Negative values of α_i lead to the oscillatory character illustrated in Example 8 above. Negative values of α_i would imply that the operators transform large values of probabilities into small values and vice versa. These properties seem undesirable to us for most problems, and moreover we have never analyzed experimental data which required negative values of the α's.

Now since $\alpha_i = 1 - a_i - b_i$ as defined in equation 1.17, we see that our condition that α_i be non-negative implies that

$$a_i + b_i \leq 1.$$

We now revise our three sets of restrictions on the parameters as follows:

I.   SLOPE-INTERCEPT FORM: $Q_i p = a_i + \alpha_i p$,

(1.27) $$0 \leq a_i \leq 1, \qquad 0 \leq \alpha_i \leq 1 - a_i.$$

II.   GAIN-LOSS FORM: $Q_i p = p + a_i(1 - p) - b_i p$,

(1.28) $$0 \leq a_i \leq 1, \qquad 0 \leq b_i \leq 1, \qquad 0 \leq a_i + b_i \leq 1.$$

III.   FIXED-POINT FORM: $Q_i p = \alpha_i p + (1 - \alpha_i)\lambda_i$,

(1.29) $$0 \leq \lambda_i \leq 1, \qquad 0 \leq \alpha_i \leq 1.$$

Figures 1.5b and 1.6b show the regions consistent with restrictions II and I just given.

The next section deals with a generalization of the matters considered up to now to more than two alternatives. This section is mathematically more formidable than the preceding material and is seldom essential to the analysis in the following chapters. It is included here for completeness.

*1.8 GENERALIZATION TO *r* ALTERNATIVES; COMBINING CLASSES CONDITION

We now return to the general case of *r* mutually exclusive and exhaustive alternatives where $r > 2$. The *r* probabilities form a column vector

$$
\mathbf{p} = \begin{bmatrix} p_1 \\ p_2 \\ \cdot \\ \cdot \\ \cdot \\ p_r \end{bmatrix},
$$

and for each of the *t* possible events we define a matrix operator

$$
\mathbf{T}_i = \begin{bmatrix} u_{11,i} & u_{12,i} & \cdots & u_{1r,i} \\ u_{21,i} & u_{22,i} & \cdots & u_{2r,i} \\ \cdot & \cdot & & \cdot \\ \cdot & \cdot & & \cdot \\ \cdot & \cdot & & \cdot \\ u_{r1,i} & u_{r2,i} & \cdots & u_{rr,i} \end{bmatrix}, \qquad (i = 1, 2, \cdots, t).
$$

In the last section, where we discussed the case of $r = 2$, we imposed the necessary condition that the probabilities of the two response classes sum to unity at all times. We now impose this condition for the general case of *r* classes of responses. We have already required that probability be conserved, that is, $\Sigma p_j = 1$, before an operator is applied. Now when we apply the operator $\mathbf{T}_i$ to the vector $\mathbf{p}$ above, we obtain the vector

$$
\mathbf{T}_i\mathbf{p} = \begin{bmatrix} u_{11,i} & u_{12,i} & \cdots & u_{1r,i} \\ u_{21,i} & u_{22,i} & \cdots & u_{2r,i} \\ \cdot & \cdot & & \cdot \\ \cdot & \cdot & & \cdot \\ \cdot & \cdot & & \cdot \\ u_{r1,i} & u_{r2,i} & \cdots & u_{rr,i} \end{bmatrix} \begin{bmatrix} p_1 \\ p_2 \\ \cdot \\ \cdot \\ \cdot \\ p_r \end{bmatrix} = \begin{bmatrix} \sum_j^r u_{1j,i}p_j \\ \sum_j^r u_{2j,i}p_j \\ \cdot \\ \cdot \\ \sum_j^r u_{rj,i}p_j \end{bmatrix} \quad \begin{array}{l} (i = 1, 2, \cdots, t), \\ (j = 1, 2, \cdots, r). \end{array}
$$

The elements of this new vector give us the *r* probabilities for the *r* response classes after the application of the operator $\mathbf{T}_i$, and so these elements must sum to unity also. This leads to the condition

$$
\sum_{k=1}^r \sum_{j=1}^r u_{kj,i}p_j = 1 \qquad (i = 1, 2, \cdots, t).
$$

This condition is a generalization of equation 1.13, which states the same condition for $r = 2$. We may interchange the order of the summations, for this merely demands rearranging terms. Thus

$$\sum_{j=1}^{r} \left(p_j \sum_{k=1}^{r} u_{kj,i} \right) = 1.$$

We then introduce the abbreviation

$$s_{j,i} = \sum_{k=1}^{r} u_{kj,i}.$$

Note that this quantity is simply the sum of the elements in the jth column of the matrix $\mathbf{T}_i$. We may now write

$$\sum_{j=1}^{r} s_{j,i} p_j = 1.$$

The original probability invariance rule of equation 1.1 was

$$\sum_{j=1}^{r} p_j = 1.$$

Now in order for both of the conditions above to hold simultaneously for all allowed values of the p_j, we must have for all values of j

$$s_{j,i} = 1, \qquad \begin{array}{l} j = 1, 2, \cdots, r \\ i = 1, 2, \cdots, t. \end{array}$$

This is readily seen if we combine those conditions:

$$\sum_{j=1}^{r} s_{j,i} p_j = \sum_{j=1}^{r} p_j.$$

We may rewrite this as

$$\sum_{j=1}^{r} (s_{j,i} - 1) p_j = 0.$$

All the p_j are non-negative and not all are zero so this last equation is satisfied only by $s_{j,i} = 1$. The meaning of the condition is simple: each column of the matrix $\mathbf{T}_i$ must sum to unity. Incidentally, for every arbitrary probability vector to be transformed into a probability vector, it is necessary that the elements of the transformation matrix be between zero and unity.

We now wish to introduce one further restriction on our operators. (This restriction is automatically fulfilled when we have only two classes of responses, as will be seen shortly.)   It arises from the arbitrariness with which we have defined the r classes of responses, and we believe that this arbitrariness is desirable.   The requirement is simply this: If r classes of responses are initially defined and if the experimenter later decides to treat any two classes in identical manner, it should be possible to combine those two classes, thereby obtaining the same results that would have been obtained had only the $r-1$ classes been defined initially.   For example, if we are describing a bar-pressing experiment, we could define three classes of responses, pressing the bar from the left, pressing the bar from the right, and not pressing the bar.   Such a distinction between the two types of bar presses might be useful for some purposes.   But if the experimenter does not make such a distinction, we should be able to combine the first two classes into a single class of bar-pressing and thereby obtain predictions equivalent to those which would have been obtained by defining only two classes in the first place.   Formally, we may think of this combining of two classes as the collapsing of an r-dimensional vector space into an $(r-1)$-dimensional vector space.

It should be further pointed out that we have implicitly assumed that our probabilities after an event were independent of the *distributions* of probabilities within classes before an event.   For example, when we have two classes of responses and two operators Q_1 and Q_2, we have probabilities p and q before and probabilities $Q_1 p$ and $\tilde{Q}_1 q$ or $Q_2 p$ and $\tilde{Q}_2 q$ afterwards.   In the various equations for $Q_i p$, nothing was said about how the probability p was distributed over various possible subclasses of response class 1.

Let us first examine the implications of this restriction for $r = 3$.   We start with three response classes and a vector

$$\mathbf{p} = \begin{bmatrix} p_1 \\ p_2 \\ p_3 \end{bmatrix} .$$

We wish to combine classes 1 and 2 to form a new class, which we label c, and to represent the vector in the collapsed space by

$$\mathbf{Cp} = \begin{bmatrix} p_c \\ 0 \\ p_3 \end{bmatrix} ,$$

requiring that

$$p_c = p_1 + p_2.$$

We denote the collapsed vector by $\mathbf{Cp}$, and this may suggest to the reader that $\mathbf{Cp}$ can be obtained by operating on $\mathbf{p}$ with a matrix $\mathbf{C}$. This is indeed true, and $\mathbf{C}$ is given by[*]

$$\mathbf{C} = \begin{bmatrix} 1 & 1 & 0 \\ 0 & 0 & 0 \\ 0 & 0 & 1 \end{bmatrix}.$$

Now if the ith event occurs, the column vector becomes, after dropping the subscripts i for simplicity,

$$\mathbf{Tp} = \begin{bmatrix} u_{11}p_1 + u_{12}p_2 + u_{13}p_3 \\ u_{21}p_1 + u_{22}p_2 + u_{23}p_3 \\ u_{31}p_1 + u_{32}p_2 + u_{33}p_3 \end{bmatrix}.$$

After collapsing the vector space by applying $\mathbf{C}$ to $\mathbf{Tp}$ we have

$$\mathbf{C(Tp)} = \begin{bmatrix} (u_{11} + u_{21})p_1 + (u_{12} + u_{22})p_2 + (u_{13} + u_{23})p_3 \\ 0 \\ u_{31}p_1 + u_{32}p_2 + u_{33}p_3 \end{bmatrix}.$$

Since the elements of this vector are to be independent of the distributions of probability within class c, the components of the vector of the last equation must not depend on p_1 and p_2 individually but only upon their sum, p_c. Hence we demand that

$$u_{31} = u_{32} = u_3.$$

This restriction assures that $u_{31}p_1 + u_{32}p_2 = u_3(p_1 + p_2) = u_3 p_c$. If we combine classes 1 and 3 (instead of 1 and 2), we obtain in a similar manner

$$u_{21} = u_{23} = u_2.$$

Or, if we combine classes 2 and 3,

$$u_{12} = u_{13} = u_1.$$

These three relations reduce our original matrix operator to the form

$$\mathbf{T} = \begin{bmatrix} u_{11} & u_1 & u_1 \\ u_2 & u_{22} & u_2 \\ u_3 & u_3 & u_{33} \end{bmatrix}.$$

---

[*] We are indebted to Gerald L. Thompson for suggesting the use of the operator $\mathbf{C}$. For a more extensive discussion see [10].

We have yet to impose the condition that the columns should separately sum to unity, which requires that

$$u_{11} + u_2 + u_3 = 1$$
$$u_1 + u_{22} + u_3 = 1$$
$$u_1 + u_2 + u_{33} = 1.$$

These relations then allow us to write

$$(1.30) \qquad \mathbf{T} = \begin{bmatrix} 1 - u_2 - u_3 & u_1 & u_1 \\ u_2 & 1 - u_1 - u_3 & u_2 \\ u_3 & u_3 & 1 - u_1 - u_2 \end{bmatrix}.$$

We may note that this matrix can be written as

$$\mathbf{T} = (1 - u_1 - u_2 - u_3) \begin{bmatrix} 1 & 0 & 0 \\ 0 & 1 & 0 \\ 0 & 0 & 1 \end{bmatrix} + \begin{bmatrix} u_1 & u_1 & u_1 \\ u_2 & u_2 & u_2 \\ u_3 & u_3 & u_3 \end{bmatrix}.$$

If we then let $\alpha = 1 - u_1 - u_2 - u_3$ and $u_j = (1 - \alpha)\lambda_j$ for $j = 1, 2, 3$, we obtain

$$\mathbf{T} = \alpha \begin{bmatrix} 1 & 0 & 0 \\ 0 & 1 & 0 \\ 0 & 0 & 1 \end{bmatrix} + \begin{bmatrix} (1 - \alpha)\lambda_1 & (1 - \alpha)\lambda_1 & (1 - \alpha)\lambda_1 \\ (1 - \alpha)\lambda_2 & (1 - \alpha)\lambda_2 & (1 - \alpha)\lambda_2 \\ (1 - \alpha)\lambda_3 & (1 - \alpha)\lambda_3 & (1 - \alpha)\lambda_3 \end{bmatrix}.$$

By Rule III, we then factor out $(1 - \alpha)$ of the last matrix and have left a matrix $\mathbf{\Lambda}$ defined by

$$(1.31) \qquad \mathbf{\Lambda} = \begin{bmatrix} \lambda_1 & \lambda_1 & \lambda_1 \\ \lambda_2 & \lambda_2 & \lambda_2 \\ \lambda_3 & \lambda_3 & \lambda_3 \end{bmatrix}.$$

We can then write $\mathbf{T}$ in the form

$$(1.32) \qquad \mathbf{T} = \alpha \mathbf{I} + (1 - \alpha)\mathbf{\Lambda}.$$

The form of $\mathbf{T}$ in this equation is a direct consequence of the combining classes restriction. When the vector $\mathbf{p}$ is multiplied by the matrix $\mathbf{\Lambda}$ we obtain a vector $\mathbf{\lambda}$ defined by

$$(1.33) \qquad \mathbf{\Lambda p} = \begin{bmatrix} \lambda_1 \\ \lambda_2 \\ \lambda_3 \end{bmatrix} = \mathbf{\lambda}.$$

Finally when we apply **T** to **p** we obtain

$$\mathbf{Tp} = \alpha\mathbf{Ip} + (1 - \alpha)\mathbf{\Lambda p}$$

(1.34)
$$= \alpha\mathbf{p} + (1 - \alpha)\mathbf{\lambda}.$$

The foregoing arguments can be directly applied when there are more than three response classes and will lead to the following general operator (cf. equation 1.32):

(1.35) $$\mathbf{T} = \alpha\mathbf{I} + (1 - \alpha)\mathbf{\Lambda},$$

where

(1.36) $$\alpha = 1 - \sum_{j=1}^{r} u_j,$$

I is the $r \times r$ identity matrix, and **Λ** is an $r \times r$ matrix given by

(1.37)
$$\mathbf{\Lambda} = \begin{bmatrix} \lambda_1 & \lambda_1 & \cdots & \lambda_1 \\ \lambda_2 & \lambda_2 & \cdots & \lambda_2 \\ \cdot & \cdot & & \cdot \\ \cdot & \cdot & & \cdot \\ \cdot & \cdot & & \cdot \\ \lambda_r & \lambda_r & \cdots & \lambda_r \end{bmatrix},$$

where the elements are defined by

(1.38) $$\lambda_j = u_j/(1 - \alpha).$$

For the jth row of the vector **Tp** we obtain the

GENERAL ELEMENT OF THE VECTOR **Tp**:

(1.39) $$p_j' = \alpha p_j + (1 - \alpha)\lambda_j.$$

Hence, we see that the new value of probability for the jth class is a linear function of p_j and not of p_k for $k \neq j$. This represents a major simplification in the form of our operators. Since the coefficient of p_j is a constant α, we can see at once that, if we combine two classes, say the jth and the kth, so that

$$p_c = p_j + p_k$$

we obtain, after applying **T**, a linear function of p_c alone for the new probability of the combined class. This, of course, is what our rule for combining classes demanded.

In the preceding development we first introduced linear operators, represented by matrices, and then imposed two conditions: (1) First we required that the operators be stochastic, which simply means that they always change probability vectors into probability vectors; this required that each column of the matrix sum to unity. (2) Then we introduced the combining classes restriction and showed that this leads to operators of the form given by equation 1.35. A considerably more general development can be made, as was first pointed out to us by L. J. Savage. It can be shown that the combining classes restriction implies that *any* stochastic operator must have the form defined by equation 1.35. Linearity need not be assumed but follows as a consequence of the combining classes condition. We refer the reader to the literature for a more precise statement of the theorem and for its proof [10].

In the preceding section we derived restrictions on our several sets of parameters for two response classes. Similar restrictions are readily obtained for *r* alternatives. First we note that $\boldsymbol{\lambda}$ is a probability vector and so its components λ_j must satisfy

$$(1.40) \qquad 0 \le \lambda_j \le 1, \quad j = 1, 2, \cdots, r.$$

The restrictions on α may be obtained from the requirement

$$(1.41) \qquad 0 \le p_j' \le 1 \quad \text{for all} \quad 0 \le p_j \le 1,$$

where p_j' is the general element of the vector **Tp**, shown by equation 1.39 above. When $p_j = 1$ we see that we must have

$$0 \le \alpha + (1 - \alpha)\lambda_j \le 1,$$

and from these inequalities we can obtain the restriction

$$\frac{-\lambda_j}{1 - \lambda_j} \le \alpha \le 1.$$

But this must hold for each value of *j* and so we must require that

$$(1.42) \qquad \operatorname*{Max}_j \left\{ \frac{-\lambda_j}{1 - \lambda_j} \right\} \le \alpha \le 1.$$

A similar condition may be obtained by letting $p_j = 0$, but the restriction so obtained will always be satisfied if inequality 1.42 is met. The smallest permissible value of α obtains when $\lambda_j = 1/r$ for $j = 1, 2, \cdots, r$ and is easily shown to be

$$(1.43) \qquad \frac{-1}{r - 1} \le \alpha.$$

This condition is necessary but ordinarily not sufficient.

As in Section 1.7 we could impose the additional condition that α is non-negative to prevent oscillation. However, L. J. Savage pointed out to us that this condition may be derived from a minor extension of the combining classes argument. If we believe in the principle of combining classes, we should be willing to break up classes of responses into arbitrarily small subclasses. And in the breaking-up of classes we would certainly want α to be invariant. But in the limit when r, the number of classes, becomes arbitrarily large, condition 1.43 leads at once to $0 \leq \alpha$. If we then conceive of combining classes we obtain the condition that α be non-negative for any finite r.

1.9 SUMMARY

In this chapter we provide the framework for the analysis contained in the remainder of this book. The basic elements of the mathematical system are a set of mutually exclusive and exhaustive alternatives A_1, A_2, $\cdots$, A_r, a vector of probabilities $p_1, p_2, \cdots, p_r$ with one component for each alternative, a set of mutually exclusive and exhaustive events $E_1, E_2, \cdots, E_t$, and a set of operators $T_1, T_2, \cdots, T_t$ corresponding to those events. The probability p_j ($j = 1, 2, \cdots, r$) represents the probability of occurrence of alternative A_j on a trial, and a trial is defined as an opportunity for choice among the r alternatives. When event E_i occurs, operator T_i is applied to the set of probabilities to yield a new set of probabilities.

The operators T_i are assumed to be linear and so are represented by matrices. For the probability invariance rule to hold, the elements of each matrix must be non-negative and each column must sum to unity. (Such matrices may be called stochastic matrices.) For more than two alternatives ($r > 2$) an additional condition on how classes can be combined is introduced in Section 1.8. Restrictions on the parameters are discussed in Section 1.7 for $r = 2$ and in Section 1.8 for $r > 2$.

In Section 1.6, a pair of *row* operators Q_i and $\tilde{Q}_i$ are introduced in order to dispense with the matrix machinery when there are only two alternatives ($r = 2$). These row operators are used throughout this book, and so the three forms of writing the operations are repeated here:

SLOPE-INTERCEPT FORM: $Q_i p = a_i + \alpha_i p$,

GAIN-LOSS FORM: $Q_i p = p + a_i(1 - p) - b_i p$,

FIXED-POINT FORM: $Q_i p = \alpha_i p + (1 - \alpha_i)\lambda_i$.

The row operator $\tilde{Q}_i$ is applied only to $q = 1 - p$, and by definition,

$$\tilde{Q}_i q = 1 - Q_i p.$$

The slope-intercept form and the fixed-point form are used mainly in Chapters 3 and 4 to simplify algebraic manipulations. The gain-loss form is used mostly in Chapter 2, where we discuss an interpretation of the operators in terms of stimulus sampling and conditioning. The fixed-point form is used in all applications in Part II.

REFERENCES

1. Bush, R. R., and Mosteller, F. A mathematical model for simple learning. *Psychol. Rev.*, 1951, **58**, 313–323.
2. Bush, R. R., and Mosteller, F. A stochastic model with applications to learning. *Annals of math. Stat.*, 1953, **24**, 559–585.
3. Uspensky, J. V. *Introduction to mathematical probability.* New York: McGraw-Hill, 1937, pp. 1–13.
4. Feller, W. *An introduction to probability theory and its applications.* New York: Wiley, 1950, pp. 1–22.
5. Davis, H. T. *The theory of linear operators.* Bloomington, Ind.: Principia Press, 1936, pp. 1–15.
6. Thurstone, L. L. *Multiple-factor analysis.* Chicago: University of Chicago Press, 1947, pp. 1–50.
7. Hull, C. L. *Principles of behavior.* New York: Appleton-Century-Crofts, 1943, p. 114.
8. Burros, R. H. Some criticisms of "A mathematical model for simple learning." *Psychol. Rev.*, 1952, **59**, 234–236.
9. Burros, R. H. The linear operator of Bush and Mosteller. *Psychol. Rev.*, 1953, **60**, 213–214.
10. Bush, R. R., Mosteller, F., and Thompson, G. L. A formal structure for multiple choice situations. *Decision Processes* (edited by R. M. Thrall, C. H. Coombs, and R. L. Davis), New York: Wiley, 1954, 99–126.

CHAPTER 2

Stimulus Sampling and Conditioning

2.1 A SET-THEORETIC APPROACH

The preceding chapter develops operators to describe the changes in the probabilities of various responses. The form of these operators is dictated chiefly by mathematical considerations—linear operators are chosen in order to make the theory more manageable. It is seen in later chapters that, even with this simplification, the theory becomes rather complicated. The only restrictions on these linear operators are those to assure that probability is conserved and to allow classes of responses to be combined in a sensible way. We omitted any explicit discussion of the stimulus aspects of the learning problem. From the point of view of psychological theory this omission may represent a serious gap. In this chapter we offer one way to narrow that gap by indicating how a stimulus model can generate the operators already presented. This model utilizes some elementary notions of mathematical set theory and is a modification of the theory developed by Estes [1]. We wish to emphasize, however, that a stimulus model is not necessary to the operator approach of Chapter 1. Rather, it is an alternative way to derive linear operators, a way we often found helpful in thinking about specific experimental problems.

The organismic responses discussed in the last chapter occur, of course, in the presence of or following various kinds of stimulation impinging on the organism. Stimulus-response psychology postulates that stimuli and responses become "connected," "associated," or "conditioned" during a learning process. Therefore it would be instructive to construct a formal model of this conditioning process—the "connecting-up" of stimuli and responses—and deduce the formulas of Chapter 1 which specify trial-by-trial changes in the probabilities of occurrence of the various possible responses.

The total environment of an organism in an experimental situation provides stimulation, but, clearly, certain parts of that environment provide more stimulation than other parts. Furthermore, some aspects of the situation may tend to evoke one response and other aspects may elicit a different response.   Therefore, psychologists have found it useful to talk about various parts of a stimulus situation: individual stimuli, stimulus elements, stimulus complexes, etc.   Certain aspects of the stimulating environment may be under experimental control whereas others may be constant or allowed to vary at random.   In any case, it is useful to break up the total stimulation into parts.   Gestalt psychologists, however, have long argued that we must be cautious in describing separate parts of the environment; patterns and relations are important in influencing behavior.   With considerations such as these in mind we shall attempt to represent the total environment of an organism by a set of elements.   These elements may be defined physically or physiologically or behaviorally, as parts or wholes, as things or relations, as stimuli or Gestalten, depending upon the problem we face.   In the following discussion we shall simply call these elements "stimuli" without further specification.   It happens that this lack of specificity gives rise to no difficulties because our final results involve neither the properties of stimulus elements nor numbers of such elements.

It has already been suggested that different parts of an organism's environment have more importance than other parts in influencing behavior.   What stimuli have how much control over a specified kind of behavior is a complex experimental problem.   Indeed it would be fair to say that this question has been and will continue to be a central problem in psychology.   To discover the stimulus units that control behavior even in the most specific situations requires technical skill, extensive empirical observations, and good intuition.   Although no mathematical model can be expected to answer these numerous quantitative questions, it is possible to describe the skeleton of the problem in mathematical terms. We would like to assign a weight or *measure* to each of the elements in the set of all stimuli.   For example, if we were to assign a measure of 2 to stimulus element s_1 and measure 4 to element s_2, we would then say that s_2 was twice as important to the organism as s_1.   (The same statement would apply if we assigned a measure of 10 to s_1 and a measure of 20 to s_2.)

Mathematicians have developed a good deal of machinery under the title of "measure theory."   To these mathematicians, measure is something applied to—a number associated with—a set.   It is a function defined on a set, satisfying the property of additivity.   This simply means that the measure of two distinct sets is the measure of one set plus the measure of the other.   In the stimulus model to be described, we shall associate with any set S of stimulus elements a measure function denoted by $\mathcal{M}(S)$.

In the preceding discussion, it was implicitly assumed that the measure

$\mathscr{M}(S)$ of a set S of stimuli was a fixed property of that S and independent of the possible responses to which S might be conditioned. The stimulus-response connections are yet to be specified in the model. It is therefore assumed that a stimulus element can exist in one and only one of a number of states. Corresponding to each class of responses, $A_1, A_2, \cdots$, there is a unique state, and so we speak of an element as "conditioned to response class A_j"; this is equivalent to saying that the stimulus element is in the state corresponding to response class A_j. Although a stimulus element can be in only one state at any instant of time, a set of elements may be partially conditioned to several response classes. (Some readers may feel that there would be stimulus elements which could not be conditioned to any of the response classes defined in a particular experimental situation. If this were so, it could be argued that such stimuli had zero measure—they would have no influence on the behavior being studied.)

In the next section we discuss some of the basic concepts of mathematical set theory, and in later sections we continue the development of the stimulus model. Our goal is to derive the linear operators postulated in Chapter 1; once we have done this we shall return to our original approach and make no further reference to the stimulus model, except for brief discussions in Sections 5.2 and 11.7. In the psychological literature, however, the reader will find other applications of similar stimulus sampling models. The authors used such a model for discussing stimulus generalization and discrimination [2], and Estes and Burke have presented a more general stimulus model to handle problems in stimulus variability [3]. In addition, Estes has applied his original model to provide a description of spontaneous recovery [4].

2.2 SUBSETS AND THEIR COMBINATIONS

In this section we present some of the elementary notions and operations of mathematical set theory (sometimes called the algebra of classes). In Fig. 2.1 we show a set S containing a number of elements represented by dots. Within S we show three subsets, U, V, and W. These are called subsets of S because all the elements of U, for example, are contained in S, but not necessarily all elements of S are contained in U. When two sets or subsets have no elements in common, such as U and W in Fig. 2.1, those subsets are said to be disjunct.

Sets may be combined in various ways, and the operations involved correspond to such operations as addition and multiplication in simple algebra. The set sum of two sets is defined as the set of all elements contained in one set or the other or both. For example, the set sum of the subsets U and V in Fig. 2.1 is the set composed of all elements belonging to either U or V or both. The set sum of U and V is usually denoted by

$U \cup V$ and is also called the *join* or *union* of U and V. The symbol $\cup$ is often called *cup*.

The set product of two sets U and V is defined as the set of elements contained in both sets and is written $U \cap V$. This set product is also called the *meet* or *intersection* of U and V; the symbol $\cap$ is read *cap*. If two sets are disjunct, such as U and W in Fig. 2.1, their set product is

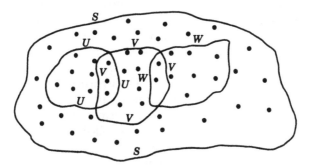

Fig. 2.1. A set S of elements, represented by dots, with three subsets, U, V, and W.

an empty set, that is, a set containing no elements, denoted by 0. So we have $U \cap W = 0$.

The union of a set U with itself is of course the set U, and the intersection of a set U with itself is also the set U. In symbols

$$U \cup U = U$$

$$U \cap U = U.$$

The complement in S of a set U is defined as the set of all elements in S but not in U. This complement is denoted by U'. From this definition we can immediately infer that

$$U \cap U' = 0$$

and

$$U \cup U' = S.$$

The first equation simply asserts that the meet (set product) of U and its complement U' is the empty set 0, that is, U and U' are disjunct. The second equation says that the union (set sum) of U and its complement in S is equal to the whole set S.

In the various set diagrams presented in this chapter, we find it convenient to represent subsets by areas, shaded in various ways, rather than by collections of dots. The position of the various dots in Fig. 2.1 was unimportant for our purposes and similarly the relative positions of the areas in other diagrams is meant to imply nothing about the positions of stimuli in an actual stimulus situation.

2.3 PROBABILITY AND STIMULUS SAMPLING

When there are just two response classes in the model, there are usually some stimulus elements conditioned to each class. Let A_1 represent the class of responses recorded by an experimenter and let the subset of stimuli conditioned to that class be denoted by C as shown in Fig. 2.2.

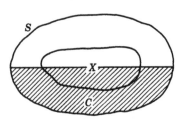

All other responses are represented by A_2 and all stimuli in S but not in C are conditioned to A_2. The entire set is denoted by S. On a particular trial or at a particular time, an organism perceives a subset, X, of the total stimuli available. This subset X is called a sample from S, and it may partly intersect the subset C conditioned to A_1, and may partly intersect the complement of C in S, that is, the subset of elements conditioned to A_2

Fig. 2.2. A stimulus set S with the subset C, which is conditioned to response A_1, and a sample X.

(unshaded in Fig. 2.2).

It is now postulated that the probability of response A_1 is

(2.1)
$$p = \frac{\mathscr{M}(X \cap C)}{\mathscr{M}(X)},$$

where $\mathscr{M}(X \cap C)$ is the measure of the intersection of X and C and $\mathscr{M}(X)$ is the measure of the sample X. Equation 2.1 asserts that the probability of response A_1 is equal to the ratio of the measure of the elements conditioned to A_1 in the sample to the measure of the entire sample. Intuitively, the probability is the relative importance of the elements conditioned to A_1 in the sample, compared with the total importance of all the elements in the sample. The value of p obtained here can be used to determine the response by analogy with a randomizing device such as the spinning disc described in Section 1.2.

We next introduce an *assumption of homogeneity*; it is related to the notion of random sampling. The assumption requires that the fraction of measure of elements conditioned to A_1 in any sample from the set S is equal to the fraction of measure of elements conditioned to A_1 in S. Hence, we have the

ASSUMPTION OF HOMOGENEITY:

(2.2)
$$p = \frac{\mathscr{M}(X \cap C)}{\mathscr{M}(X)} = \frac{\mathscr{M}(C)}{\mathscr{M}(S)}.$$

A fluid model gives us an easy interpretation of this assumption. Suppose that the total situation is represented by a vessel containing two fluids which do not chemically interact but are completely miscible.   For discussion let the substances be water and alcohol.   The weight of the water corresponds to the measure of the subset conditioned to A_2, and the weight of the alcohol corresponds to the measure of the subset C conditioned to A_1.   The subset X corresponds to a jigger full of the mixture, and of course, if the fluids are well mixed, the fractional weight

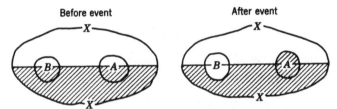

Before event   After event

Fig. 2.3.   A sample X of stimuli from the entire set before and after the event. A and B are two subsamples of X. The shaded portion represents the elements conditioned to A_1.

of alcohol in a jigger full is much the same as that in the whole highball. Following the same sort of assumption, a standard method of determining the hemoglobin content of the blood proceeds by assuming that the mixture is homogeneous.   Twenty cubic millimeters of blood are withdrawn from the patient, and the concentration of hemoglobin in the sample is determined.   It is assumed that the concentration obtained is the concentration of hemoglobin in all the patient's blood as it passes by the point of withdrawal.

2.4   DEDUCTION OF THE OPERATORS

Changes in probabilities are described by changes in the measure of elements conditioned to A_1; we next postulate a procedure for describing these changes.   It is assumed that a subset A of the sample X becomes completely conditioned to A_1, whereas another subset B of X becomes completely conditioned to A_2 responses, after an event has occurred. This situation is illustrated in Fig. 2.3.   Since an element can be conditioned to only one class of responses, A and B are disjunct, that is, have no elements in common.   Furthermore, the measures of A and B depend upon what event has just occurred.   After this reconditioning has taken place the sample is returned to the population S.   In terms of the fluid model discussed in the preceding section, the procedure corresponds to this: Two jiggers, A and B, are removed from the highball mixture; the jiggers need not be of the same size.   Jigger A is emptied, and an equivalent weight of pure alcohol is returned to the highball; for jigger B an equivalent weight of water is returned.

Initially the measure of C is $\mathcal{M}(C)$. An event will change this measure an amount $\Delta\mathcal{M}(C)$. The part of A which was not in C has been added to the part of S conditioned to A_1; its measure is $\mathcal{M}(A) - \mathcal{M}(A \cap C)$. The part of B which was in C has been removed from the part of S conditioned to A_1, and its measure is $\mathcal{M}(B \cap C)$. Hence we have

$$\Delta\mathcal{M}(C) = \mathcal{M}(A) - \mathcal{M}(A \cap C) - \mathcal{M}(B \cap C)$$

(2.3)
$$= \mathcal{M}(A)\left[1 - \frac{\mathcal{M}(A \cap C)}{\mathcal{M}(A)}\right] - \mathcal{M}(B)\frac{\mathcal{M}(B \cap C)}{\mathcal{M}(B)}$$

provided that neither denominator vanishes. We now extend our assumption of homogeneity so that

(2.4)
$$\frac{\mathcal{M}(A \cap C)}{\mathcal{M}(A)} = \frac{\mathcal{M}(B \cap C)}{\mathcal{M}(B)} = \frac{\mathcal{M}(C)}{\mathcal{M}(S)} = p.$$

Hence, equation 2.3, representing the change in the measure of C, becomes

(2.5)
$$\Delta\mathcal{M}(C) = \mathcal{M}(A)(1 - p) - \mathcal{M}(B)p.$$

We then define

(2.6)
$$a = \frac{\mathcal{M}(A)}{\mathcal{M}(S)}, \qquad b = \frac{\mathcal{M}(B)}{\mathcal{M}(S)},$$

so that after dividing equation 2.5 through by $\mathcal{M}(S)$ we write

(2.7)
$$\Delta p = \frac{\Delta\mathcal{M}(C)}{\mathcal{M}(S)} = a(1 - p) - bp.$$

The new value of probability is then

(2.8)
$$Qp = p + \Delta p = p + a(1 - p) - bp.$$

This equation is the gain-loss form given by equation 1.19 of the last chapter and so we have deduced from the stimulus model the operators in Chapter 1. The definitions (2.6) of the parameters a and b in terms of sets of stimulus elements may provide readers with an additional interpretation of the meaning of those parameters. As we saw in equation 2.5, the increment $\Delta\mathcal{M}(C)$ in the measure of elements conditioned to A_1 had two parts. The first part, $\mathcal{M}(A)(1 - p)$, represents a gain in the measure of elements conditioned to A_1 and is parallel to the gain term, $a(1 - p)$, in the gain-loss form of Qp. Similarly the part, $\mathcal{M}(B)p$, represents a loss in measure and is parallel to the loss term bp.

*2.5   EXTENSION TO r RESPONSE CLASSES; HOMOGENEITY AND COMBINING CLASSES

The foregoing theory of stimulus sampling and conditioning was developed in terms of two classes of responses, but the extension to r classes is straightforward. Again we shall exhibit correspondence between the operators of Chapter 1 and the set-theoretic model. At some point in the process, let S have r subsets, C_j, $(j = 1, 2, \cdots, r)$ conditioned to the r response classes. A sample X is drawn, and the probability of response A_j is

$$(2.9) \qquad p_j = \frac{\mathscr{M}(X \cap C_j)}{\mathscr{M}(X)} = \frac{\mathscr{M}(C_j)}{\mathscr{M}(S)} \qquad (j = 1, 2, \cdots, r).$$

We then let X contain r disjunct subsamples, U_j, which become conditioned to the response classes indicated by their subscripts when an event occurs. An $(r + 1)$st class in X consists of the residual elements not contained in the U_j; these residual elements are left unchanged in the reconditioning process. If we are describing the changes in the measure of elements conditioned to the jth class, the measure of the elements in U_j and not in C_j are added to the measure of C_j, that is, there is an increment, $\mathscr{M}(U_j) - \mathscr{M}(U_j \cap C_j)$. Moreover, the measure of elements in any other subset U_k which are also in C_j will be subtracted. This leads to decrements, $\mathscr{M}(U_k \cap C_j)$, from all subsets except U_j, that is, for all $k \neq j$. The change in the measure of C_j is then

$$\Delta\mathscr{M}(C_j) = \mathscr{M}(U_j) - \mathscr{M}(U_j \cap C_j) - \sum_{k \neq j} \mathscr{M}(U_k \cap C_j)$$

(2.10)

$$= \mathscr{M}(U_j) - \sum_{k=1}^{r} \mathscr{M}(U_k \cap C_j).$$

Our assumption of homogeneity becomes

$$(2.11) \qquad \frac{\mathscr{M}(U_k \cap C_j)}{\mathscr{M}(U_k)} = \frac{\mathscr{M}(C_j)}{\mathscr{M}(S)} = p_j \qquad (k = 1, 2, \cdots, r).$$

We then define

$$(2.12) \qquad u_k = \frac{\mathscr{M}(U_k)}{\mathscr{M}(S)} \qquad (k = 1, 2, \cdots, r).$$

Combining the last two equations then leads to

$$(2.13) \qquad \mathscr{M}(U_k \cap C_j) = p_j \mathscr{M}(U_k) = u_k p_j \mathscr{M}(S).$$

Inserting these relations in equation 2.10 expressing the change in the measure of C_j, and then dividing through by $\mathscr{M}(S)$, we obtain

$$(2.14) \qquad \Delta p_j = u_j - \sum_{k=1}^{r} u_k p_j.$$

The new probability of the jth class is

$$(2.15) \qquad p_j + \Delta p_j = (1 - \sum_{k=1}^{r} u_k) p_j + u_j = \alpha p_j + u_j$$

$$= \alpha p_j + (1 - \alpha)\lambda_j,$$

where as before we have $\alpha = 1 - \sum_{k=1}^{r} u_k$ and $u_j = (1 - \alpha)\lambda_j$.

This result is the general element of the vector **Tp** given in Chapter 1, expression 1.39. It is worth noting that expression 1.39 was developed as a consequence of our rule for combining classes; without that restriction, it would have been much more complicated. Therefore we see that our rule for combining classes automatically falls out of this generalization of Estes' stimulus theory.

2.6 SUMMARY

In this chapter we have deduced the event operators, first introduced in Chapter 1, from a simple model of stimulus sampling and conditioning. This model is described in terms of mathematical set theory. It is assumed that any stimulus situation can be represented by a set of abstract elements and that elements exist in one of r states, each state corresponding to a class of responses. In these terms, a conditioning process involves changing the states of elements in the set. The probability of a response class is defined in terms of the states of elements in a sample drawn on a trial. After an assumption of homogeneity in the sampling process is introduced, the operators of Chapter 1 are deduced from the model. In Section 2.5 it is shown that the operators deduced from the set-theoretic model for r response classes agree with those which result from the combining classes restriction given in Section 1.8.

REFERENCES

1. Estes, W. K.   Toward a statistical theory of learning. *Psychol. Rev.*, 1950, **57**, 94–107.
2. Bush, R. R., and Mosteller, F.   A model for stimulus generalization and discrimination. *Psychol. Rev.*, 1951, **58**, 413–423.
3. Estes, W. K., and Burke, C. J.   A theory of stimulus variability in learning. *Psychol. Rev.*, 1953, **60**, 276–286.
4. Estes, W. K., personal communication.

CHAPTER 3

Sequences of Events

3.1 INTRODUCTION

Let us review briefly the basic framework of the mathematical system. We have a set of *alternatives* A_j, which are to be identified with classes of responses, and a set of probability variables p_j corresponding to those alternatives. In the simplest case, which is of major interest in the following chapters, there are two alternatives A_1 and A_2 with probabilities of occurrence p and $1 - p$, respectively. As described in Chapter 1, these probabilities are to be changed from time to time by mathematical operators which correspond to occurrences of *events*. These events in the mathematical system are to be identified with various kinds of events in the real world which influence the behavior of an organism, events which produce learning. These events may be almost entirely at the disposal of an experimenter; for example, they may correspond to rewards and punishments. On the other hand, events which alter behavior may be partially controlled by the organism whose behavior is being studied. An occurrence of a response and whatever stimulus changes it may produce may constitute an event, or perhaps a reward may be contingent upon the occurrence of a particular response.

In nearly all learning experiments, we are less interested in the effect of a single event than in the cumulative effect of a sequence of events. In some experiments, a single event occurs repeatedly on a series of trials. An example is the Graham-Gagné runway experiment [1] discussed in Chapter 14; a rat finds a reward at the end of the runway on each experimental trial. In other experiments, an event such as reward may occur periodically; food might follow every fourth emission of a particular response. Such a schedule of reinforcement has been called a *fixed ratio* schedule [2, 3]. In this example, we would consider a rewarded response one type of event and an unrewarded response another type of event. Thus *fixed ratio reinforcement* involves a systematic sequence of two types of events. In still other kinds of reinforcement schedules, called *random ratio*, reward occurs a fixed proportion of the time, but in a more or less random sequence of rewards and non-rewards [3]. Strict randomness is seldom used, but randomization within blocks of trials is common.

Because effects of events are represented by the application of operators and because the events occur repeatedly, we are concerned with sequences of operators that are applied to the probability variables. In this chapter we analyze various kinds of sequences of operators, sequences that correspond to sequences of events in common use by the experimental psychologist. Before analyzing these operator sequences in detail we first discuss the structure of some simple sequences of events to focus attention upon one important property: the degree of dependence of a given event upon earlier events in the sequence. There are three sequences of events described: a random sequence, a somewhat dependent sequence, and a completely systematic sequence. Corresponding to these three sequences there are operator sequences, and results for these various cases are obtained. Finally, it turns out that such operator sequences are related to three kinds of experiments: (1) where the experimenter controls the occurrence of events, (2) where the subject controls the events, and (3) where the experimenter and subject together control the events.

In connection with one of the event sequences described, we present an elementary discussion of the theory of Markov chains and how they are related to the general model. This discussion is closely related to an exposition by Miller [4]. (The reader may also wish to consult two papers by Miller and Frick [5, 6] for somewhat different analyses of the sequential properties of learning data.)

3.2 THE STRUCTURE OF SOME ELEMENTARY SEQUENCES

This section introduces some easy sequences that illustrate differing degrees of dependency. By dependency we mean the amount of information earlier members of a sequence provide about later members.

Suppose there are two types of events *A* and *B*, and that we record in temporal order a sequence of fifty-one events. Below are examples of three types of sequences that might be observed. If they are examined, certain important similarities and differences will be discovered. The sequences are to be read from left to right; they are given in groups of five for convenience in reading, so the separations have no meaning.

SEQUENCE 1

ABABB   ABABA   BAABB   AAAAA   AABBB
AAAAA   BAAAB   BABAA   AABAA   AAABA   A

SEQUENCE 2

AABAA ABAAA AABAA BAABA BABAB
ABABA ABAAA BAABA BAAAA BABAB A

SEQUENCE 3

AABAA BAABA ABAAB AABAA BAABA
ABAAB AABAA BAABA ABAAB AABAA B

Examination of these three sequences reveals that if we disregard the arrangement and merely count the number of type *B* events, each sequence has 17 (one-third) *B*'s.   In the first sequence the *B*'s occur in a haphazard order.   In the second sequence no *B*'s occur together.   In the third sequence, not only do no *B*'s occur together, but given two *A*'s in succession, a *B* is bound to follow; and just as inevitably a *B* will be followed by two *A*'s.   These three sequences illustrate three of the many possible degrees of dependence that sequences can have.

The first sequence was constructed by letting the probability of an *A* on each trial be 2/3.   Then a long sequence was put together with the aid of a random number table, and we lifted out the first consecutive sequence of length 51 that had 17 *B*'s.

The second sequence was constructed by starting with an *A* and letting the probability of a *B* following an *A* be 0.5, and the probability of a *B* following a *B* be zero.   Then such a sequence was constructed with random numbers, and the first consecutive sequence of length 51 containing 17 *B*'s was chosen for display.

The third sequence was arranged with a completely systematic pattern of two *A*'s followed by a *B*.   If we know the first two elements of the third sequence and the rule of formation we can state the entry in the 1175th position (to choose a position quite far from the starting point) without difficulty, and if we know the last two elements that have occurred we can always predict the next element.

Sequence 2 has in common with sequence 3 the property that if the last element was a *B* we know the next element is an *A*.   But the resemblance stops here.   The next element in sequence 2 has a 50–50 chance of being a *B*, and we are still certain of the next element in sequence 3.   And knowing only the first two elements of sequence 2 we would at best be able to say that the probability of an *A* on the 1175th trial is about 2/3.

Sequences 1 and 2 have these properties in common.   If they were extended infinitely, knowing the early elements helps us very little in predicting elements far from the start.   But knowing an element of sequence 2 we can make a differential prediction of the next element, depending on which element has just occurred; if a *B* has just occurred an *A* is next, if an *A* has just occurred *A* and *B* have an equal chance of occurring in the next position. But for long* sequences of the type given by sequence 1 above, knowledge of the previous value is of no real assistance—each trial is independent of the previous one, and all that can be said is that the probability of an A on the next trial is 2/3.

One way of analyzing sequences (sometimes called time series) arises from considering what length of run of values contributes to accurate prediction of the next entry. In our sequence 1, with a knowledge of the rule of formation, complete knowledge of the sequence up to the 1174th entry does not assist us in our prediction of the 1175th. In sequence 2, knowledge of the last entry was helpful, but knowledge of the entire previous sequence is no better than knowledge of the last entry. Sequence 3 illustrates a case where knowledge of the last two entries improves the prediction, but it differs from the others in that knowledge of any two adjacent previous entries, along with their entry number, gives complete information about any entry in the sequence. Actually knowledge of one B and its entry number is adequate, but two A's are needed. We can, of course, construct situations where the last three, the last four, etc., members of a sequence provide all the information there is for predicting the next entry. This is a common approach to time series analysis.

When previous entries provide no information about the next entry other than about the long-run proportions of A's and B's, we are said to have independent trials (example: sequence 1). When the last entry provides all the available information about the next entry we have what is commonly called a Markov chain [7] (example: sequence 2). In Section 3.11 we have more to say about Markov chains.

3.3 REPETITIVE APPLICATION OF A SINGLE OPERATOR Q_i

If we should have a sequence of identical events, we need to compute the effect of a repetitive application of a single operator. In this section we consider only two response classes, A_1 and A_2, with probabilities p and $1 - p$, respectively. The single event E_i is then represented by the row operator Q_i, which we write in the fixed point form of equation 1.20:

$$(3.1) \qquad Q_i p = \alpha_i p + (1 - \alpha_i)\lambda_i.$$

If the probability of response A_1 before event E_i occurs is p, then $Q_i p$ is the probability of that response after E_i has occurred. Hence, if there is

---

* We say "long" because in a finite sequence constructed to have a *fixed* proportion of A's and B's some information is available from knowledge of the values that have previously occurred. This is like sampling without replacement from a finite urn of known composition. We wish to sample with replacement and to avoid introducing this special point into the discussion.

another occurrence of E_i, the operator Q_i must be applied to the probability $Q_i p$. We accomplish this second application of Q_i by letting $(Q_i p)$ be the operand instead of p in equation 3.1. We have

$$(3.2) \qquad Q_i(Q_i p) = \alpha_i(Q_i p) + (1 - \alpha_i)\lambda_i.$$

We then replace $(Q_i p)$ on the right side of this equation with its equivalent given by equation 3.1 and have

$$(3.3) \qquad \begin{aligned} Q_i(Q_i p) &= \alpha_i[\alpha_i p + (1 - \alpha_i)\lambda_i] + (1 - \alpha_i)\lambda_i \\ &= \alpha_i^2 p + (1 - \alpha_i^2)\lambda_i. \end{aligned}$$

It is convenient to denote this double application of Q_i to p by $Q_i^2 p$. Now if event E_i occurs for a third time we need to compute $Q_i^3 p$:

$$(3.4) \qquad \begin{aligned} Q_i^3 p = Q_i(Q_i^2 p) &= \alpha_i(Q_i^2 p) + (1 - \alpha_i)\lambda_i \\ &= \alpha_i[\alpha_i^2 p + (1 - \alpha_i^2)\lambda_i] + (1 - \alpha_i)\lambda_i \\ &= \alpha_i^3 p + (1 - \alpha_i^3)\lambda_i. \end{aligned}$$

The forms of $Q_i p$, $Q_i^2 p$, and $Q_i^3 p$ suggest that the general form of the result for any number n of applications is

$$(3.5) \qquad Q_i^n p = \alpha_i^n p + (1 - \alpha_i^n)\lambda_i.$$

Assuming that this is the correct form for n applications we can show by mathematical induction* that it is correct for $n + 1$, because

$$(3.6) \qquad \begin{aligned} Q_i(Q_i^n p) &= \alpha_i(Q_i^n p) + (1 - \alpha_i)\lambda_i \\ &= \alpha_i[\alpha_i^n p + (1 - \alpha_i^n)\lambda_i] + (1 - \alpha_i)\lambda_i \\ &= \alpha_i^{n+1} p + (1 - \alpha_i^{n+1})\lambda_i. \end{aligned}$$

The fact that the right side of equation 3.6 is identical to the right side of equation 3.5 except that n has been replaced by $n + 1$ proves that conjecture 3.5 is correct.

---

* For discussion of mathematical induction see G. Birkhoff and S. MacLane [8]. We are using what they call the "principle of finite induction" which they give as follows (with notation slightly changed): "Let there be associated with each positive integer k a proposition $P(k)$ which is either true or false. If, firstly, $P(1)$ is true and secondly, for all n, $P(n)$ implies $P(n + 1)$, then $P(k)$ is true for all positive integers k." (This quotation is used with the permission of the Macmillan Company.)

Induction proofs are particularly good when we have a good guess about the correct answer and want to test its truth or falsity. In our problem we know $P(1)$ is true, because here $P(1)$ is the definition shown in equation 3.1. The $P(n)$ we have chosen is shown in equation 3.5. In equation 3.6 we have shown that we can go from an arbitrary number to the next higher by legitimate operations and still get the same formula; and this completes the proof.

When α_i is less than unity in absolute value, we see that, as n tends to infinity, then α_i^n tends to zero.   Hence the asymptote of $Q_i^n p$ is λ_i:

$$(3.7) \qquad \lim_{n \to \infty} Q_i^n p = \lambda_i \qquad (-1 < \alpha_i < 1).$$

(Read: the limiting value of $Q_i^n p$ as n tends to infinity is λ_i.)

It will be recalled that when λ_i was first introduced in Section 1.6 it was shown that $Q_i \lambda_i = \lambda_i$, and so λ_i was called the *fixed point* of the operator Q_i. We have just shown that λ_i is also the *limit point* of the operator Q_i; that is, λ_i is the point which $Q_i^n p$ approaches as n gets large for any value of p.

The results of this section say that, if the same event occurs on every trial, it is possible to compute the response probability on any trial and that this probability stabilizes at some fixed value.   In Fig. 3.1 we give an example of the values of $Q_i^n p$ plotted against the trial number n.

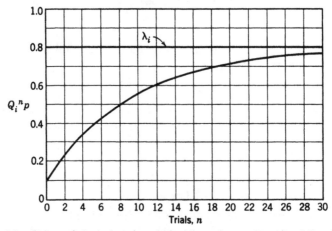

Fig. 3.1.   Values of $Q_i^n p$ plotted against trial number n.   Equation 3.5 and the values $p = 0.1$, $\lambda_i = 0.8$, and $\alpha_i = 0.9$ were used in plotting the curve.

Although the derivation in this section was in terms of response probabilities and event operators, this was an unnecessary restriction.   Elsewhere we shall have occasion to use these mathematical results in other connections, and so we now indicate the more general result.   If we have a recursive formula of the form

$$(3.8) \qquad x_{n+1} = \beta x_n + (1 - \beta)\mu,$$

the solution is

$$(3.9) \qquad x_n = \beta^n x_0 + (1 - \beta^n)\mu.$$

Equation 3.8 is analogous to equation 3.1 above, and solution 3.9 is analogous to solution 3.5. When the magnitude of β is less than unity, x_n tends to μ as n approaches infinity.

*3.4 ALTERNATIVE APPROACHES

In understanding the operators it may help some readers to realize that in finding the general form of $Q_i{}^n p$ we are actually solving a difference equation of the form

$$(3.10) \qquad Q_i{}^{n+1}p - \alpha_i Q_i{}^n p = (1 - \alpha_i)\lambda_i.$$

Equation 3.10 is linear with constant coefficients, and represents quite a simple case. Such equations are handled in full generality by C. Jordan [9], whose book the reader may consult for more details. Because we have solved this equation by induction, the difference equation derivation is not given here.

Another approach to this problem is through a differential equation approximation. If we think of $y = Q_i{}^n p$, then $y + \Delta y = Q_i{}^{n+1}p$, where the independent variable is n. When we go from n to $n + 1$ the increment in n is Δn (which happens to be unity). Now our recursion relation can be written

$$(3.11) \qquad Q_i{}^{n+1}p = \alpha_i Q_i{}^n p + (1 - \alpha_i)\lambda_i,$$

and, subtracting $Q_i{}^n p$ from both sides, we have

$$(3.12) \qquad Q_i{}^{n+1}p - Q_i{}^n p = (1 - \alpha_i)(\lambda_i - Q_i{}^n p),$$

or

$$(3.13) \qquad \frac{(y + \Delta y) - y}{\Delta n} = \frac{\Delta y}{\Delta n} = (1 - \alpha_i)(\lambda_i - y).$$

If we think of n as continuous, we may approximate this last equation by replacing the ratio $\dfrac{\Delta y}{\Delta n}$ by the derivative $\dfrac{dy}{dn}$. In this way we obtain the differential equation

$$(3.14) \qquad \frac{dy}{dn} = (1 - \alpha_i)(\lambda_i - y),$$

which has the solution

$$(3.15) \qquad y = e^{-(1-\alpha_i)n}p + (1 - e^{-(1-\alpha_i)n})\lambda_i.$$

This equation gives an approximation to $Q_i{}^n p$ and has the same form as the exact formula, equation 3.5. Where we formerly had $\alpha_i{}^n$ we now have the exponential. When $(1 - \alpha_i)$ is small we can take α_i as an approximation to the exponential $e^{-(1-\alpha_i)}$. Hence we conclude that the solution of the differential equation is a reasonable approximation to the exact formula for $Q_i^n p$ when $(1 - \alpha_i)$ is small compared to unity. The approximate formula does yield the correct values when $n = 0$ and when n tends to infinity.

We have included the foregoing discussion of the differential equation approach mainly because it has been a commonly used device in psychological literature [see, for example, 1, 10, 11]. However, in linear problems in this book we have found it better and easier to solve the difference equation directly.

*3.5   REPETITIVE APPLICATION OF MATRIX OPERATOR T

This section generalizes the results of the last section to any number r of alternatives when the combining classes restriction of Section 1.8 is imposed. According to equation 1.32 the matrix operator $\mathbf{T}_i$ has the form

$$(3.16) \quad \mathbf{T}_i = \alpha_i \begin{bmatrix} 1 & 0 & \cdots & 0 \\ 0 & 1 & \cdots & 0 \\ \cdot & \cdot & & \cdot \\ \cdot & \cdot & & \cdot \\ \cdot & \cdot & & \cdot \\ 0 & 0 & \cdots & 1 \end{bmatrix} + (1 - \alpha_i) \begin{bmatrix} \lambda_{i1} & \lambda_{i1} & \cdots & \lambda_{i1} \\ \lambda_{i2} & \lambda_{i2} & \cdots & \lambda_{i2} \\ \cdot & \cdot & & \cdot \\ \cdot & \cdot & & \cdot \\ \cdot & \cdot & & \cdot \\ \lambda_{ir} & \lambda_{ir} & \cdots & \lambda_{ir} \end{bmatrix}$$

$$= \alpha_i \mathbf{I} + (1 - \alpha_i)\boldsymbol{\Lambda}_i.$$

In other words, $\mathbf{T}_i$ is a linear combination (or weighted sum) of two matrices, the identity matrix $\mathbf{I}$ and the matrix $\boldsymbol{\Lambda}_i$. We now wish to apply the matrix operator $\mathbf{T}_i$ to the probability vector $\mathbf{p}$. One application will yield a vector $\mathbf{T}_i\mathbf{p}$. We then apply $\mathbf{T}_i$ to $\mathbf{T}_i\mathbf{p}$, that is, multiply the vector $\mathbf{T}_i\mathbf{p}$ by the matrix $\mathbf{T}_i$ to get a vector $\mathbf{T}_i(\mathbf{T}_i\mathbf{p})$. But according to the associativity law for matrix multiplication, discussed in Section 1.5, we get the same result by first multiplying $\mathbf{T}_i$ by $\mathbf{T}_i$ and then multiplying $\mathbf{p}$ by this product, namely, $\mathbf{T}_i(\mathbf{T}_i\mathbf{p}) = (\mathbf{T}_i\mathbf{T}_i)\mathbf{p}$. More generally, if we wish to apply $\mathbf{T}_i$ repeatedly, say n times, to $\mathbf{p}$, we may do so by computing $(\mathbf{T}_i\mathbf{T}_i \cdots \mathbf{T}_i)\mathbf{p}$ where there are n matrices $\mathbf{T}_i$ in the parentheses. The matrix product $(\mathbf{T}_i\mathbf{T}_i \cdots \mathbf{T}_i)$ is denoted by $\mathbf{T}_i^n$, and it is called the nth *power* or *iterate* of the matrix $\mathbf{T}_i$. Our task then is to compute the powers of $\mathbf{T}_i$.

We first compute the second power or square of $\mathbf{T}_i$:

$$(3.17) \quad \mathbf{T}_i^2 = \mathbf{T}_i\mathbf{T}_i = [\alpha_i\mathbf{I} + (1 - \alpha_i)\boldsymbol{\Lambda}_i][\alpha_i\mathbf{I} + (1 - \alpha_i)\boldsymbol{\Lambda}_i].$$

In computing the indicated multiplication we usually must be careful about the order of the factors in a product because we saw in Section 1.5 that matrix multiplication is generally *non-commutative*.   In the present case there is no trouble.   Thus,

$$(3.18) \quad \mathbf{T}_i^2 = \alpha_i^2 \mathbf{I}^2 + \alpha_i(1 - \alpha_i)\mathbf{I}\mathbf{\Lambda}_i + \alpha_i(1 - \alpha_i)\mathbf{\Lambda}_i\mathbf{I} + (1 - \alpha_i)^2 \mathbf{\Lambda}_i^2.$$

This result is simplified by making a number of observations.   First, the square of $\mathbf{I}$ yields just $\mathbf{I}$; in fact $\mathbf{I}^n = \mathbf{I}$ for $n > 0$.   Second, since $\mathbf{I}$ is the identity matrix, $\mathbf{I}\mathbf{\Lambda}_i = \mathbf{\Lambda}_i\mathbf{I} = \mathbf{\Lambda}_i$.   Finally we compute the square of $\mathbf{\Lambda}_i$:

$$\mathbf{\Lambda}_i^2 =
\begin{bmatrix}
\lambda_{i1} & \lambda_{i1} & \cdots & \lambda_{i1} \\
\lambda_{i2} & \lambda_{i2} & \cdots & \lambda_{i2} \\
\cdot & \cdot & & \cdot \\
\cdot & \cdot & & \cdot \\
\cdot & \cdot & & \cdot \\
\lambda_{ir} & \lambda_{ir} & \cdots & \lambda_{ir}
\end{bmatrix}
\begin{bmatrix}
\lambda_{i1} & \lambda_{i1} & \cdots & \lambda_{i1} \\
\lambda_{i2} & \lambda_{i2} & \cdots & \lambda_{i2} \\
\cdot & \cdot & & \cdot \\
\cdot & \cdot & & \cdot \\
\cdot & \cdot & & \cdot \\
\lambda_{ir} & \lambda_{ir} & \cdots & \lambda_{ir}
\end{bmatrix}$$

(3.19)

$$=
\begin{bmatrix}
\lambda_{i1}\sum_j \lambda_{ij} & \lambda_{i1}\sum_j \lambda_{ij} & \cdots & \lambda_{i1}\sum_j \lambda_{ij} \\
\lambda_{i2}\sum_j \lambda_{ij} & \lambda_{i2}\sum_j \lambda_{ij} & \cdots & \lambda_{i2}\sum_j \lambda_{ij} \\
\cdot & \cdot & & \cdot \\
\cdot & \cdot & & \cdot \\
\cdot & \cdot & & \cdot \\
\lambda_{ir}\sum_j \lambda_{ij} & \lambda_{ir}\sum_j \lambda_{ij} & \cdots & \lambda_{ir}\sum_j \lambda_{ij}
\end{bmatrix}$$

But from Section 1.8 we know that

$$(3.20) \qquad\qquad \sum_{j=1}^{r} \lambda_{ij} = 1,$$

that is, that the vector $\mathbf{\lambda}_i$ is a probability vector, and so its elements λ_{ij} sum to unity.   Hence we have the result that

$$(3.21) \qquad\qquad \mathbf{\Lambda}_i^2 = \mathbf{\Lambda}_i.$$

Equation 3.18 now becomes

$$(3.22) \quad
\begin{aligned}
\mathbf{T}_i^2 &= \alpha_i^2 \mathbf{I} + 2\alpha_i(1 - \alpha_i)\mathbf{\Lambda}_i + (1 - \alpha_i)^2 \mathbf{\Lambda}_i \\
&= \alpha_i^2 \mathbf{I} + (1 - \alpha_i^2)\mathbf{\Lambda}_i.
\end{aligned}$$

The computation of higher powers of T_i follows very easily now. By the same procedure we get

(3.23)
$$T_i^3 = \alpha_i^3 I + (1 - \alpha_i^3)\Lambda_i.$$

An induction proof like the one used in Section 3.3 gives

(3.24)
$$T_i^n = \alpha_i^n I + (1 - \alpha_i^n)\Lambda_i.$$

Finally, when we apply T_i^n to the vector $\mathbf{p}$ we obtain

(3.25)
$$T_i^n \mathbf{p} = \alpha_i^n \mathbf{p} + (1 - \alpha_i^n)\lambda_i,$$

and this, of course, is exactly the form that might have been conjectured from our work with the Q_i. However, had we not imposed the combining classes condition, we would not have obtained this simple generalization. The vector λ_i may now be interpreted as a limit vector of the operator T_i, provided that $|\alpha_i| < 1$, since $T_i^n \mathbf{p}$ tends to λ_i as n becomes large.

3.6 COMMUTATIVITY OF THE OPERATORS, Q_1 AND Q_2

In later sections we handle more complicated sequences than those given thus far, but first we inquire about the commutativity of the operators Q_1 and Q_2. For example, if Q_1 is applied when reinforcement is given and Q_2 is applied when reinforcement is withheld, would their order make any difference? Would it matter whether the reinforcements were scattered randomly through the sequence or all bunched together at the beginning or at the end? If it does not matter, the operators should commute. If the order makes a difference, the operators do not commute. (Such questions were introduced in Sections 1.4 and 1.5.) First we apply Q_1, then Q_2, using the fixed-point form of the operators:

(3.26)
$$Q_2 Q_1 p = Q_2(Q_1 p) = \alpha_2[\alpha_1 p + (1 - \alpha_1)\lambda_1] + (1 - \alpha_2)\lambda_2$$
$$= \alpha_1 \alpha_2 p + \alpha_2(1 - \alpha_1)\lambda_1 + (1 - \alpha_2)\lambda_2.$$

On the other hand, applying Q_2 first and then Q_1 gives

(3.27)
$$Q_1 Q_2 p = \alpha_2 \alpha_1 p + \alpha_1(1 - \alpha_2)\lambda_2 + (1 - \alpha_1)\lambda_1.$$

The difference between these two results is

(3.28)
$$(Q_1 Q_2 - Q_2 Q_1)p = (1 - \alpha_1)(1 - \alpha_2)(\lambda_1 - \lambda_2).$$

Now if this difference is zero, the operators Q_1 and Q_2 commute, and we obtain a considerable simplification in our theory. For example, all sequences of these operators which contain m applications of Q_1 and n of Q_2 terminate in the same value of probability if the order of the events is irrelevant.

From the right side of equation 3.28 we see that Q_1 and Q_2 commute if any of the following three conditions holds:

$$(a)\ \alpha_1 = 1;$$

$$(b)\ \alpha_2 = 1;$$

$$(c)\ \lambda_1 = \lambda_2.$$

Condition c implies that $Q_1{}^n p$ and $Q_2{}^n p$ tend to the same limits as n gets large; this case will be of particular interest in Chapters 8 and 11. Conditions a and b imply that either $Q_1 p = p$ or $Q_2 p = p$, respectively, that is, that one of the two operators is the identity operator. In Chapter 8 we examine the mathematical consequences of such a condition, and in Chapter 10 we apply this case in handling some rote learning data.

*3.7   COMMUTATIVITY OF THE MATRIX OPERATORS T_1 AND T_2

We now generalize the results of the last section to obtain the conditions for the commutativity of two matrix operators T_1 and T_2. We write these operators in the forms

(3.29)
$$T_1 = \alpha_1 I + (1 - \alpha_1)\Lambda_1,$$
$$T_2 = \alpha_2 I + (1 - \alpha_2)\Lambda_2.$$

We first compute the product $T_2 T_1$:

(3.30)
$$T_2 T_1 =$$
$$\alpha_1\alpha_2 I + \alpha_2(1 - \alpha_1)\Lambda_1 + \alpha_1(1 - \alpha_2)\Lambda_2 + (1 - \alpha_1)(1 - \alpha_2)\Lambda_2\Lambda_1.$$

We have used the fact that $I\Lambda_i = \Lambda_i I = \Lambda_i$. Similarly,

(3.31)
$$T_1 T_2 =$$
$$\alpha_1\alpha_2 I + \alpha_1(1 - \alpha_2)\Lambda_2 + \alpha_2(1 - \alpha_1)\Lambda_1 + (1 - \alpha_1)(1 - \alpha_2)\Lambda_1\Lambda_2.$$

We then take the difference of these two product matrices:

$$(3.32)\qquad T_1 T_2 - T_2 T_1 = (1 - \alpha_1)(1 - \alpha_2)(\Lambda_1\Lambda_2 - \Lambda_2\Lambda_1).$$

Now as a consequence of the fact that each column of both Λ_1 and Λ_2 sums to unity, it is easy to show that

$$(3.33)\qquad\qquad\qquad \Lambda_1\Lambda_2 = \Lambda_1,$$

and

$$(3.34)\qquad\qquad\qquad \Lambda_2\Lambda_1 = \Lambda_2.$$

Therefore,

$$(3.35) \qquad \mathbf{T}_1\mathbf{T}_2 - \mathbf{T}_2\mathbf{T}_1 = (1 - \alpha_1)(1 - \alpha_2)(\mathbf{\Lambda}_1 - \mathbf{\Lambda}_2).$$

If this expression is equal to a matrix containing all zeros, $\mathbf{T}_1$ and $\mathbf{T}_2$ commute. As in the last section we have three cases:

$\qquad\qquad$ (a) $\alpha_1 = 1$, which implies $\mathbf{T}_1 = \mathbf{I}$,

$\qquad\qquad$ (b) $\alpha_2 = 1$, which implies $\mathbf{T}_2 = \mathbf{I}$,

$\qquad\qquad$ (c) $\mathbf{\Lambda}_1 = \mathbf{\Lambda}_2$.

This last case requires that

$$(3.36) \qquad \lambda_{1j} = \lambda_{2j}, \qquad j = 1, 2, \cdots, r.$$

And this in turn implies that the limit vectors $\mathbf{\lambda}_1$ and $\mathbf{\lambda}_2$ are identical.

3.8　THE SYSTEMATIC SEQUENCE $(Q_2{}^v Q_1{}^u)^n$

With certain kinds of schedules of reinforcement, two classes of events such as reward and non-reward occur in a systematic sequence such as that illustrated by sequence 3 in Section 3.2. For example, an experimenter may want to reward three out of every ten trials on the average but may not want to allow very long runs without rewards. So he may use the same sequence on each block of ten trials, giving reward on the third, fifth, and tenth trials in each block, for example. We represent such a repeated sequence of events by a systematic sequence of operators Q_1 and Q_2, and it is straightforward to compute the effect of such a sequence on the appropriate probability variables.

In this section we illustrate such systematic sequences of operators by analyzing in detail a rather simple sequence. Suppose event E_1 occurs u times followed by v occurrences of event E_2, and that this sequence is repeated n times. An example is the sequence

$$E_1E_2E_2 \qquad E_1E_2E_2 \qquad E_1E_2E_2 \qquad E_1E_2E_2 \qquad E_1E_2E_2.$$

Here we may take the subsequence $E_1E_2E_2$ as basic. Event E_1 occurs $u = 1$ times, event E_2 occurs $v = 2$ times, and the subsequence occurs $n = 5$ times. If we write out the sequence of operators operating on p corresponding to the above sequence we have

$$Q_2Q_2Q_1Q_2Q_2Q_1Q_2Q_2Q_1Q_2Q_2Q_1Q_2Q_2Q_1p$$

$$(3.37) \qquad\qquad\qquad = Q_2{}^2Q_1Q_2{}^2Q_1Q_2{}^2Q_1Q_2{}^2Q_1Q_2{}^2Q_1p$$

$$= (Q_2{}^2Q_1)^5 p.$$

Note that the operators are written in the reverse order of the occurrence of the events. The most recent operator is farthest from p when written in the extended form at the left.

The general operator of the type given above is $(Q_2{}^v Q_1{}^u)^n$, or the subscripts may be reversed if Q_2 is applied first. We would like a general expression for this systematic type of operator. A general expression can be derived by making successive use of our previous result (equation 3.5) obtained by applying Q_1 successively many times. First let us evaluate $Q_2{}^v Q_1{}^u p$:

$$Q_2{}^v Q_1{}^u p = \alpha_2{}^v(Q_1{}^u p) + (1 - \alpha_2{}^v)\lambda_2$$

(3.38)
$$= \alpha_2{}^v[\alpha_1{}^u p + (1 - \alpha_1{}^u)\lambda_1] + (1 - \alpha_2{}^v)\lambda_2$$

$$= \alpha_1{}^u \alpha_2{}^v p + \alpha_2{}^v(1 - \alpha_1{}^u)\lambda_1 + (1 - \alpha_2{}^v)\lambda_2.$$

Note that this result is linear in p and so could be written as a new operator

(3.39)
$$Q_{u,v} p = \alpha_{u,v} p + (1 - \alpha_{u,v})\lambda_{u,v},$$

where

$$\alpha_{u,v} = \alpha_1{}^u \alpha_2{}^v,$$

(3.40)
$$\lambda_{u,v} = \frac{\alpha_2{}^v(1 - \alpha_1{}^u)\lambda_1 + (1 - \alpha_2{}^v)\lambda_2}{1 - \alpha_1{}^u \alpha_2{}^v}.$$

We apply $Q_{u,v}$ to p a total of n times, and we get

(3.41)
$$Q_{u,v}^n p = \alpha_{u,v}^n p + (1 - \alpha_{u,v}^n)\lambda_{u,v}.$$

If $\alpha_{u,v}$ is not unity, the asymptotic result is, of course, $\lambda_{u,v}$ as defined above. It is worth noting the behavior of $\lambda_{u,v}$ when u or v becomes large. First, if u is fixed and v tends to infinity, we have

(3.42)
$$\lim_{v \to \infty} \lambda_{u,v} = \lambda_2, \qquad \alpha_2 \neq 1.$$

On the other hand, if v is fixed and u tends to infinity, we get

(3.43)
$$\lim_{u \to \infty} \lambda_{u,v} = \alpha_2{}^v \lambda_1 + (1 - \alpha_2{}^v)\lambda_2, \qquad \alpha_1 \neq 1.$$

The asymmetry in this pair of results comes from the fact that Q_2 is applied after Q_1 in each block, and that the more recent applications usually have relatively greater effects on the final outcome.

In Table 3.1 we have tabulated the asymptote for an example where $u = 1$, and v takes on values $0, 1, 2, \cdots, 9, \infty$, with $\lambda_1 = 0.75$, $\lambda_2 = 0.10$, $\alpha_1 = 0.6$, $\alpha_2 = 0.9$. Clearly the asymptote starts at 0.75, the asymptote of $Q_1{}^n p$, and gradually decreases to the asymptote of $Q_2{}^n p$ as the number of Q_2 operations occurring after each Q_1 operation increases.

TABLE 3.1

Asymptotic values of $(Q_2^v Q_1)^n p$ for
$\lambda_1 = 0.75$, $\lambda_2 = 0.10$, $\alpha_1 = 0.6$, and $\alpha_2 = 0.9$.

v	Asymptote
0	0.75
1	0.61
2	0.51
3	0.44
4	0.38
5	0.34
6	0.30
7	0.27
8	0.25
9	0.23
∞	0.10

Using the approach outlined above it is possible to examine the effects of any sequence of operators that is repeated over and over again, and generalization of these results to more than two events is straightforward. When the combining classes restriction of Section 1.8 is employed, a further generalization to more than two classes of responses is also straightforward though somewhat tedious.

3.9 EXPERIMENTER-CONTROLLED EVENTS

Sequence 1 in Section 3.2 is an example of a sequence of two events, A and B, which occur in random order but with a fixed proportion of A's and B's. Such a "random ratio" sequence of rewards and non-rewards might be used in an actual experiment. In those experiments we would consider reward of a particular response, when it occurred, to be one event (E_1) and non-reward of that response, when it occurred, another event (E_2). If we further assumed that no other events influenced the animals' behavior we could apply the mathematical machinery of this section to describe such experiments.

Another illustration is the Brunswik T-maze experiment [12]. In the Introduction we described a special case of that experiment in which a rat always was rewarded for turning right and never rewarded for turning left. We assumed that a rewarded right turn and an unrewarded left turn had the same effect on the probability p of a right turn on the next trial. Hence we implicitly assumed that a rewarded right turn and an unrewarded left turn constituted a single event class E_1 with associated operator Q_1. Now in the Brunswik experiment, turning right was rewarded a fixed proportion, π_1, of trials when the rat went right, and turning left was rewarded some other fixed proportion, π_2, of trials when the animal went left. Suppose we make an assumption similar to the one described above: Assume that an unrewarded right turn and a rewarded left turn have the same effect on p and so constitute an event class E_2 with operator Q_2. (In Section 3.13 we present a more general analysis which does not require these strong assumptions which may or may not be appropriate. The position here is that, if one cared to make these assumptions, the analysis of this section would be applicable.) Furthermore, let $\pi_1 + \pi_2 = 1$; this is not true in all Brunswik-type experiments, but it has often been true. For example, with one group of rats Brunswik chose $\pi_1 = 0.75$ and $\pi_2 = 0.25$. Now let us see what the probability of event E_1 is under the above assumptions. The probability of a rewarded right turn is $p\pi_1$, and the probability of an unrewarded left turn is $q(1 - \pi_2) = (1 - p)\pi_1$. Hence the probability of E_1 is $p\pi_1 + (1 - p)\pi_1 = \pi_1$. Similarly, it is easy to show that the probability of E_2 is $\pi_2 = 1 - \pi_1$. The four possible response-outcome pairs are summarized below:

Response	Outcome	Operator	Probability of Occurrence
right turn	reward	Q_1	$p\pi_1$
right turn	non-reward	Q_2	$p(1 - \pi_1) = p\pi_2$
left turn	reward	Q_2	$q\pi_2 = (1 - p)\pi_2$
left turn	non-reward	Q_1	$q(1 - \pi_2) = (1 - p)\pi_1$

An experimenter may actually go farther than we have indicated. He may decide on π_1 of the trials to place food in the right box and no food in the left box (event E_1) and on $(1 - \pi_1)$ of the trials to place food on the left and no food on the right (event E_2). Thus on no trials will food be placed on both sides, and on no trials will food be absent from both sides.

In these illustrations the probabilities π_1 and π_2 are fixed by the experimenter. Hence we call this case the *case of experimenter-controlled events*. The mathematical problem is to compute the effect of applying operator Q_1 with fixed probability π_1 and operator Q_2 with fixed probability $\pi_2 = 1 - \pi_1$. This is the task of the present section.

Consider two operators Q_1 and Q_2 for two events; suppose that they are applied successively with fixed probabilities π_1 and π_2, respectively, where $\pi_1 + \pi_2 = 1$. If this is done it no longer makes sense to speak of an organism's probability p of making response A_1 at the end of n events, unless by this we mean that we are told the actual sequence of events that did occur after the chance sequence had been determined. For example, in the T-maze example described above, one animal might obtain five occurrences of event E_1 in five trials and another might obtain only two such occurrences, depending upon the precise sequence of events for each animal. At the end of five trials there are $2^5 = 32$ possible values of p that an organism might have for a fixed initial value. So rather than ask what the value of p is at the end of a number of trials, we ordinarily would inquire about the average probability $\bar{p}$ of the response for a large number of organisms. This is not the result for an average organism.

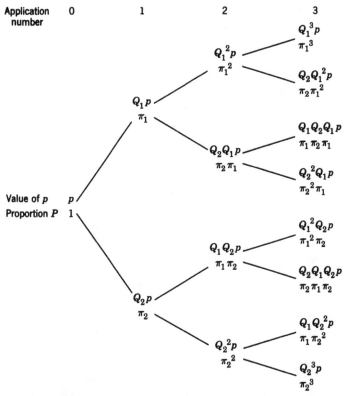

Fig. 3.2. The successive splits that a large group of animals goes through after three events when the probability of applying operator Q_1 is π_1, given an operator is to be applied (experimenter-controlled events). The proportions in the groups written beneath the p value of the group have been written to parallel the Q's rather than in the simplest form—thus under $Q_2 Q_1 Q_2 p$ is written $\pi_2 \pi_1 \pi_2$ rather than $\pi_1 \pi_2^2$.

There usually is no average organism in the sense that that organism would ever assume the mean probability except for $\pi_1 = 1$ or 0.

Let us start with an initial value p. The proportion of animals with Q_1 applied first is π_1, whereas that with Q_2 applied is π_2. Thus there are now two groups of animals that have been treated alike up to this point, the Q_1 group and the Q_2 group, and their sizes are the proportions π_1 and π_2, respectively. After the next event there are usually four groups. The proportion in the $Q_1{}^2$ group is $\pi_1{}^2$, that for the Q_1Q_2 group is $\pi_1\pi_2$, that for the Q_2Q_1 group is $\pi_1\pi_2$, and that for the $Q_2{}^2$ group is $\pi_2{}^2$. The tree diagram (Fig. 3.2) shows the sequence of events together with their probabilities for the first three applications of the operators.

From an examination of the tree diagram it should be clear that the proportion of animals in a given group is known as soon as the number of times each operator is applied is known. In particular, if Q_1 has been applied u times, and Q_2 has been applied v times, the proportion in the group is $\pi_1{}^u\pi_2{}^v$; this proportion applies to every rearrangement of the operators with u applications of Q_1 and v of Q_2.

We are interested in the average fraction of organisms that will make response A_1 after n applications of operators. The situation after n applications is that there are 2^n groups at most. The groups are homogeneous in the sense that all animals in a group have the same probability of response A_1, but different groups have in general different probabilities. If we think of the groups as ordered after a particular number of applications of the operators, we can call p_ν the probability of response A_1 in the νth group and P_ν the proportion that the group is of the total population. The average contribution of the νth group to the total that would make response A_1 is just the product $p_\nu P_\nu$; and therefore the average fraction $\bar{p}_n$ for the whole population is the sum of such products over all the groups in the population, that is,

$$(3.44) \qquad \bar{p}_n = \sum_{\nu=1}^{2^n} p_\nu P_\nu.$$

This result is just the mean of the distribution of p_ν.

Let us calculate some means after successive applications of the operators, using the slope-intercept form of the operators,

$$(3.45) \qquad \bar{p}_0 = p \times 1 = p,$$

$$(3.46) \qquad \begin{aligned} \bar{p}_1 &= \pi_1 Q_1 p + \pi_2 Q_2 p \\ &= \pi_1(a_1 + \alpha_1 p) + \pi_2(a_2 + \alpha_2 p) \\ &= (\pi_1 a_1 + \pi_2 a_2) + (\pi_1 \alpha_1 + \pi_2 \alpha_2)p. \end{aligned}$$

The fact that $\bar{p}_1$ can be written as a linear function of p suggests the slope-intercept form:

$$(3.47) \qquad \bar{p}_1 = \bar{a} + \bar{\alpha}p,$$

where

$$\bar{a} = \pi_1 a_1 + \pi_2 a_2,$$

(3.48)

$$\bar{\alpha} = \pi_1 \alpha_1 + \pi_2 \alpha_2.$$

Note that $\bar{a}$ and $\bar{\alpha}$ are the averages of the a's and α's of the operators when weighted by their frequencies of occurrence. This way of writing $\bar{p}_1$ suggests in turn the conjecture:

(3.49) $$\bar{p}_n = \bar{\alpha}^n p + (1 - \bar{\alpha}^n)\bar{\lambda},$$

where

(3.50) $$\bar{\lambda} = \frac{\bar{a}}{1 - \bar{\alpha}}.$$

To verify this conjecture let us call the probability of the νth group on the nth trial $p_{\nu n}$ and the proportion of the population in the νth group $P_{\nu n}$. On the $(n + 1)$st application of the operators this group will split into two groups, with new values of p and proportions as follows:

New Values of p	New Proportions
$Q_1 p_{\nu n} = a_1 + \alpha_1 p_{\nu n}$	$\pi_1 P_{\nu n}$
$Q_2 p_{\nu n} = a_2 + \alpha_2 p_{\nu n}$	$\pi_2 P_{\nu n}$

To get $\bar{p}_{n+1}$ we must multiply the new values of p by the new proportions and then sum over all the groups:

$$\bar{p}_{n+1} = \sum_{\nu=1}^{2^n} [\pi_1 P_{\nu n}(a_1 + \alpha_1 p_{\nu n}) + \pi_2 P_{\nu n}(a_2 + \alpha_2 p_{\nu n})]$$

(3.51)

$$= (\pi_1 a_1 + \pi_2 a_2) \sum_\nu P_{\nu n} + (\pi_1 \alpha_1 + \pi_2 \alpha_2) \sum_\nu p_{\nu n} P_{\nu n}.$$

Naturally the sum of $P_{\nu n}$ over ν is unity, because it represents the total size of the group, and by definition

(3.52) $$\sum_\nu p_{\nu n} P_{\nu n} = \bar{p}_n,$$

leading to the difference equation

(3.53) $$\bar{p}_{n+1} = \pi_1 a_1 + \pi_2 a_2 + (\pi_1 \alpha_1 + \pi_2 \alpha_2)\bar{p}_n$$

$$= \bar{a} + \bar{\alpha}\bar{p}_n.$$

We already know the solution to this difference equation, because it was obtained in Section 3.3. The solution is given by the conjecture 3.49, and so we have verified the conjecture.

These results for fixed probabilities of applying the operators show that for the purpose of obtaining means we could define an expected operator $\bar{Q}$ by

(3.54) $$\bar{Q}p = \bar{a} + \bar{\alpha}p$$

that behaves just like the Q's (cf. Section 6.4).

3.10 EXPERIMENTER-CONTROLLED EVENTS WITH t OPERATORS AND r ALTERNATIVES

The preceding results can be extended to any number, t, of events immediately.   We merely define

(3.55)
$$\bar{a} = \sum_{i=1} \pi_i a_i,$$

$$\bar{\alpha} = \sum_{i=1}^{t} \pi_i \alpha_i.$$

Equation 3.53 then gives the correct mean probabilities.

The analysis can also be extended to any number of response classes. A matrix operator corresponds to each event, and these operators generate distributions of probability vectors.   We give the detailed discussion of this extension in Section 4.4.

3.11 SIMPLE MARKOV CHAINS

In Section 3.2 of this chapter we discussed three kinds of elementary sequences of events.   One of these, sequence 3, was completely systematic —two A's and a B, two A's and a B, etc.   In Section 3.8 we analyzed the effects of such sequences of operators Q_1 and Q_2.   Sequence 1 of Section 3.2, on the other hand, involved a random order of A's and B's but with a fixed proportion of each.   In the preceding two sections we discussed such random sequences of operators.   There remains, then, a need to discuss applications of the operators with dependent probabilities such as those illustrated by sequence 2 of Section 3.2.   In that example, the probability of a B following an A was 0.5, whereas the probability of a B following a B was zero.   We pointed out that this type of sequence was the simplest type of Markov chain, and we want to expand this notion in this section.   The application of Markov chain theory to psychology has been discussed by Miller [4].

In any Markov chain involving two events, A and B, we must know three quantities: an initial probability and two transition probabilities. The initial probability of event A is denoted by $p_0(A)$, and the initial probability of event B is $p_0(B)$.   One of these events necessarily starts the sequence, and so we have

$$(3.56) \qquad p_0(A) + p_0(B) = 1.$$

As soon as the first event occurs, the next event is chosen according to the two conditional probabilities. If event A has just occurred, the probability of an A is $p(A|A)$ and the probability of a B is $p(B|A)$. (The notation $p(x|y)$ is read "the probability of x, given that y has occurred.") If event B had occurred, the probability of an A is $p(A|B)$ and the probability of a B is $p(B|B)$. We must have, of course,

$$(3.57) \qquad p(A|A) + p(B|A) = 1,$$

$$(3.58) \qquad p(A|B) + p(B|B) = 1.$$

A total of six probabilities has been introduced, but the preceding three equations reduce the number of independent variables to three.

If we know the three numbers $p_0(A)$, $p(A|A)$ and $p(A|B)$ we completely specify the Markov process. If we know what event occurs at any point in the process, we know the probability of an A occurring next. However, we ask another question: What is the probability $p_n(A)$ on trial n if we do not know the events on earlier trials? This question acquires more meaning if we think of the population of all possible sequences that are generated by the specified rules. The fraction of this population that begins with event A is $p_0(A)$; of these sequences, a fraction $p(A|A)$ will have an event A in the second position (trial $n = 1$). The fraction of the total population of sequences which begin with event B is $p_0(B)$ whereas the fraction of these that have an A on trial $n = 1$ is $p(A|B)$. Thus the total fraction of sequences with an A on trial $n = 1$ is

$$(3.59) \qquad p_1(A) = p_0(A)p(A|A) + p_0(B)p(A|B).$$

For example, if $p_0(A) = 0.2$, $p(A|A) = 0.7$, and $p(A|B) = 0.5$, then

$$p_1(A) = (0.2)(0.7) + (0.8)(0.5) = 0.54.$$

Therefore, we see that the unconditional probability of an A on trial n may be interpreted as the fraction of the total population of sequences that have an A on trial n. For example, on trial $n = 2$ we have

$$(3.60) \qquad p_2(A) = p_1(A)p(A|A) + p_1(B)p(A|B).$$

We have already found that $p_1(A)$ for our numerical example was 0.54. Thus, for that example,

$$p_2(A) = (0.54)(0.7) + (0.46)(0.5) = 0.608.$$

More generally, on trial $n + 1$,

$$(3.61) \qquad \begin{aligned} p_{n+1}(A) &= p_n(A)p(A|A) + p_n(B)p(A|B) \\ &= p_n(A)p(A|A) + [1 - p_n(A)] p(A|B) \\ &= p(A|B) + [p(A|A) - p(A|B)] p_n(A). \end{aligned}$$

Hence, we obtain a linear difference equation in the $p_n(A)$. If we let

$$p(A|B) = a,$$

(3.62)

$$p(A|A) - p(A|B) = \alpha,$$

we have

(3.63)
$$p_{n+1}(A) = a + \alpha p_n(A).$$

This linear equation is an example of difference equation 3.8 encountered in Section 3.3. The solution is

(3.64)
$$p_n(A) = \alpha^n p_0(A) + (1 - \alpha^n) p_\infty(A),$$

where

$$p_\infty(A) = \frac{a}{1 - \alpha} = \frac{p(A|B)}{1 - p(A|A) + p(A|B)}$$

(3.65)

$$= \frac{p(A|B)}{p(B|A) + p(A|B)}.$$

This result is quite well known [4]. We see that, if $-1 < \alpha < +1$, then as $n \to \infty$, the quantity α^n tends to zero, and so $p_n(A)$ tends to $p_\infty(A)$. In the example with $p_0(A) = 0.2$, $p(A|A) = 0.7$, $p(A|B) = 0.5$,

$$p_\infty(A) = \frac{0.5}{0.3 + 0.5} = \frac{5}{8} = 0.625.$$

We note that $p_2(A) = 0.608$, so that the limit is very closely approximated in this case after a very few trials. The speed of convergence, of course, depends on the magnitude of $p(A|A) - p(A|B)$, and in the present case this is 0.2, and powers of 0.2 tend to zero rapidly.

To return to the general problem, suppose that the starting proportions are the same as the asymptotic proportions, that is,

(3.66)
$$p_0(A) = p_\infty(A) = \frac{p(A|B)}{p(B|A) + p(A|B)}.$$

If we substitute this in equation 3.64 we see that

(3.67)
$$p_n(A) = p_0(A).$$

Thus, when the starting condition $p_0(A)$ is adjusted properly, the sequences have the same proportion of A events on every trial. The reason for the trial-to-trial change in the more general case is that the starting proportions differ from the asymptotic proportions, and this discrepancy can be adjusted only gradually.

Newman found a simple example of a Markov chain in the written Samoan language [13]. If a consonant is called a B event, and a vowel an A event, the sequence of vowels and consonants in Samoan can be represented by a Markov chain with $p(B|B) = 0$ and $p(B|A) = 0.49$. Thus consonants never follow consonants in written Samoan. By a happy accident, sequence 2 of Section 3.2 provides a close approximation to a sequence of vowels and consonants in Samoan writing. There $p(B|B) = 0$ and $p(B|A) = 0.5$. The asymptotic proportion of vowels in Samoan writing would, according to equation 3.65, have the value $1/(0.49 + 1) = 0.671$, or about 2/3.

Simple Markov chains of the type just described are seldom used for constructing schedules of reinforcements, but were used, for example, by Hake and Hyman in a prediction experiment [14]. We do not analyze a Markov sequence of the operators Q_1 and Q_2 in this chapter. However, a more common experimental procedure leads to a considerably more complicated Markov chain which is the subject of the next section.

3.12 SUBJECT-CONTROLLED EVENTS

We now consider a problem that is of major importance to most of the remainder of this book. Suppose that we equate a response occurrence with an event in the mathematical system. For two alternatives A_1 and A_2 we simply let event E_1 be an occurrence of response A_1 and also let event E_2 be an occurrence of response A_2. We suggest two possible rationales for making these equivalences. First, the Guthrian school of association theorists postulates that conditioning is a consequence of the contiguity between stimulus and response. From this position, we conclude that the mere occurrence of a response and the stimulus change it produces alter future behavior, and thus, in our language, constitute an event. The second rationale arises from the common experimental procedure of making a reward contingent upon the occurrence of a particular response. The simple T-maze experiment described in the Introduction is an example. If the rat turns right, reward is always found; if the rat turns left, reward is never presented. The two responses are turning right (A_1) and turning left (A_2). Event E_1 in the mathematical system is identified with turning right and finding reward, and event E_2 is identified with turning left and finding no reward. Note that the event includes the response and the *outcome* (reward or no reward), but the outcome of a trial is completely determined by the response which is made in this experiment. Hence we can say that E_1 occurs whenever A_1 occurs and that E_2 occurs whenever A_2 occurs. For this reason we refer to this case as the *case of subject-controlled events*.

When the foregoing equivalence between responses and events is assumed, the probability of event E_1 on trial n is no longer constant; it is equal to the probability p_n of response A_1. Moreover, the conditional probability of event E_1 on trial $n + 1$, given that E_1 has occurred on trial n, is not even constant; it is equal to $Q_1 p_n$, and this probability will ordinarily depend upon n. Similarly, the conditional probability of event E_1 on trial $n + 1$, given that E_2 occurred on trial n, is $Q_2 p_n$. Hence the simple type of Markov chain described in the last section is not appropriate for the present problem; the conditional probabilities are not constant. We can, however, make the conditional probabilities constant by choosing the operators Q_1 and Q_2 in a very special way. If we let $\alpha_1 = \alpha_2 = 0$, we have $Q_1 p = \lambda_1$ and $Q_2 p = \lambda_2$, that is, the probability of an A_1 following an A_1 is λ_1, and the probability of an A_1 following an A_2 is λ_2. For this special choice of the operators we do have a simple two-state Markov chain of the type described in the last section. The process is completely defined by the two conditional probabilities, λ_1 and λ_2, and the initial probability, p_0, of an A_1 response. In this sense, then, the simple two-state Markov chain is a special case of our general mathematical system. Except for this special choice of the parameters of the operators, a two-state Markov chain is inadequate for our present problem, but we now show that a more general Markov process is appropriate.

In the theory of Markov chains, we talk about a number of *states* and the probabilities of the system being in these states. In Section 3.11 we let these states be events A and B and so we had but two states. The Markov chains discussed in this book have transition probabilities that are the same on every trial. (More general Markov chains having time-dependent transition probabilities are not considered, though this complication might be an asset to a learning model. Feller discusses these more general Markov chains [7].) The number of states may be greater than two and, in fact, may be infinite. The learning process referred to in the preceding paragraph may be considered a Markov process, provided we *let the possible values of the response probability p represent the states of the system.* We shall try to make this notion more explicit. Ordinarily we like to number the states or give them letters, A, B, C, $\cdots$. But we shall label a particular state by specifying a value of p, the probability of occurrence of response A_1 (or of event E_1). Since p can usually assume any value between zero and unity, an infinite number of states is possible. On any trial, however, only two states can be reached from the state p, namely, $Q_1 p$ and $Q_2 p$. The transition probabilities to all other states are zero. The Markov condition that transition probabilities remain constant requires only that those transition probabilities be completely determined by the existence of the system in any given state. In our problem, the conditional probability of state Q_1p given state p is simply p, the conditional probability of state Q_2p given state p is $1-p$, and the conditional probability of all other states is zero.

The characteristic feature of a Markov process is the *independence of path*. By this is meant that the transition probabilities from a particular state do not depend upon how that particular state was reached. The mathematical process described above is Markovian because the conditional probabilities of all other states, given state p, do not depend upon how state p was achieved. This may be said in another way: Given that the probability of response A_1 is p_n on trial n, then the probabilities of the various possible values of p_{n+1} do not depend in any way upon $p_{n-1}, p_{n-2}, \cdots, p_0$.

Since the probability of applying Q_1 to p_n to obtain p_{n+1} is p_n and the probability of applying Q_2 to p_n is $1-p_n$, we have a branching process similar to the one described in Section 3.9. As before, we give a diagram showing the results of the first three applications of the operators (Fig. 3.3). As in previous sections we are interested in the average p value at the end of n applications of the operators. Let us compute $\bar{p}_0$ and $\bar{p}_1$:

$$(3.68) \qquad\qquad \bar{p}_0 = p,$$

$$(3.69) \qquad \bar{p}_1 = pQ_1p + qQ_2p = a_1p + \alpha_1p^2 + a_2q + \alpha_2pq.$$

And substituting $q = 1 - p$ in this equation, we obtain

$$(3.70) \qquad \bar{p}_1 = a_2 + (a_1 - a_2 + \alpha_2)p + (\alpha_1 - \alpha_2)p^2.$$

This is the first time we have met a $\bar{p}$ that was expressed in any way other than linearly in the initial value of p. It is a great blow because the nice methods we have used no longer work. We note that when $\alpha_1 = \alpha_2$ the p^2 term vanishes, and we do have a linear expression in p. This special case will be handled in detail in Chapter 5. Here we consider only the general case when $\alpha_1 \neq \alpha_2$. Suppose for a moment that we could use the expression on the right of equation 3.70 to generate $\bar{p}_2$, $\bar{p}_3$, etc., just as we did for fixed probabilities of application of the operators in Section 3.9. Then the conjectured equation would be

$$(3.71) \qquad \bar{p}_{n+1} = a_2 + (a_1 - a_2 + \alpha_2)\bar{p}_n + (\alpha_1 - \alpha_2)\bar{p}_n^2.$$

Furthermore,

$$\bar{p}_{n+2} = a_2 + (a_1 - a_2 + \alpha_2)\bar{p}_{n+1} + (\alpha_1 - \alpha_2)\bar{p}_{n+1}^2$$

$$(3.72) \qquad = a_2 + (a_1 - a_2 + \alpha_2)[a_2 + (a_1 - a_2 + \alpha_2)\bar{p}_n + (\alpha_1 - \alpha_2)\bar{p}_n^2]$$
$$+ (\alpha_1 - \alpha_2)[a_2 + (a_1 - a_2 + \alpha_2)\bar{p}_n + (\alpha_1 - \alpha_2)\bar{p}_n^2]^2.$$

Without simplifying this expression further it is clear that $\bar{p}_2$ will have

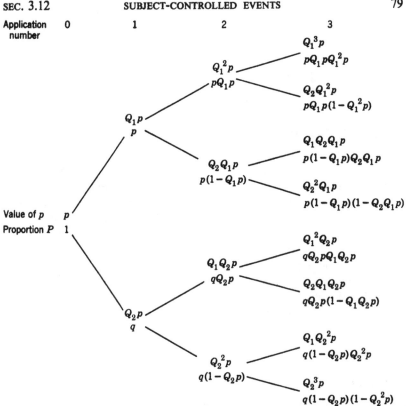

Fig. 3.3. This diagram shows the successive splits that a large number of organisms would go through on successive applications of the operators for the case of subject-controlled events. Both the p value of the group and the proportional size of the group are given at each stage of the operation. The probability of applying Q_1 is equal to the p value of the group.

a term involving the fourth power of the starting value of p, unless $\alpha_1 = \alpha_2$. Now let us consider what happens when we actually try to compute $\bar{p}_2$, so as to compare with the above conjecture. From the tree diagram, Fig. 3.3, we have expected values to compute by multiplying the p values of the groups by their probabilities and summing. We notice that the p value in general involves only a linear expression in p for any group, but the probability of occurrence contains as high a power of p as the number of times the operator has been applied. Therefore the highest power of p in $\bar{p}_2$ is clearly the highest power that can be obtained by multiplying a linear expression in p by a quadratic in p; thus the highest possible power is p^3. We have already shown that an attempt to use the conjectured equation 3.71 must lead to terms in p^4, and therefore the conjecture does not give the correct answer. This is unfortunate because it complicates matters tremendously. Furthermore, the case under discussion is of interest in many learning situations, so that we shall have to treat it more extensively, and give particular study to a number of special cases as well as investigate methods of handling the general case. Indeed, the later chapters of Part I are mainly devoted to this investigation.

3.13 EXPERIMENTER-SUBJECT-CONTROLLED EVENTS

We have previously discussed two important cases. In Section 3.9 we described the case of experimenter-controlled events, for which the event probabilities are fixed. In Section 3.12 we introduced the case of subject-controlled events, for which the event probabilities were identical to the response probabilities. We now combine these ideas as follows. We identify an event E_i with the occurrence of an alternative A_j and an occurrence of a particular outcome O_k. We consider two possible alternatives and s possible outcomes ($k = 1, 2, \cdots s$), and we define a set of constant conditional probabilities π_{jk}, which specify the conditional probability of outcome O_k, given that alternative A_j has occurred. The s outcomes are mutually exclusive and exhaustive and so

$$(3.73) \qquad \sum_{k=1}^{s} \pi_{jk} = 1, \qquad j = 1, 2.$$

Instead of the subscript i on the operator Q, we use the double subscript jk; we have the operators Q_{jk} defined by

$$(3.74) \qquad Q_{jk}p = \alpha_{jk}p + (1 - \alpha_{jk})\lambda_{jk}.$$

This operator is applied when alternative A_j occurs and is followed by outcome O_k. Hence the probability of application of Q_{1k} is $p\pi_{1k}$ and the probability of application of Q_{2k} is $(1 - p)\pi_{2k}$. Since the conditional probabilities π_{jk} are determined by the experimenter and because p and $1 - p$ are the response probabilities of the subject, we call this case the *case of experimenter-subject-controlled events*.

Particularly in Chapter 13 will we be interested in two alternatives A_1 and A_2 and two outcomes O_1 and O_2. We then have four operators Q_{jk} as given below:

	Alternative	Outcome	Operation	Probability of Application
	A_1	O_1	$Q_{11}p = \alpha_{11}p + (1 - \alpha_{11})\lambda_{11}$	$p\pi_{11}$
	A_1	O_2	$Q_{12}p = \alpha_{12}p + (1 - \alpha_{12})\lambda_{12}$	$p\pi_{12}$
(3.75)	A_2	O_1	$Q_{21}p = \alpha_{21}p + (1 - \alpha_{21})\lambda_{21}$	$(1 - p)\pi_{21}$
	A_2	O_2	$Q_{22}p = \alpha_{22}p + (1 - \alpha_{22})\lambda_{22}$	$(1 - p)\pi_{22}$

We have the relations

(3.76)

$$\pi_{11} + \pi_{12} = 1,$$

$$\pi_{21} + \pi_{22} = 1.$$

In Chapter 4 we consider the sequences of probability values generated by such operators.

3.14 SUMMARY

This chapter describes various possible sequences of events and the resulting sequences of operators. Three types of simple sequences, one with independent random elements, one with elements dependent on the preceding entry, and a completely systematic sequence, are introduced in Section 3.2. These are later used as prototypes for operator sequences. In Section 3.3 the effects of a repetitive application of a single operator Q_i are analyzed; the following formula is developed for computing the response probability on every trial:

(3.5) $$Q_i{}^n p = \alpha_i{}^n p + (1 - \alpha_i{}^n)\lambda_i.$$

The analogous problem for matrix operators T_i is solved in Section 3.5.

The commutativity of two operators is considered in Sections 3.6 and 3.7. It is shown that two event operators commute—their order of application is irrelevant—if and only if one operator is an identity operator or the operators have equal limit points. A particular kind of systematic sequence, one with repeated cycles of u events of one class and v events of another, is analyzed in Section 3.8; the response probabilities are computed for each trial from a formula similar to equation 3.5 shown above.

When the event occurrences are uncertain, that is, when only probability statements about the event on a given trial can be made, the analysis becomes more complicated. We introduced three such cases which will be used for describing various kinds of learning data: (1) experimenter-controlled events, where the event probabilities have fixed values, π_1 and π_2 (Sections 3.9, 3.10), (2) subject-controlled events, where the responses are considered to be events and so the event probabilities change (Section 3.12), and (3) experimenter-subject-controlled events, where the events correspond to response-outcome pairs (Section 3.13). For the case of experimenter-controlled events, the following formulas are developed for computing the mean response probabilities $\bar{p}_n$ on each trial:

(3.49) $$\bar{p}_n = \bar{\alpha}^n p + (1 - \bar{\alpha}^n)\bar{\lambda},$$

where p is the initial response probability, where

$$\bar{\alpha} = \pi_1\alpha_1 + \pi_2\alpha_2,$$

(3.48)

$$\bar{a} = \pi_1 a_1 + \pi_2 a_2,$$

and

(3.50)
$$\bar{\lambda} = \frac{\bar{a}}{1 - \bar{\alpha}}.$$

(The event probabilities are π_1 and π_2.) This analysis does not generalize to the other two cases which are treated in more detail in later chapters.

The case of subject-controlled events is related to the theory of Markov chains as discussed in Section 3.12, but this discussion is preceded by an elementary exposition of Markov processes in Section 3.11. It is pointed out that when the α_i are zero, the case of subject-controlled events gives a simple type of Markov chain, and that otherwise that case leads to a Markov chain with an infinite number of states.

REFERENCES

1. Graham, C., and Gagné, R. M. The acquisition, extinction, and spontaneous recovery of a conditioned operant response. *J. exp. Psychol.*, 1940, **26**, 251–280.
2. Jenkins, W. O., McFann, H., and Clayton, F. L. A methodological study of extinction following aperiodic and continuous reinforcement. *J. compar. physiol. Psychol.*, 1950, **43**, 155–167.
3. Jenkins, W. O., and Stanley, J. C., Jr. Partial reinforcement: a review and critique. *Psychol. Bull.*, 1950, **47**, 193–234.
4. Miller, G. A. Finite Markov processes in psychology. *Psychometrika*, 1952, **17**, 149–167.
5. Miller, G. A., and Frick, F. C. Statistical behavioristics and sequences of responses. *Psychol. Rev.*, 1949, **56**, 311–324.
6. Frick, F. C., and Miller, G. A. A statistical description of operant conditioning. *Amer. J. Psychol.*, 1951, **64**, 20–36.
7. Feller, W. *An introduction to probability theory and its applications.* New York: Wiley, 1950, Chapter 15.
8. Birkhoff, G., and MacLane, S. *A survey of modern algebra.* New York: Macmillan, 1941, pp. 11–13.
9. Jordan, C. *Calculus of finite differences.* New York: Chelsea Publishing Co., 1947, second edition, pp. 558–559.
10. Estes, W. K. Toward a statistical theory of learning. *Psychol. Rev.*, 1950, **57**, 94–107.
11. Bush, R. R., and Mosteller, F. A mathematical model for simple learning. *Psychol. Rev.*, 1951, **58**, 313–323.
12. Brunswik, E. Probability as a determiner of rat behavior. *J. exp. Psychol.*, 1939, **25**, 175–197.
13. Newman, E. B. The pattern of vowels and consonants in various languages. *Amer. J. Psychol.*, 1951, **64**, 369–379.
14. Hake, H. W., and Hyman, R. Perception of the statistical structure of a random series of binary symbols. *J. exp. Psychol.*, 1953, **45**, 64–74.

CHAPTER 4

Distributions of Response Probabilities

4.1 INTRODUCTION

Suppose that a large group of organisms is run through an experiment for many trials. Furthermore, assume that different organisms make different responses on each trial and that different things happen to them. Under these circumstances the organisms will not all have the same probability of response A_1 on a given trial. What percentage of organisms is more likely to make response A_1 than response A_2? Or what percentage of the organisms is at least 90 percent sure to make response A_1? What is the average probability of response A_1? Questions such as these are considered in this chapter.

In the preceding chapter we described various kinds of sequences of events and computed the effects of the corresponding sequences of operators. However, in Sections 3.9 through 3.13 we encountered problems in which the event occurrence at any position in the sequence was uncertain. At most we knew the probability that the event was of a particular kind such as E_1. As a result for a large group of organisms a new *distribution* of response probabilities is obtained after each trial. For two response classes we had a set of possible response probabilities p_{vn} for trial n. The values of the index v were just labels for the possible response probabilities. Furthermore, we introduced a set of probabilities P_{vn} which specified the likelihood that the various values of p_{vn} would occur on trial n. In other words, P_{vn} is the probability of occurrence of the response probability p_{vn} on trial n. In order to avoid the awkward phrase "probability of a probability" we call the p_{vn} "p values" in this chapter. Thus P_{vn} is the probability of the p value p_{vn}. When we speak of a distribution of p values, p_{vn} is the variable of the distribution and P_{vn} is the density function of p_{vn}. Frequently we call P_{vn} a proportion (of organisms) since it may help the reader to think of a very large population of identical organisms undergoing the same learning process. Then P_{vn} is the proportion of this large population that has the p value p_{vn} on trial n.

We considered three main cases in Chapter 3:

1. Experimenter-controlled events,
2. Subject-controlled events,
3. Experimenter-subject-controlled events.

When the events are experimenter-controlled, they occur with fixed probabilities π_i as determined by the experimenter, whereas for subject-controlled events, event E_i necessarily occurs when alternative A_i occurs. For the more general problem of experimenter-subject-controlled events, the conditional probabilities of various outcomes, given response A_j, were fixed by the experimenter; but the probability of alternative A_1 being chosen by the subject was p_n on trial n. Each alternative-outcome pair was considered to be an event and, as usual, an operator was associated with an event. In this chapter we again consider these three cases.

In Fig. 3.2 we wrote out the exact probability distribution (for trials 0, 1, 2, and 3) that would be obtained by an infinite population of subjects with two experimenter-controlled events. On trial n there are at most 2^n groups of subjects who have had different experiences since the experiment began; therefore, writing out the complete details even for n as small as 10 would usually involve 1024 groups. Since the explicit statement of all these 2^n probabilities is most tedious for all but the smallest n, we would like to have some general rules to calculate what portion of the subjects will have values of p between p_v and $p_v + \Delta p_v$ on trial n. That is to say, we would like to know the probability distribution on successive trials. The best information, of course, would be an explicit formula for the distribution function. Failing this we could describe the distribution through its moments.

In this chapter recurrence relations are derived for the raw moments of the distribution function after n trials. The derivation is given first for experimenter-controlled events, and the recurrence relations are solved for the mean and variance of the distributions. Next, the corresponding recurrence relations for subject-controlled events are developed and extended to experimenter-subject-controlled events. Unfortunately we are not able to solve for the moments explicitly in these latter two cases, but the recurrence formulas are used in later chapters to discuss special problems and to develop some useful approximations.

This may be a good place to emphasize again the importance of studying the general properties of any model more deeply than specific applications may require. Only by understanding a model very generally is it possible to have a good "feel" for what it will do and how it will behave in new situations, and only in this way do we get a good notion about both its scope and its limitations. Therefore in this chapter and some others in

Part I the reader should not expect results that will be particularly useful in analyzing a specific experiment. Rather the hope is that this material will improve the reader's understanding just as it has the authors'.

Heretofore we have assumed that sequences of p values began at some single point p_0. We have seen how distributions of p values are generated by the operators even with a single initial probability. Some readers may have felt that this restriction to a single value of p_0 is undesirable—different organisms may come into an experiment with different initial probabilities, that is, we might reasonably expect an *initial distribution* of p values. Nevertheless, most of our analysis will be in terms of a single value of p_0. The generalization to an arbitrary initial distribution is straightforward but involves some mathematical complications we do not care to introduce. Most of the results in this chapter, however, are valid for an arbitrary distribution of initial p values.

In Section 4.7 we present some theorems about the asymptotic distribution of p values when the events are experimenter controlled and subject controlled. The proofs have been omitted. Finally, in Section 4.8 we discuss the lengths of runs of one response or another.

4.2 DEFINITION OF MOMENTS

The reader who is familiar with moments of distributions will find nothing new in this section; we suggest that he skip to Section 4.3.

If we have a set of discrete probabilities $P_\nu \geq 0$, $\nu = 1, 2, \cdots$, finite or infinite in number, each associated with a number p_ν, then the mth raw moment of the distribution of p's is defined to be

$$(4.1) \qquad V_m = \sum_\nu p_\nu{}^m P_\nu,$$

where, of course,

$$(4.2) \qquad V_0 = \sum_\nu P_\nu = 1.$$

Thus the mth raw moment is the average or mean of the mth power of the variable whose distribution is under study. For example, the first raw moment, V_1, is the mean of the distribution. It is often convenient to define *moments about the mean* instead of the V's which are *moments about the origin*. This can readily be done in terms of the raw moments. The mth moment about the mean is defined as

$$(4.3) \qquad \mu_m = \sum_\nu (p_\nu - V_1)^m P_\nu.$$

In words, the mth moment about the mean is the average of the mth power of the deviation of the variable from its own mean. When this form of the μ's is expanded we see that

$$\mu_0 = \sum_\nu P_\nu = 1,$$

$$\mu_1 = \sum_\nu (p_\nu - V_1)P_\nu = \sum_\nu p_\nu P_\nu - \sum_\nu V_1 P_\nu = V_1 - V_1 = 0,$$

(4.4)

$$\mu_2 = \sum_\nu (p_\nu - V_1)^2 P_\nu = \sum_\nu (p_\nu^2 - 2V_1 p_\nu + V_1^2)P_\nu = V_2 - 2V_1^2 + V_1^2$$

$$= V_2 - V_1^2.$$

In a similar way, the reader can verify that the third moment about the mean is

(4.5) $$\mu_3 = V_3 - 3V_2 V_1 + 2V_1^3.$$

In order to obtain a general formula for μ_m in terms of the raw moments, we can use the well-known binomial expansion of $(x - y)^m$:

(4.6) $$(x - y)^m = \sum_{u=0}^{m} (-1)^u \binom{m}{u} x^{m-u} y^u,$$

where $\binom{m}{u}$ is the binomial coefficient equal to $m!/[(m - u)!u!]$. For example $(x - y)^2 = x^2 - 2xy + y^2$ and $(x - y)^3 = x^3 - 3x^2 y + 3xy^2 - y^3$. Using this binomial expansion in equation 4.3, we have

$$\mu_m = \sum_\nu \left\{ \sum_{u=0}^{m} (-1)^u \binom{m}{u} p_\nu^{m-u} V_1^u \right\} P_\nu.$$

By interchanging the order of summation (this amounts to rearranging terms) we have

$$\mu_m = \sum_{u=0}^{m} \left\{ (-1)^u \binom{m}{u} V_1^u \sum_\nu p_\nu^{m-u} P_\nu \right\}$$

(4.7)

$$= \sum_{u=0}^{m} \left\{ (-1)^u \binom{m}{u} V_1^u V_{m-u} \right\}.$$

For $m = 0, 1, 2,$ and 3, this formula gives results which agree with our earlier formulas.

Classically for any distribution the second moment about the mean is called the variance of that distribution, and when it is used in statistical problems the notation σ^2 usually replaces μ_2. The square root of μ_2, or σ, is called the standard deviation. It is a measure of the spread of a distribution as opposed to V_1, the mean, which is a measure of location. Changing only V_1 moves the whole distribution, but changing σ spreads the distribution out or pulls it together.   We note that σ^2 is always positive or zero because it is a sum of squares with positive weights.   Because p_ν is between 0 and 1, σ^2 is always finite for our problems, as are all other moments.   This may be seen in equation 4.1; $p_\nu{}^m$ is at most unity for all values of p_ν and m, and, since the P_ν sum to unity, we see that $0 \leq V_m \leq 1$ for all m.

It is not our intention to expand on this notion of moments in great detail, because they are discussed in any elementary statistics book. One important point that may not be mentioned in elementary textbooks is that knowledge of all the moments usually determines the cumulative distribution function, just as the distribution function completely determines the moments.   However it is usually not so easy to get the distribution from the moments as it is to get the moments from the distribution.

If we order the groups according to their p values so that $p_{\nu+1} > p_\nu$ for all ν, the cumulative distribution is given by

$$(4.8) \qquad F(p) = \sum_{\nu=1}^{t} P_\nu, \qquad t = \text{largest } \nu \text{ such that } p_\nu \leq p.$$

This means that as p increases, F takes a jump each time p comes to a p_ν with positive P_ν, then stays the same until the next such jump.

EXAMPLE: $p_1 = 0.2,\, p_2 = 0.6,\, P_1 = 0.3,\, P_2 = 0.7$.   Then

$$F(p) = \begin{cases} 0 & -\infty \leq p < 0.2 \\ 0.3 & 0.2 \leq p < 0.6 \\ 1.0 & 0.6 \leq p \leq \infty \end{cases}$$

In words, the cumulative distribution gives the probability that the variable is less than or equal to a specified value.

Sometimes it is convenient to think of the p's as having a distribution without specifying whether it is discrete, continuous, or both at once. Then we write the cumulative as either

$$F(p) \quad \text{or} \quad \int_{-\infty}^{p} dF(p'),$$

where the prime on p is used to distinguish the variable of integration from the upper limit of the integral.   Although such a procedure can be rigorously justified, it suffices for our purpose to remind the reader that an integral is just the limit of a sum.   Therefore if he prefers to think of sums instead of integrals, not only should there be no difficulty, but he will be essentially correct.

We proceed now to find moments of the distribution for the several cases discussed in Sections 3.9 through 3.13.

4.3 MOMENTS FOR TWO EXPERIMENTER-CONTROLLED EVENTS

When the probabilities of applying Q_1 and Q_2 are π_1 and $\pi_2 = 1 - \pi_1$, respectively, we have already seen (Section 3.9) that the probability of applying any sequence of Q's is exactly $\pi_1{}^u \pi_2{}^v$, where u and v are the numbers of times Q_1 and Q_2 have been applied, respectively. If we regard p_{vn}, $v = 1, 2, \cdots, 2^n$, as the p value for one of the 2^n groups of organisms available after the nth random application of the operators to a single initial probability p_0, and, if the proportion in the group is P_{vn}, on the next trial this group splits into two parts as follows:

	New p Value	New Proportion
(4.9)	$Q_1 p_{vn} = a_1 + \alpha_1 p_{vn}$	$\pi_1 P_{vn}$
	$Q_2 p_{vn} = a_2 + \alpha_2 p_{vn}$	$\pi_2 P_{vn}.$

The mth raw moment on trial $n + 1$ will be called $V_{m,n+1}$. It can be evaluated by summing the product of the mth power of the new p values with the new proportions in the manner shown below:

$$V_{m,n+1} = \sum_{v=1}^{2^n} (Q_1 p_{vn})^m \pi_1 P_{vn} + \sum_{v=1}^{2^n} (Q_2 p_{vn})^m \pi_2 P_{vn}$$

(4.10)

$$= \pi_1 \sum_v (a_1 + \alpha_1 p_{vn})^m P_{vn} + \pi_2 \sum_v (a_2 + \alpha_2 p_{vn})^m P_{vn}.$$

Using the binomial expansion of equation 4.6, we obtain

$$V_{m,n+1} =$$

$$\pi_1 \sum_v \sum_u \binom{m}{u} a_1{}^{m-u} \alpha_1{}^u p_{vn}{}^u P_{vn} + \pi_2 \sum_v \sum_u \binom{m}{u} a_2{}^{m-u} \alpha_2{}^u p_{vn}{}^u P_{vn}.$$

After interchanging the order of summation we can sum on v, remembering the definition of moments given in equation 4.1. We get the

RECURRENCE FORMULA FOR THE MOMENTS:

$$(4.11) \quad V_{m,n+1} = \pi_1 \sum_u \binom{m}{u} a_1{}^{m-u} \alpha_1{}^u V_{u,n} + \pi_2 \sum_u \binom{m}{u} a_2{}^{m-u} \alpha_2{}^u V_{u,n}.$$

Thus the mth moment of the distribution on trial $n + 1$ depends on all the moments up to the mth on trial n.

We already know the mean of the distribution of p values on trial n from equation 3.49; knowledge of the spread of the distribution, that is, its variance, would be useful. Then we could know under what circumstances the final distribution of p values is tightly clustered about its mean. Writing out the formulas for the first two raw moments on trial $n + 1$ gives

$$V_{1,n+1} = \pi_1 a_1 + \pi_2 a_2 + (\pi_1 \alpha_1 + \pi_2 \alpha_2)V_{1,n}$$

(4.12)
$$V_{2,n+1} = \pi_1 a_1{}^2 + \pi_2 a_2{}^2 + 2(\pi_1 a_1 \alpha_1 + \pi_2 a_2 \alpha_2)V_{1,n}$$
$$+ (\pi_1 \alpha_1{}^2 + \pi_2 \alpha_2{}^2)V_{2,n}.$$

The formula for the second moment can be written more conveniently as

(4.13)
$$V_{2,n+1} = C_0 + C_1 V_{1,n} + C_2 V_{2,n},$$

where

$$C_0 = \pi_1 a_1{}^2 + \pi_2 a_2{}^2,$$

(4.14)
$$C_1 = 2(\pi_1 a_1 \alpha_1 + \pi_2 a_2 \alpha_2),$$
$$C_2 = \pi_1 \alpha_1{}^2 + \pi_2 \alpha_2{}^2.$$

The solution of the first of equations 4.12 is given by equation 3.49 if we replace $\bar{p}_n$ with $V_{1,n}$, p with $V_{1,0}$, and $\bar{\lambda}$ with $V_{1,\infty}$. We obtain then the

EXPLICIT FORMULA FOR THE MEAN:

(4.15)
$$V_{1,n} = V_{1,\infty} - (V_{1,\infty} - V_{1,0})\bar{\alpha}^n.$$

From equations 3.48 and 3.50 we get the

ASYMPTOTIC FORMULA FOR THE MEAN:

(4.16)
$$V_{1,\infty} = \bar{\lambda} = \frac{\bar{a}}{1 - \bar{\alpha}},$$

where

$$\bar{a} = \pi_1 a_1 + \pi_2 a_2,$$

(4.17)
$$\bar{\alpha} = \pi_1 \alpha_1 + \pi_2 \alpha_2.$$

With this evaluation of $V_{1,n}$ (equation 4.15), substituted into equation 4.13, we get

(4.18) $\quad V_{2,n+1} = (C_0 + C_1 V_{1,\infty}) - C_1(V_{1,\infty} - V_{1,0})\bar{\alpha}^n + C_2 V_{2,n}.$

Writing out equation 4.18 for a few values of n quickly gives us the clue to the solution of this difference equation. Let

$$C_0{}' = C_0 + C_1 V_{1,\infty},$$

(4.19)
$$C_1{}' = -C_1(V_{1,\infty} - V_{1,0}),$$

so that

(4.20)
$$V_{2,n+1} = C_0{}' + C_1{}'\bar{\alpha}^n + C_2 V_{2,n}.$$

We know by definition that

$$V_{2,0} = V_{1,0}{}^2 = p_0{}^2;$$

therefore

$$V_{2,1} = C_0' + C_1' + C_2 p_0{}^2,$$

and

$$V_{2,2} = C_0' + C_1'\bar{\alpha} + C_2(C_0' + C_1' + C_2 p_0{}^2)$$
$$= C_0'(1 + C_2) + C_1'(\bar{\alpha} + C_2) + C_2{}^2 p_0{}^2.$$

Similarly, we can easily show that

$$V_{2,3} = C_0'(1 + C_2 + C_2{}^2) + C_1'(\bar{\alpha}^2 + C_2\bar{\alpha} + C_2{}^2) + C_2{}^3 p_0{}^2.$$

Continuing in this way, we can develop the general formula

(4.21) $$V_{2,n} = C_0' \sum_{u=0}^{n-1} C_2{}^u + C_1' \sum_{u=0}^{n-1} \bar{\alpha}^{n-1-u} C_2{}^u + C_2{}^n p_0{}^2,$$

which can be proved correct by finishing the mathematical induction. (See footnote in Section 3.3.) Provided that $C_2 \neq 1$ and $\bar{\alpha} \neq C_2$, we can perform the summations and get the

EXPLICIT FORMULA FOR THE SECOND MOMENT:

(4.22)
$$V_{2,n} = C_0' \frac{(1 - C_2{}^n)}{1 - C_2} + C_1' \frac{(\bar{\alpha}^n - C_2{}^n)}{\bar{\alpha} - C_2} + C_2{}^n p_0{}^2$$

$$= (C_0 + C_1 V_{1,\infty}) \frac{1 - C_2{}^n}{1 - C_2} - C_1(V_{1,\infty} - p_0) \frac{\bar{\alpha}^n - C_2{}^n}{\bar{\alpha} - C_2} + C_2{}^n p_0{}^2.$$

From the definition of C_2 (equation 4.14) we see that $0 \leq C_2 \leq 1$ since $\pi_1 + \pi_2 = 1$ and $0 \leq \alpha_1{}^2 \leq 1$ and $0 \leq \alpha_2{}^2 \leq 1$. Also, from definition 4.17 we see that $-1 \leq \bar{\alpha} \leq 1$. When $0 \leq C_2 < 1$ and $-1 < \bar{\alpha} < 1$ much of this result tends to zero when $n \to \infty$ and we have the

ASYMPTOTIC FORMULA FOR THE SECOND MOMENT:

(4.23) $$V_{2,\infty} = \frac{C_0 + C_1 V_{1,\infty}}{1 - C_2}.$$

This result can also be obtained more directly by setting $V_{2,\infty+1} = V_{2,\infty}$ in the second of equations 4.12 and solving for $V_{2,\infty}$. Asymptotically, the variance is

(4.24) $$\sigma_\infty{}^2 = V_{2,\infty} - V_{1,\infty}^2 = \frac{C_0 + C_1 V_{1,\infty}}{1 - C_2} - V_{1,\infty}^2.$$

Let us consider an example with parameter values $a_1 = 0.3$, $\alpha_1 = 0.6$, $a_2 = 0.01$, $\alpha_2 = 0.9$, $\pi_1 = \pi_2 = 0.5$.

$$V_{1,\infty} = \frac{0.5(0.31)}{1 - 0.5(1.5)} = 0.62, \quad C_0 = 0.04505, \quad C_1 = 0.189, \quad C_2 = 0.585.$$

For this example the asymptotic mean is 0.62, and the asymptotic standard deviation is $\sigma_\infty = 0.08$.

The results of this section, together with a reasonable amount of computation, can tell us a good bit about the behavior of the probability distribution on successive trials. In addition to the moments, however, it might be valuable to have the percentage points of the cumulative distribution in the asymptotic case. For any finite number of applications of the operators the cumulative can be computed with the help of diagrams like Fig. 3.2. However, the labor shortly becomes prohibitive because 2^n increases so rapidly. For the asymptotic distribution we can use the theorem presented in Section 4.7.

*4.4 MOMENTS FOR t EXPERIMENTER-CONTROLLED EVENTS

In this section we generalize the analysis of the preceding section to any number t of events. For two response classes the generalization is direct. If the probability of application of Q_i is π_i, where

(4.25)
$$\sum_{i=1}^{t} \pi_i = 1,$$

equation 4.10 is replaced by

$$V_{m,n+1} = \sum_{i=1}^{t} \pi_i \Big\{ \sum_{\nu=1}^{t^n} (Q_i p_{\nu n})^m P_{\nu n} \Big\}$$

(4.26)
$$= \sum_{i=1}^{t} \pi_i \Big\{ \sum_{\nu=1}^{t^n} (a_i + \alpha_i p_{\nu n})^m P_{\nu n} \Big\}.$$

Using the binomial expansion, interchanging the order of summation of ν and u, and using the definitions of the moments, we get

(4.27)
$$V_{m,n+1} = \sum_{i=1}^{t} \pi_i \Big\{ \sum_{u=0}^{m} \binom{m}{u} a_i^{m-u} \alpha_i^{u} V_{u,n} \Big\}.$$

Computations similar to those in Section 4.3 give for the means

(4.28)
$$V_{1,n} = V_{1,\infty} - (V_{1,\infty} - p_0)\bar{\alpha}^n,$$

and, for the second raw moments,

(4.29)

$$V_{2,n} = (C_0 + C_1 V_{1,\infty}) \frac{1 - C_2^n}{1 - C_2} - C_1(V_{1,\infty} - p_0) \frac{(\bar{\alpha}^n - C_2^n)}{\bar{\alpha} - C_2} + C_2^n p_0^2,$$

where

(4.30)

$$V_{1,\infty} = \frac{\bar{a}}{1 - \bar{\alpha}}, \qquad C_0 = \sum_{i=1}^{t} \pi_i a_i^2,$$

$$\bar{a} = \sum_{i=1}^{t} \pi_i a_i, \qquad C_1 = 2 \sum_{i=1}^{t} \pi_i a_i \alpha_i,$$

$$\bar{\alpha} = \sum_{i=1}^{t} \pi_i \alpha_i, \qquad C_2 = \sum_{i=1}^{t} \pi_i \alpha_i^2.$$

The preceding analysis can be extended to more than two response classes by using the matrix operators of Section 1.8. If event E_i occurs, the operator,

(4.31)
$$\mathbf{T}_i = \alpha_i \mathbf{I} + (1 - \alpha_i)\mathbf{\Lambda}_i,$$

is applied to the response probability vector. Event E_i occurs with probability π_i, and

(4.32)
$$\sum_{i=1} \pi_i = 1,$$

since we assume that precisely one event occurs on each trial. The remaining problem is to specify something about the distributions of probabilities which obtain.

We are necessarily involved in a multivariate distribution of probability values on each trial. Such a distribution on any trial will have moments and cross-product moments, but we shall restrict our attention to the means of the marginal distributions. There will be t^n possible sequences leading up to trial n; the νth sequence has probability of occurrence $P_{\nu n}$, and the probability vector generated by that sequence is $\mathbf{p}_{\nu n}$. We then define a *vector of marginal means*

(4.33)
$$\mathbf{V}_{1,n} = \sum_{\nu=1}^{t^n} P_{\nu n}\mathbf{p}_{\nu n}.$$

This vector on the next trial is given by

(4.34)
$$\mathbf{V}_{1,n+1} = \sum_{\nu=1}^{t^n} P_{\nu n}\{ \sum_{i=1}^{t} \pi_i \mathbf{T}_i \mathbf{p}_{\nu n}\}.$$

When we apply an operator $\mathbf{T}_i$ to $\mathbf{p}_{\nu n}$ we get

(4.35)
$$\mathbf{T}_i \mathbf{p}_{\nu n} = \alpha_i \mathbf{p}_{\nu n} + (1 - \alpha_i)\mathbf{\lambda}_i,$$

where λ_i is the limit vector of T_i. Using this in the expression for $V_{1,n+1}$, we have

(4.36)
$$V_{1,n+1} = \sum_{\nu=1}^{t^n} P_{\nu n} \{ \sum_{i=1}^{t} \pi_i [\alpha_i p_{\nu n} + (1 - \alpha_i)\lambda_i] \}.$$

After interchanging the order of summation, using the above definition for $V_{1,n}$ and the fact that the $P_{\nu n}$ sum to unity, we have

(4.37)
$$V_{1,n+1} = \{ \sum_{i=1}^{t} \pi_i \alpha_i \} V_{1,n} + \sum_{i=1}^{t} \pi_i (1 - \alpha_i)\lambda_i.$$

As before we let

(4.38) .
$$\bar{\alpha} = \sum_{i=1}^{t} \pi_i \alpha_i,$$

and also define an average limit vector by

(4.39)
$$\bar{\lambda} = \frac{\sum_{i=1}^{t} \pi_i (1 - \alpha_i)\lambda_i}{\sum_{i=1}^{t} \pi_i (1 - \alpha_i)} = \frac{\sum_{i=1}^{t} \pi_i (1 - \alpha_i)\lambda_i}{1 - \bar{\alpha}}.$$

We can then write

(4.40)
$$V_{1,n+1} = \bar{\alpha} V_{1,n} + (1 - \bar{\alpha})\bar{\lambda}.$$

This vector difference equation may be solved at once by the method used in Section 3.5 to yield

(4.41)
$$V_{1,n} = \bar{\alpha}^n V_{1,0} + (1 - \bar{\alpha}^n)\bar{\lambda}.$$

Provided only that $|\bar{\alpha}| < 1$, the asymptotic vector of marginal means is

(4.42)
$$V_{1,\infty} = \bar{\lambda}.$$

These results are closely related to those obtained by Neimark [1]. The main difference is that in the problems she treated, special restrictions on the α_i and λ_i were appropriate. (See Chapter 13 for discussion of experiment.)

4.5 MOMENTS FOR TWO SUBJECT-CONTROLLED EVENTS

We consider next the special case of variable probability when there are two possible responses and the probability of application of Q_1 is identical with the p value at the time. We proceed in a manner quite the same as that used for experimenter-controlled events.

If $p_{\nu n}$ is the p value for the νth group of organisms after n applications of the operators, if P_{vn} is the size of the vth group expressed as a proportion of the whole population, and if the operators are applied one more time, we get the following table:

	New p Values	New Proportion
(4.43)	$Q_1 p_{vn} = a_1 + \alpha_1 p_{vn}$	$p_{vn} P_{vn}$
	$Q_2 p_{vn} = a_2 + \alpha_2 p_{vn}$	$(1 - p_{vn}) P_{vn}.$

Note that the only difference between this table and equation 4.9 is the replacement of the constants π_1 and π_2 by the variables p_{vn} and $1 - p_{vn}$ in the calculation of the new proportions. The mth moment on the $(n + 1)$st trial is the sum of the mth powers of the p values weighted by the group sizes over all the groups. Thus

$$(4.44) \qquad V_{m,n+1} = \sum_{v=1}^{2^n} (Q_1 p_{vn})^m p_{vn} P_{vn} + \sum_{v=1}^{2^n} (Q_2 p_{vn})^m (1 - p_{vn}) P_{vn}$$

$$= \sum_v (a_1 + \alpha_1 p_{vn})^m p_{vn} P_{vn} + \sum_v (a_2 + \alpha_2 p_{vn})^m (1 - p_{vn}) P_{vn}.$$

(In the previous section we factored out the π's at this point, but the p's cannot be factored out.) Using the binomial expansion of equation 4.6, we have

$$V_{m,n+1} = \sum_v \sum_{u=0}^m \binom{m}{u} a_1^{m-u} \alpha_1^u p_{vn}^{u+1} P_{vn} - \sum_v \sum_{u=0}^m \binom{m}{u} a_2^{m-u} \alpha_2^u p_{vn}^{u+1} P_{vn}$$

$$(4.45)$$

$$+ \sum_v \sum_{u=0}^m \binom{m}{u} a_2^{m-u} \alpha_2^u p_{vn}^u P_{vn}.$$

In view of the definition of moments, we can sum on v in each of the three sums on the right of the last equation to get the

RECURRENCE FORMULA FOR THE MOMENTS:

$$V_{m,n+1} = \sum_{u=0}^m \binom{m}{u} a_1^{m-u} \alpha_1^u V_{u+1,n} - \sum_{u=0}^m \binom{m}{u} a_2^{m-u} \alpha_2^u V_{u+1,n}$$

$$(4.46)$$

$$+ \sum_{u=0}^m \binom{m}{u} a_2^{m-u} \alpha_2^u V_{u,n}.$$

This equation can be written in a more convenient form if we collect together the terms in $V_{1,n}$, $V_{2,n}$, etc., on the right side. We need to determine a set of coefficients $C_{m,u}$ which allows us to write

$$(4.47) \qquad V_{m,n+1} = \sum_{u=0}^{m+1} C_{m,u} V_{u,n}.$$

The problem, therefore, is to discover the correct form of the coefficients.

We first change the dummy index from u to v in the first two sums on the right of equation 4.46 to get

$$(4.48) \qquad \begin{aligned} V_{m,n+1} = &\sum_{v=0}^{m} \binom{m}{v} \{a_1{}^{m-v}\alpha_1{}^{v} - a_2{}^{m-v}\alpha_2{}^{v}\} V_{v+1,n} \\ &+ \sum_{u=0}^{m} \binom{m}{u} a_2{}^{m-u}\alpha_2{}^{u} V_{u,n}. \end{aligned}$$

Then we let $v = u - 1$ and have

$$(4.49) \qquad \begin{aligned} V_{m,n+1} = &\sum_{u=1}^{m+1} \binom{m}{u-1} \{a_1{}^{m-u+1}\alpha_1{}^{u-1} - a_2{}^{m-u+1}\alpha_2{}^{u-1}\} V_{u,n} \\ &+ \sum_{u=0}^{m} \binom{m}{u} a_2{}^{m-u}\alpha_2{}^{u} V_{u,n}. \end{aligned}$$

We then separate out the terms for $u = 0$ in the last sum and for $u = m + 1$ in the first sum and combine the other terms to get

$$(4.50) \qquad \begin{aligned} V_{m,n+1} = &a_2{}^{m} V_{0,n} + \sum_{u=1}^{m} \left\{ \binom{m}{u-1}(a_1{}^{m-u+1}\alpha_1{}^{u-1} - a_2{}^{m-u+1}\alpha_2{}^{u-1}) \right. \\ &\left. + \binom{m}{u} a_2{}^{m-u}\alpha_2{}^{u} \right\} V_{u,n} + (\alpha_1{}^{m} - \alpha_2{}^{m}) V_{m+1,n}. \end{aligned}$$

Therefore the coefficients $C_{m,u}$ are

$$(4.51) \quad C_{m,u} = \begin{cases} a_2{}^{m} & (u = 0) \\ \binom{m}{u-1}(a_1{}^{m-u+1}\alpha_1{}^{u-1} - a_2{}^{m-u+1}\alpha_2{}^{u-1}) + \binom{m}{u} a_2{}^{m-u}\alpha_2{}^{u} & \\ & (u = 1, 2, \cdots, m) \\ \alpha_1{}^{m} - \alpha_2{}^{m} & (u = m + 1). \end{cases}$$

For the mean on trial $n + 1$ we have

(4.52) $$V_{1,n+1} = C_{10} + C_{11}V_{1,n} + C_{12}V_{2,n},$$

where

(4.53) $$\begin{aligned} C_{10} &= a_2 \\ C_{11} &= a_1 - a_2 + \alpha_2 \\ C_{12} &= \alpha_1 - \alpha_2. \end{aligned}$$

For the second raw moment we have

(4.54) $$V_{2,n+1} = C_{20} + C_{21}V_{1,n} + C_{22}V_{2,n} + C_{23}V_{3,n},$$

where

(4.55) $$\begin{aligned} C_{20} &= a_2{}^2 \\ C_{21} &= a_1{}^2 - a_2{}^2 + 2a_2\alpha_2 \\ C_{22} &= 2(a_1\alpha_1 - a_2\alpha_2) + \alpha_2{}^2 \\ C_{23} &= \alpha_1{}^2 - \alpha_2{}^2. \end{aligned}$$

The result just obtained in equations 4.47 and 4.51 is a general recurrence relation for all the raw moments. It has one unfortunate aspect. It will be noted that the mth moment on the $(n + 1)$st trial depends on all the moments through the $(m + 1)$st on the nth trial. If we start with a particular value of p_0 at the zeroth trial and try to trace through, say, the mean for the first six trials, then at the first trial we need $V_{1,0}$ and $V_{2,0}$. At the second trial we need $V_{1,1}$ and $V_{2,1}$, which in turn implies that we need $V_{3,0}$. At the third trial $V_{1,2}$ and $V_{2,2}$ generate $V_{1,1}$, $V_{2,1}$, and $V_{3,1}$, which in turn require $V_{1,0}$, $V_{2,0}$, $V_{3,0}$, and $V_{4,0}$. So to get $V_{1,6}$ we require $V_{7,0}$. This dependence on more and more early moments is tedious, not because of the difficulty of computing $V_{m,0}$—which is a slight task—but because substitutions must be made into longer and longer equations, and more and more equations. To continue the example of $V_{1,6}$, we need a total of twenty-eight evaluations (if we regard obtaining a quantity like $V_{3,4}$ as one evaluation). The quantities needed are

$$\begin{aligned} V_{m,0} &\qquad m = 1, 2, \cdots, 7 \\ V_{m,1} &\qquad m = 1, 2, \cdots, 6 \\ V_{m,2} &\qquad m = 1, 2, \cdots, 5 \\ V_{m,3} &\qquad m = 1, 2, 3, 4 \\ V_{m,4} &\qquad m = 1, 2, 3 \\ V_{m,5} &\qquad m = 1, 2 \\ V_{m,6} &\qquad m = 1. \end{aligned}$$

All these are quite apart from the values of the coefficients $C_{m,u}$ that must be computed.   Except for small numbers of trials then, or for special problems that are treated later, the moments are awkward to evaluate. However, it may sometimes be valuable to evaluate them for small n in some specific cases to see how the curves for $V_{1,n}$ begin.   It is probably about the same amount of trouble to compute the moments from the successive applications of the recurrence moment formulas as it is to compute the p values and probabilities for the 2^n groups at the mth application of the operators and then compute the moments from their definition.   With either procedure, rounding errors are cumulative and must be watched carefully.

Even though we do not do much with the general recurrence formula for the moments here, it is the starting point of our investigations of special problems treated in later chapters.

4.6 MOMENTS FOR EXPERIMENTER-SUBJECT-CONTROLLED EVENTS

We here consider the problem introduced in Section 3.13 and compute the moments for the probability distribution.   We restrict the discussion to two alternatives, A_1 and A_2, with probabilities p and $1 - p$, respectively, but allow s possible outcomes O_k.   The conditional probability of outcome O_k, given alternative A_j, is π_{jk}.   The operators, Q_{jk}, are defined by

$$Q_{jk}p = \alpha_{jk}p + (1 - \alpha_{jk})\lambda_{jk}$$

(4.56)
$$= a_{jk} + \alpha_{jk}p.$$

The probability of application of Q_{1k} is $p\pi_{1k}$, and the probability of application of Q_{2k} is $(1 - p)\pi_{2k}$.   Furthermore,

(4.57)
$$\sum_{k=1}^{s} \pi_{jk} = 1, \qquad j = 1, 2.$$

On trial n there are $(2s)^n$ possible values of p.   The values of p are denoted by p_{vn}, and the probability of the sequence up to p_{vn} is denoted by P_{vn}. Hence the mth raw moment on trial $n + 1$ is

(4.58)
$$
\begin{aligned}
V_{m,n+1} = \sum_{v=1}^{(2s)^n} &\{ p_{vn}P_{vn} \sum_{k=1}^{s} \pi_{1k}(Q_{1k}p_{vn})^m \\
&+ (1 - p_{vn})P_{vn} \sum_{k=1}^{s} \pi_{2k}(Q_{2k}p_{vn})^m \} \\
= \sum_{v=1}^{(2s)^n} &\{ p_{vn}P_{vn} \sum_{k=1}^{s} \pi_{1k}[a_{1k} + \alpha_{1k}p_{vn}]^m \\
&+ (1 - p_{vn})P_{vn} \sum_{k=1}^{s} \pi_{2k}[a_{2k} + \alpha_{2k}p_{vn}]^m \}.
\end{aligned}
$$

As before, we expand the expressions in square brackets by the binomial expansion and get the general

RECURRENCE FORMULA FOR THE MOMENTS:

$$
(4.59) \quad V_{m,n+1} = \sum_{k=1}^{s} \sum_{u=0}^{m} \binom{m}{u} \{ \pi_{1k} a_{1k}{}^{m-u} \alpha_{1k}{}^{u} V_{u+1,n} \\
+ \pi_{2k} a_{2k}{}^{m-u} \alpha_{2k}{}^{u} (V_{u,n} - V_{u+1,n}) \}.
$$

In particular, for the means we have

$$
V_{1,n+1} = \sum_{k=1}^{s} \{ \pi_{1k}[a_{1k}V_{1,n} + \alpha_{1k}V_{2,n}] \\
+ \pi_{2k}[a_{2k}(1 - V_{1,n}) + \alpha_{2k}(V_{1,n} - V_{2,n})] \}
$$

$$
(4.60) \quad = \sum_{k=1}^{s} \pi_{2k} a_{2k} \\
+ [\sum_{k=1}^{s} (\pi_{1k}a_{1k} - \pi_{2k}a_{2k} + \pi_{2k}\alpha_{2k})] V_{1,n} \\
+ [\sum_{k=1}^{s} (\pi_{1k}\alpha_{1k} - \pi_{2k}\alpha_{2k})] V_{2,n}.
$$

For compactness in writing we define

$$
(4.61) \quad \bar{a}_j = \sum_{k=1}^{s} \pi_{jk}a_{jk}, \quad j = 1, 2, \\
\bar{\alpha}_j = \sum_{k=1}^{s} \pi_{jk}\alpha_{jk}, \quad j = 1, 2,
$$

and so we can then write

$$
(4.62) \quad V_{1,n+1} = \bar{a}_2 + (\bar{a}_1 - \bar{a}_2 + \bar{\alpha}_2)V_{1,n} + (\bar{\alpha}_1 - \bar{\alpha}_2)V_{2,n}.
$$

The similarity of this result to equation 4.52 may be noted.

4.7 THEOREMS ABOUT THE p-VALUE DISTRIBUTIONS

In this section we state several theorems about the distributions of p values. Rigorous proofs of these theorems are long and require advanced methods, and so are not included.

Some important theorems and references to related mathematical papers are given by Karlin [5].

TRAPPING THEOREM FOR $r = 2$. This theorem specifies an interval within which p values are ultimately trapped. First, consider two event operators having unique limit points and non-negative α's, that is, operators for which $0 \leq \alpha_i < 1$. The theorem asserts that if the p value of any sequence is ever between the two limit points the sequence forever remains there; this part of the theorem is intuitively obvious, for neither operator can take a p value outside the interval. The theorem also asserts that, when the sequence is outside the interval, it will later enter that interval with probability 1, provided that the probability of application of each operator is never zero. For more than two event operators with unique limit points and non-negative α's, the same statements apply to the interval between the largest and smallest limit points. Similar theorems have been proved [2] without the restriction to non-negative α's, but we are not concerned with negative α's in this book. When only one operator has a unique limit point, that is, an α less than unity, all sequences will approach that limit point with probability 1 (see Section 8.5).

TRAPPING THEOREM FOR $r \geq 3$. For more than two response classes, a similar trapping theorem applies to the *convex union* of the operator limit points. (The convex union of a set of points is the smallest convex set containing those points.) For example, when $r = 3$, the probability vectors can be represented by points in a plane. If there are three operators with distinct limit points, the convex union of those three points is the triangle formed by connecting the points with straight lines, and the theorem says that all sequences of probability vectors will be trapped in this convex union. A similar theorem applies when negative α's are allowed [2].

ASYMPTOTIC DISTRIBUTION THEOREM. For two response classes and experimenter-controlled events, it has been shown that an asymptotic distribution exists, that is, that the p-value distributions on trials n and $n + 1$ may be made as close to one another as desired by making n sufficiently large. Furthermore, this distribution is independent of the initial distribution of p values. For two subject-controlled events, Harris has shown [3] that the same theorem is true provided that the α_i are non-negative and the absolute value of the difference between the two limit points is less than unity. When the limits are zero and unity, special cases arise that are discussed in Chapter 7.

ERGODIC THEOREM FOR SINGLE SEQUENCES. Whenever an asymptotic distribution of p values exists, single sequences of p values possess an important property. Such a single sequence traces out a population of p values which form a distribution. In the limit as $n \to \infty$, this distribution is almost certainly equivalent to the distribution from all possible sequences. In Section 6.3, we develop a computation scheme, based upon this theorem, for approximating the form of the asymptotic distribution.

4.8 LENGTH OF RUNS

In this section we are concerned with the expected or average length of a run of responses of one kind or another. Suppose we have a long sequence of A_1 and A_2 responses and we inquire about the average length of a run of A_1 responses or A_2 responses or both. For example, if the sequence is $A_1A_1A_2A_1A_1A_1A_2A_2A_2A_2A_2$, we observe two runs of A_1 responses, one of length two and one of length three so that the mean length of A_1 runs is 2.5. Similarly we observe two runs of A_2 responses, one of length one and one of length five, giving a mean length of A_2 runs of 3.0.

If the probability p of an A_1 response is constant it is a simple matter to compute the mean or expected length of a run. The solution to this problem is well known [4], but we develop it here as an illustration. Let there be an A_1 response on some trial n preceded by an A_2 response on trial $n - 1$; then an A_1 run begins on trial n. The probability that the run is of length one is $(1 - p)$ since this is the probability of an A_2 response on trial $n + 1$. The probability that the run is of length two is $p(1 - p)$, that is, the probability that an A_1 occurs on trial $n + 1$ and an A_2 occurs on trial $n + 2$. More generally, the probability that the run is of length v is $p^{v-1}(1 - p)$. For convenience in computations we introduce a *random variable* R_1 which represents the length of a run of A_1 responses. It has possible values of $1, 2, \cdots, v, \cdots$, and the probability that R_1 has the value v is

$$(4.63) \qquad Pr\{R_1 = v\} = p^{v-1}(1 - p).$$

The mean length of an A_1 run is simply the expected value of R_1:

$$(4.64) \qquad \begin{aligned} E(R_1) &= \sum_{v=1}^{\infty} v Pr\{R_1 = v\} = \sum_{v=1}^{\infty} v p^{v-1}(1 - p) \\ &= \sum_{v=1}^{\infty} v p^{v-1} - \sum_{v=1}^{\infty} v p^{v}. \end{aligned}$$

If we write out a few terms in these sums we quickly see how to evaluate. When $p < 1$ we may re-arrange terms as follows (the series is absolutely convergent):

$$(4.65) \qquad \begin{aligned} E(R_1) &= (1 + 2p + 3p^2 + \cdots) - (p + 2p^2 + 3p^3 + \cdots) \\ &= 1 + p + p^2 + p^3 + \cdots. \end{aligned}$$

For $p < 1$ this last series is simply the expansion of $(1 - p)^{-1}$; the reader may verify this by long division. Hence

$$(4.66) \qquad E(R_1) = \frac{1}{1 - p}.$$

If $p = 1/2$ for example, as in a coin-flipping experiment, $E(R_1) = 2$, that is, the average length of a run of heads is 2. Or, if $p = 1/6$ is the probability of a four in a roll of a die, the expected length of a run of fours is $E(R_1) = (1 - 1/6)^{-1} = 1.2$.

When the probability p_n of an A_1 response on trial n changes from trial to trial, we can compute the probability of runs of various lengths in certain special cases. Suppose again that an A_2 response occurs on trial $n - 1$ and an A_1 on trial n. The probability that the run is of length one is $1 - p_{n+1}$; the probability of a run of length two is $p_{n+1}(1 - p_{n+2})$, etc. We introduce a random variable $R_{1,n}$ denoting the length of a run of A_1 responses beginning on trial n, and write

$$
\begin{aligned}
Pr\{R_{1,n} = \nu\} &= p_{n+1}p_{n+2} \cdots p_{n+\nu-1}(1 - p_{n+\nu}) \\
(4.67) \qquad &= \left\{ \prod_{K=1}^{\nu-1} p_{n+K} \right\} (1 - p_{n+\nu}) \\
&= \prod_{K=1}^{\nu-1} p_{n+K} - \prod_{K=1}^{\nu} p_{n+K} \\
&\qquad\qquad (\nu = 2, 3, \cdots).
\end{aligned}
$$

We find it convenient to define a quantity $\tau_{n,\nu}$ by

$$
\begin{aligned}
\tau_{n,\nu} &= \prod_{K=1}^{\nu} p_{n+K} \qquad \nu = 1, 2, \cdots . \\
(4.68) \qquad & \\
\tau_{n,0} &= 1.
\end{aligned}
$$

With this definition we can write equation 4.67 in the form

$$(4.69) \qquad Pr\{R_{1,n} = \nu\} = \tau_{n,\nu-1} - \tau_{n,\nu}, \qquad \nu = 1, 2, \cdots .$$

The expected value of $R_{1,n}$ is the mean length of run beginning on trial n. It is

$$
\begin{aligned}
E(R_{1,n}) &= \sum_{\nu=1}^{\infty} \nu Pr\{R_{1,n} = \nu\} \\
(4.70) \qquad &= \sum_{\nu=1}^{\infty} \nu(\tau_{n,\nu-1} - \tau_{n,\nu}) \\
&= \sum_{\nu=1}^{\infty} \nu\tau_{n,\nu-1} - \sum_{\nu=1}^{\infty} \nu\tau_{n,\nu}.
\end{aligned}
$$

By writing out a few terms we see that

$$
\begin{aligned}
E(R_{1,n}) &= (\tau_{n,0} + 2\tau_{n,1} + 3\tau_{n,2} + \cdots) \\
(4.71) \qquad &\qquad - (\tau_{n,1} + 2\tau_{n,2} + 3\tau_{n,3} + \cdots) \\
&= \tau_{n,0} + \tau_{n,1} + \tau_{n,2} + \cdots \\
&= \sum_{\nu=0}^{\infty} \tau_{n,\nu}.
\end{aligned}
$$

We have assumed that the series is absolutely convergent.

For any particular sequence of response probabilities, p_n, if we can compute the products $\tau_{n,\nu}$ of equation 4.68, we can obtain the distribution function and the expected run length. For the case considered previously, when p_n is constant, $\tau_{n,\nu}$ reduces to p^ν and $E(R_{1,n}) = 1/(1 - p)$ as obtained before. In addition to the mean run length we may also wish to obtain its variance. Thus we first compute the expected value of $R_{1,n}^2$, and then use the formula

$$(4.72) \qquad \sigma^2(R_{1,n}) = E(R_{1,n}^2) - [E(R_{1,n})]^2.$$

We have then

$$(4.73) \qquad
\begin{aligned}
E(R_{1,n}^2) &= \sum_{\nu=1}^{\infty} \nu^2 Pr\{R_{1,n} = \nu\} \\
&= \sum_{\nu=1}^{\infty} \nu^2(\tau_{n,\nu-1} - \tau_{n,\nu}).
\end{aligned}$$

Again, we write out a few terms:

$$(4.74) \qquad
\begin{aligned}
E(R_{1,n}^2) &= (\tau_{n,0} + 4\tau_{n,1} + 9\tau_{n,2} + 16\tau_{n,3} + \cdots) \\
&\quad - (\tau_{n,1} + 4\tau_{n,2} + 9\tau_{n,3} + \cdots) \\
&= \tau_{n,0} + 3\tau_{n,1} + 5\tau_{n,2} + 7\tau_{n,3} + \cdots.
\end{aligned}$$

We may write this as a sum of two series:

$$(4.75) \qquad
\begin{aligned}
E(R_{1,n}^2) &= \tau_{n,0} + \tau_{n,1} + \tau_{n,2} + \tau_{n,3} + \cdots \\
&\quad\quad + 2\tau_{n,1} + 4\tau_{n,2} + 6\tau_{n,3} + \cdots \\
&= \sum_{\nu=0}^{\infty} \tau_{n,\nu} + 2 \sum_{\nu=0}^{\infty} \nu\tau_{n,\nu} \\
&= 2 \sum_{\nu=0}^{\infty} \nu\tau_{n,\nu} + E(R_{1,n}).
\end{aligned}$$

Therefore, the variance is

$$(4.76) \qquad \sigma^2(R_{1,n}) = 2 \sum_{\nu=0}^{\infty} \nu\tau_{n,\nu} + E(R_{1,n}) - [E(R_{1,n})]^2.$$

Thus we see again that, if we can compute the products $\tau_{n,\nu}$, we can use these in obtaining further information about length of runs. For the previously discussed case of fixed probabilities, for which $\tau_{n,\nu} = p^\nu$, we have

$$(4.77) \qquad \sum_{\nu=0}^{\infty} \nu\tau_{n,\nu} = \sum_{\nu=0}^{\infty} \nu p^\nu = p + 2p^2 + 3p^3 + \cdots.$$

This series is well known, and the reader may easily verify by the binomial expansion of $(1 - p)^{-2}$ that

$$(4.78) \qquad \sum_{\nu=0}^{\infty} \nu p^{\nu} = \frac{p}{(1-p)^2}.$$

The variance is then

$$(4.79) \qquad \sigma^2(R_{1,n}) = \frac{2p}{(1-p)^2} + \frac{1}{(1-p)} - \frac{1}{(1-p)^2} = \frac{p}{(1-p)^2}.$$

We now apply the preceding analysis to a more complicated case, that discussed in Section 3.3, for which a single operator Q_1 is applied repeatedly. We then obtain $p_{n+K} = Q_1^{n+K} p_0$ for the probability of an A_1 response on trial $n + K$. We further assume that λ_1, the limit point of Q_1, is zero so that

$$(4.80) \qquad p_{n+K} = \alpha_1^{n+K} p_0.$$

Equation 4.68 then gives for the product $\tau_{n,\nu}$,

$$(4.81) \qquad \begin{aligned} \tau_{n,\nu} &= \prod_{K=1}^{\nu} \alpha_1^{n+K} p_0 = p_0^{\nu} \alpha_1^{n} \prod_{K=1}^{\nu} \alpha_1^{K} \\ &= p_0^{\nu} \alpha_1^{n} \{(\alpha_1)(\alpha_1^2)(\alpha_1^3) \cdots (\alpha_1^{\nu})\} \qquad \nu \neq 0. \end{aligned}$$

The product in the brackets is α_1 raised to a power which is the sum of the first ν integers. It is well known (and easily verified) that

$$(4.82) \qquad 1 + 2 + 3 + \cdots + \nu = \frac{\nu(\nu + 1)}{2}.$$

Thus

$$(4.83) \qquad \tau_{n,\nu} = p_0^{\nu} \alpha_1^{n} \alpha_1^{\nu(\nu+1)/2} \qquad \nu \neq 0.$$

The expected run length of equation 4.71 is then

$$(4.84) \qquad E(R_{1,n}) = \alpha_1^{n} \sum_{\nu=1}^{\infty} \alpha_1^{\nu(\nu+1)/2} p_0^{\nu} + 1.$$

This sum cannot be evaluated except by numerical means, but in Table A at the end of the book we give values of the function

$$(4.85) \qquad \Phi(\alpha, \beta) = \sum_{\nu=0}^{\infty} \alpha^{\nu(\nu+1)/2} \beta^{\nu}.$$

Thus, for our present problem, we have

$$(4.86) \qquad E(R_{1,n}) = \alpha_1^{n} \Phi(\alpha_1, p_0) + (1 - \alpha_1^{n}).$$

For known values of α_1 and p_0, we may readily compute $E(R_{1,n})$ by using Table A.

The variance of the lengths of A_1 runs when equation 4.80 gives the probabilities is

(4.87) $\sigma^2(R_{1,n}) = 2 \sum_{\nu=0}^{\infty} \nu p_0{}^\nu \alpha_1{}^n \alpha_1{}^{\nu(\nu+1)/2} + E(R_{1,n}) - [E(R_{1,n})]^2.$

Also in Table A we give the function

(4.88) $$\Psi(\alpha, \beta) = \sum_{\nu=0}^{\infty} \nu \alpha^{\nu(\nu+1)/2} \beta^\nu,$$

and so we may write

(4.89) $$\sigma^2(R_{1,n}) = 2\alpha_1{}^n \Psi(\alpha_1, p_0) + E(R_{1,n}) - [E(R_{1,n})]^2.$$

For known values of α_1 and p_0 we can compute this variance by using Table A.

The mean and variance of the distribution of run lengths can be easily computed only in certain special cases. We have illustrated the procedure when p_n is constant and when $p_n = \alpha^n p_0$. Unfortunately, most other cases are rather complicated and require tedious numerical computations. We shall have occasion in Chapters 8 and 11 to use some of the results of this section. We summarize these results by the following formulas:

$$E(R_{1,n}) = \sum_{\nu=0}^{\infty} \tau_{n,\nu}$$

$$\sigma^2(R_{1,n}) = 2 \sum_{\nu=0}^{\infty} \nu \tau_{n,\nu} + E(R_{1,n}) - [E(R_{1,n})]^2$$

$$\tau_{n,\nu} = \prod_{K=1}^{\nu} p_{n+K}, \quad \nu = 1, 2, \cdots,$$

$$\tau_{n,0} = 1.$$

For runs of A_2 instead of A_1 responses, we need a random variable $R_{2,n}$ representing the length of an A_2 run beginning on trial n. Its expected value and variance are

(4.90) $$E(R_{2,n}) = \sum_{\nu=0}^{\infty} \tilde{\tau}_{n,\nu},$$

(4.91) $$\sigma^2(R_{2,n}) = 2 \sum_{\nu=0}^{\infty} \nu \tilde{\tau}_{n,\nu} + E(R_{2,n}) - [E(R_{2,n})]^2,$$

where

(4.92) $$\tilde{\tau}_{n,\nu} = \prod_{K=1}^{\nu} q_{n+K}, \quad \tau_{n,0} = 1.$$

The q's are the A_2 response probabilities.

4.9 SUMMARY

Distributions of response probabilities produced by various types of event sequences are analyzed in this chapter; recurrence formulas for the moments are developed for all cases discussed. For experimenter-controlled events, the recurrence formulas for the means and second raw moments are solved to yield explicit formulas in Sections 4.3 and 4.4. These derivations are restricted to two response classes, but in Section 4.4 the analysis of the means is extended to any number r of responses. For subject-controlled and experimenter-subject-controlled events, discussed in Sections 4.5 and 4.6, respectively, recurrence formulas for the moments are derived but not solved to give explicit formulas.

Several theorems about the distributions of p values are stated without proof in Section 4.7. These theorems are important in studying the properties of the mathematical system but have little direct relevance to applications of the system. Finally, in Section 4.8, we discuss the distributions of run lengths and derive expressions to be used in Chapters 8 and 11.

REFERENCES

1. Neimark, E. D. *Effects of type of non-reinforcement and number of alternative responses in two verbal conditioning situations.* Ph.D. thesis, Indiana University, 1953.
2. Bush, R. R., Mosteller, F., and Thompson, G. L. A formal structure for multiple choice situations, *Decision processes* (edited by R. M. Thrall, C. H. Coombs, and R. L. Davis), New York: Wiley, 1954, 99–126.
3. Harris, T. E. Personal communication and abstract: *Annals of math. Stat.*, 1952 **23**, 141.
4. Feller, W. *An introduction to probability theory and its applications.* New York: Wiley, 1950, pp. 56–59.
5. Karlin, S. Some random walks arising in learning models I. *Pacific J. of Math.*, 1953, **3**, 725–756.

CHAPTER 5

The Equal Alpha Condition

5.1 INTRODUCTION

In a number of experimental designs, the psychologist introduces a certain kind of complementarity between two events. Roughly speaking, event E_1 has the same effect on response A_1 as event E_2 has on response A_2. For example, a subject may be asked to predict which of two light bulbs will be illuminated on each trial. The responses are the two possible predictions, and the events are the turning on of the light bulbs. We would expect that the illumination of the left bulb would influence the probability of predicting "left" in the same way as the illumination of the right bulb would alter the probability of predicting "right." In other words, if the roles of the two events were interchanged and the response labels were reversed, the basic design of the experiment would not be altered.

Another example of this kind of symmetry is the Brunswik T-maze experiment first described in the Introduction. We would expect that a rewarded right turn would have the same effect on p, the probability of turning right, as a rewarded left turn would have on q, the probability of turning left. Again, "right" and "left" could be interchanged without changing the design. Indeed, in many such experiments, half of the subjects are trained with one choice "favorable" and the other half with the opposite choice "favorable."

The symmetry or complementarity which exists in the design of experiments such as those just mentioned should not be confused with position preferences or initial tendencies to make one response more than the other. These preferences may be described by taking the initial response probabilities different from 0.5, whereas the symmetry has to do with the event operators.

Within the framework of our general model, we can make the symmetry notion described above precise. Consider two responses A_1 and A_2, with probabilities p and $q = 1 - p$, and two events E_1 and E_2 with operators Q_1 and Q_2 defined by

(5.1)
$$Q_1p = \alpha_1 p + (1 - \alpha_1)\lambda_1,$$
$$Q_2p = \alpha_2 p + (1 - \alpha_2)\lambda_2.$$

It is convenient to replace the second equation by one involving the complementary operator $\tilde{Q}_2$ introduced in Section 1.6. We have

(5.2)
$$\tilde{Q}_2q = 1 - Q_2p$$
$$= 1 - \alpha_2 p - (1 - \alpha_2)\lambda_2.$$

Replacing p with $1 - q$ and re-arranging, we get

(5.3)
$$\tilde{Q}_2q = \alpha_2 q + (1 - \alpha_2)(1 - \lambda_2).$$

The symmetry requirement is as follows: We want Q_1p to be the same function of p as $\tilde{Q}_2q$ is of q. This is accomplished by letting

(5.4)
$$\alpha_2 = \alpha_1,$$

(5.5)
$$1 - \lambda_2 = \lambda_1.$$

The first equation says that the slope parameters are equal—this is called the equal alpha condition—and the second equation says that the two operators Q_1 and Q_2 have complementary limit points.

The symmetric effect of two events in many experiments strongly suggests the equal alpha condition as described above, but there is another motivation for investigating consequences of this condition. In Sections 4.5 and 4.6 we encountered some serious *mathematical* problems which become greatly simplified when the equal alpha condition is imposed. The recurrence formulas for the moments for subject-controlled and experimenter-subject-controlled events could not be solved for general explicit formulas; the one exception to this arises when the α_i are equal. We consider this major mathematical simplification reason enough for studying the model with the equal alpha condition imposed.

With experimenter-controlled events, the equal alpha condition leads to only minor algebraic simplifications; in Sections 4.3 and 4.4 we obtained explicit formulas for the means and variances without using the equal alpha restriction. Nevertheless, we examine the consequences of the equal alpha condition for experimenter-controlled events, mainly because the results are used in Chapter 13 to analyze some data.

As mentioned above, the main mathematical simplifications result from imposing the equal alpha condition when the events are subject controlled or experimenter-subject controlled. However, in the analysis of experiments for which those cases seem most appropriate, the equal alpha condition seems least appropriate, as we shall see in Chapter 13. The equal alpha condition implies that reward and non-reward have "equal but opposite" effects on behavior; such an assumption is seldom warranted, either by psychological theory or by data. In spite of such objections, the equal alpha assumption leads to a useful "base-line" model for many experiments, that is, the data can be compared with this base-line model in much the same way that we compare results with a null hypothesis in routine statistical analyses.

5.2 IMPLICATIONS IN THE SET-THEORETIC MODEL

We now examine what the equal alpha condition means in terms of the theory of stimulus conditioning presented in Chapter 2. In Section 2.4 we showed that the parameters a and b in the operators for two response classes could be considered as the relative measures of two subsamples, A and B, of stimuli from the sample X. After a response occurred, the subsample A became conditioned to response A_1, and the subsample B became conditioned to response A_2. By definition, we have

$$(5.6) \qquad \alpha_i = 1 - a_i - b_i, \qquad i = 1, 2,$$

and so we see that the condition $\alpha_1 = \alpha_2$ is equivalent to

$$(5.7) \qquad a_1 + b_1 = a_2 + b_2.$$

For two events, E_1 and E_2, we must distinguish between a sample X_1 and its subsamples A_1 and B_1 which correspond to event E_1, and a sample X_2 with its subsamples A_2 and B_2 which correspond to event E_2. The equal alpha restriction is then equivalent to

$$(5.8) \qquad \mathscr{M}(A_1) + \mathscr{M}(B_1) = \mathscr{M}(A_2) + \mathscr{M}(B_2),$$

or, since A_i and B_i are disjunct,

$$(5.9) \qquad \mathscr{M}(A_1 \cup B_1) = \mathscr{M}(A_2 \cup B_2),$$

where the symbol $\cup$ indicates the set sum or union. Now the stimulus set $A_1 \cup B_1$ is the total set of stimuli available for reconditioning when event E_1 occurs. Similarly, $A_2 \cup B_2$ is the set available for reconditioning when event E_2 occurs. Equation 5.9 shows that the measure of the elements available for reconditioning is the same for both events.

5.3 EXPERIMENTER-CONTROLLED EVENTS

For several experiments discussed in Chapter 13, we would be willing to identify the events in the model with certain stimulus changes which are scheduled in advance by the experimenter. For example, a pre-arranged sequence of appearances and non-appearances of a light is used in some such experiments. As a result, the case of experimenter-controlled events is applicable. When we further assume that $\alpha_1 = \alpha_2 = \alpha$ for problems involving two events, the computation of the means and variances of the response probability distributions is quite simple. In this section we develop the necessary formulas for such computations.

In Section 4.3 we derive explicit formulas for the mean and variance of the p-value distributions for two experimenter-controlled events. When $\alpha_1 = \alpha_2 = \alpha$, equations 4.15, 4.16, and 4.17 give the

EXPLICIT FORMULA FOR THE MEANS:

$$(5.10) \qquad V_{1,n} = V_{1,\infty} - (V_{1,\infty} - V_{1,0})\alpha^n,$$

and the

ASYMPTOTIC FORMULA FOR THE MEANS:

$$(5.11) \qquad V_{1,\infty} = \pi_1\lambda_1 + \pi_2\lambda_2.$$

The formulas for the second raw moments, equations 4.22 and 4.23, are also simplified when $\alpha_1 = \alpha_2 = \alpha$. Furthermore, the variances, defined by

$$(5.12) \qquad \sigma_n^2 = V_{2,n} - V_{1,n}^2,$$

can be computed from the following simple formulas which result when $\alpha_1 = \alpha_2 = \alpha$:

EXPLICIT FORMULA FOR THE VARIANCES:

$$(5.13) \qquad \sigma_n^2 = \sigma_\infty^2 - (\sigma_\infty^2 - \sigma_0^2)\alpha^{2n},$$

ASYMPTOTIC FORMULA FOR THE VARIANCE:

$$(5.14) \qquad \sigma_\infty^2 = \frac{1-\alpha}{1+\alpha}[(\pi_1\lambda_1^2 + \pi_2\lambda_2^2) - V_{1,\infty}^2].$$

For many applications, as we see in Chapters 7 and 13, we are also willing to assume that $\lambda_1 = 1$ and $\lambda_2 = 0$. Then we have

$$(5.15) \qquad V_{1,\infty} = \pi_1,$$

$$(5.16) \qquad \sigma_\infty^2 = \frac{1-\alpha}{1+\alpha}\pi_1(1-\pi_1).$$

The first of this pair of equations shows that the asymptotic mean probability of response A_1 is equal to the probability of occurrence of event E_1. This conclusion is consistent with a rather large class of experimental data described in Chapter 13.

When there are more than two events, similar formulas are obtained as indicated by the analysis in Section 4.4. For two response classes and

t experimenter-controlled events with $\alpha_i = \alpha$ ($i = 1, 2, \cdots, t$), the foregoing explicit formulas for the means and variances are still correct, but the asymptotic formulas become

(5.17) $$V_{1,\infty} = \sum_{i=1}^{t} \pi_i \lambda_i,$$

(5.18) $$\sigma_\infty^2 = \frac{1-\alpha}{1+\alpha} \left\{ \sum_{i=1}^{t} \pi_i \lambda_i^2 - V_{1,\infty}^2 \right\}.$$

When we have more than two response classes (and t events) we can compute a vector of marginal means, as shown in Section 4.4. These vectors $V_{1,n}$ are given by

(5.19) $$\mathbf{V}_{1,n} = \mathbf{V}_{1,\infty} - (\mathbf{V}_{1,\infty} - \mathbf{V}_{1,0})\alpha^n,$$

(5.20) $$\mathbf{V}_{1,\infty} = \sum_{i=1}^{t} \pi_i \lambda_i,$$

where λ_i is the limit vector corresponding to event E_i. In Chapter 13 we use these results for analyzing data obtained by Neimark in a three-choice situation.

5.4 THE DISTRIBUTION MEANS FOR TWO SUBJECT-CONTROLLED EVENTS

When the events are subject controlled, that is, when Q_1 is applied to p_n with probability p_n and Q_2 is applied with probability $1 - p_n$, explicit formulas for the mean and variance were not obtained in Chapter 4. However, when $\alpha_1 = \alpha_2 = \alpha$, major simplifications obtain, and so we now derive the desired explicit formulas.

The recurrence formula for the means is given by equations 4.52 and 4.53:

(5.21) $$V_{1,n+1} = a_2 + (a_1 - a_2 + \alpha_2)V_{1,n} + (\alpha_1 - \alpha_2)V_{2,n}.$$

This equation cannot be solved to obtain an explicit formula for $V_{1,n}$ in terms of $V_{1,0}$, a_1, a_2, α_1, and α_2 because of the term in $V_{2,n}$. However, when we let $\alpha_1 = \alpha_2 = \alpha$, this term disappears, and we have the

RECURRENCE FORMULA FOR THE MEANS:

(5.22) $$V_{1,n+1} = a_2 + (a_1 - a_2 + \alpha)V_{1,n}.$$

This is a linear difference equation of the type we first encountered in Section 3.3, and it can be solved immediately. However, it is instructive to note that the right side of this equation is given by a new operator $\bar{Q}$ applied to $V_{1,n}$, that is,

(5.23) $$\bar{Q}V_{1,n} = a_2 + (a_1 - a_2 + \alpha)V_{1,n}.$$

This operator $\bar{Q}$ has the same form as the operators Q_i when they are written in the slope-intercept form

(5.24) $$Q_i p = a_i + \alpha_i p.$$

The operand is $V_{1,n}$ instead of p, the intercept is a_2 instead of a_i, and the slope is $(a_1 - a_2 + \alpha)$ instead of α_i, but mathematically $\bar{Q}$ has the same form as Q_1 and Q_2.

The operator $\bar{Q}$ may be obtained in another way as already suggested by the analysis in Section 3.12. Suppose we were to compute a weighted average of the operations $Q_1 V_{1,n}$ and $Q_2 V_{1,n}$ with weights $V_{1,n}$ and $1 - V_{1,n}$, respectively. We may easily verify that this will yield the expression on the right side of equation 5.23:

(5.25)
$$
\begin{aligned}
V_{1,n} Q_1 V_{1,n} &+ (1 - V_{1,n}) Q_2 V_{1,n} \\
&= V_{1,n}(a_1 + \alpha V_{1,n}) + (1 - V_{1,n})(a_2 + \alpha V_{1,n}) \\
&= a_2 + (a_1 - a_2 + \alpha) V_{1,n},
\end{aligned}
$$

and so we have

(5.26) $$\bar{Q} V_{1,n} = V_{1,n} Q_1 V_{1,n} + (1 - V_{1,n}) Q_2 V_{1,n}.$$

This means that $\bar{Q}$ can be considered a weighted mean or expectation of the operators Q_1 and Q_2 and so we call $\bar{Q}$ the *expected operator*.

The mean $V_{1,n}$ can be obtained by applying the expected operator $\bar{Q}$ to the initial mean n times:

(5.27) $$V_{1,n} = \bar{Q}^n V_{1,0}.$$

The results of Section 3.3 may be used directly if we first write $\bar{Q}$ in its fixed point form. We will call the slope β, that is,

(5.28) $$\beta = a_1 - a_2 + \alpha,$$

and we will denote the fixed point by $V_{1,\infty}$. Therefore the fixed point form of $\bar{Q}$ is

(5.29) $$\bar{Q} V_{1,n} = \beta V_{1,n} + (1 - \beta) V_{1,\infty}.$$

Comparison of this result with the slope-intercept form of $\bar{Q}$, equation 5.23, shows that

$$(1 - \beta) V_{1,\infty} = a_2.$$

Thus we have, for $\beta \neq 1$, the

ASYMPTOTIC FORMULA FOR THE MEAN:

(5.30) $$V_{1,\infty} = \frac{a_2}{1 - \beta} = \frac{a_2}{1 - a_1 + a_2 - \alpha} = \frac{\lambda_2}{1 - \lambda_1 + \lambda_2}.$$

Moreover, from equation 3.9 we see that $\tilde{Q}$ may be applied n times to $V_{1,0}$ to give the

EXPLICIT FORMULA FOR THE MEANS:

$$(5.31) \qquad V_{1,n} = \tilde{Q}^n V_{1,0} = \beta^n V_{1,0} + (1 - \beta^n) V_{1,\infty}.$$

This is the desired equation for the distribution means.

Before we develop the corresponding explicit formula for the second raw moments we wish to point out a rather interesting property of the means, namely, that they behave *as if* they were the means associated with a simple two-state Markov chain with constant conditional probabilities. Suppose we introduce the parameter b_1 of the gain-loss form of Q_1. It is defined by equation 1.17, that is,

$$(5.32) \qquad b_1 = 1 - a_1 - \alpha.$$

Substituting this in recurrence formula 5.22, we have

$$(5.33) \qquad V_{1,n+1} = a_2 + (1 - a_2 - b_1) V_{1,n}.$$

We note that the parameter α has been eliminated from this equation and only the parameters a_2 and b_1 remain; that is to say, the means behave the same way no matter what value α has. It is instructive to consider $\tilde{Q}_1$ which is applied to $q = 1 - p$. We have

$$(5.34) \qquad \tilde{Q}_1 q = 1 - Q_1 p = 1 - a_1 - \alpha(1 - q);$$

or, using 5.32, we have

$$(5.35) \qquad \tilde{Q}_1 q = b_1 + \alpha q.$$

The operator Q_2 may be written

$$(5.36) \qquad Q_2 p = a_2 + \alpha p.$$

We then note that $\tilde{Q}_1 q$ is the conditional probability $Pr\{A_2|A_1\}$, and $Q_2 p$ is the conditional probability $Pr\{A_1|A_2\}$. As pointed out in Section 3.12, if we take $\alpha = 0$ these two conditional probabilities are constant:

$$(5.37) \qquad \begin{aligned} Pr\{A_2|A_1\} &= b_1, \qquad (\alpha = 0), \\ Pr\{A_1|A_2\} &= a_2, \qquad (\alpha = 0), \end{aligned}$$

and we have a simple two-state Markov chain. The analogue of equation 3.61 is then

$$(5.38) \qquad V_{1,n+1} = a_2 + (1 - a_2 - b_1) V_{1,n},$$

but this is precisely the result we had in equation 5.33 without the restriction that $\alpha = 0$. This is not surprising because we saw in equation 5.33 that the means $V_{1,n}$ do not depend on α; the means are the same for $\alpha = 0$ as for any other value of α. Thus we conclude that the distribution means for the equal alpha case behave as if we had $\alpha = 0$ and hence as if we had the constant conditional probabilities of equations 5.37. In Fig. 5.1 we illustrate how the means vary with trial number for $b_1 = a_2 = 0.1$.

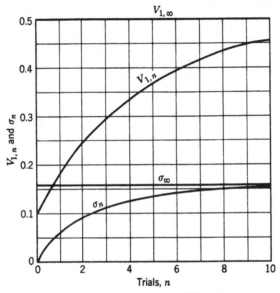

Fig. 5.1. Showing the curve of mean probability, $V_{1,n}$ vs. n for ten trials, and the standard deviation, σ_n. Equations 5.31 and 5.41 were used with $a_1 = 0.3$, $a_2 = 0.1$, $\alpha = 0.6$, $V_{1,0} = 0.1$, $\sigma_0 = 0$.

Although the value of α is irrelevant for computing the means, in the next section we show that the variances of the distributions do depend on α.

5.5 THE DISTRIBUTION VARIANCES FOR TWO SUBJECT-CONTROLLED EVENTS

It is often useful to compute moments of the distribution higher than the first, and this can be done when $\alpha_1 = \alpha_2 = \alpha$. Equation 4.46 for the mth raw moment on trial $n + 1$ becomes

$$(5.39) \quad V_{m,n+1} = \sum_{u=0}^{m} \binom{m}{u} \{a_1^{m-u}\alpha^u V_{u+1,n} + a_2^{m-u}\alpha^u(V_{u,n} - V_{u+1,n})\}.$$

In this sum we may separate out the terms for $u = m$ and obtain

$$(5.40) \quad V_{m,n+1}$$
$$= \sum_{u=0}^{m-1} \binom{m}{u} \{a_1^{m-u}\alpha^u V_{u+1,n} + a_2^{m-u}\alpha^u(V_{u,n} - V_{u+1,n})\} + \alpha^m V_{m,n}.$$

We now note that the highest moment on the right side is $V_{m,n}$, and so we express $V_{m,n+1}$ in terms of $V_{0,n} = 1$, $V_{1,n}$, $\cdots$, $V_{m,n}$. These equations may be solved explicitly for $m = 1, 2, \cdots$ to yield expressions for the raw moments in terms of the trial number n and the parameters.

We solved for the means $V_{1,n}$ in the preceding section so we now proceed to solve for the second raw moments. For $m = 2$ we have, from equation 5.40, the

RECURRENCE FORMULA FOR THE SECOND MOMENTS:

$$
\begin{aligned}
V_{2,n+1} &= a_1{}^2 V_{1,n} + a_2{}^2 (1 - V_{1,n}) \\
&\quad + 2\{a_1\alpha V_{2,n} + a_2\alpha(V_{1,n} - V_{2,n})\} + \alpha^2 V_{2,n} \\
&= a_2{}^2 + (a_1{}^2 - a_2{}^2 + 2a_2\alpha)V_{1,n} + \alpha(2a_1 - 2a_2 + \alpha)V_{2,n}.
\end{aligned}
$$
(5.41)

We then introduce the abbreviations

$$
\begin{aligned}
B_1 &= a_1{}^2 - a_2{}^2 + 2a_2\alpha, \\
B_2 &= \alpha(2a_1 - 2a_2 + \alpha),
\end{aligned}
$$
(5.42)

so that we may write

$$
V_{2,n+1} = a_2{}^2 + B_1 V_{1,n} + B_2 V_{2,n}.
$$
(5.43)

From the preceding section we know that

$$
V_{1,n} = V_{1,\infty} - (V_{1,\infty} - V_{1,0})\beta^n.
$$
(5.44)

When we substitute this in the above expression for $V_{2,n+1}$ we have

$$
V_{2,n+1} = (a_2{}^2 + B_1 V_{1,\infty}) - B_1(V_{1,\infty} - V_{1,0})\beta^n + B_2 V_{2,n}.
$$
(5.45)

We solved a difference equation of precisely this form in Section 4.3 (cf. equations 4.18 and 4.22), and so we can write down the solution immediately. We then get the

EXPLICIT FORMULA FOR THE SECOND MOMENTS:

$$
V_{2,n} = (a_2{}^2 + B_1 V_{1,\infty}) \frac{1 - B_2{}^n}{1 - B_2} - B_1(V_{1,\infty} - V_{1,0}) \frac{\beta^n - B_2{}^n}{\beta - B_2} + B_2{}^n V_{2,0},
$$
(5.46)

where $V_{1,0}$ and $V_{2,0}$ are the initial mean and second raw moment, respectively. When $-1 < B_2 < 1$, $-1 < \beta < 1$, and $\beta \neq B_2$, we have in the limit as $n \to \infty$ the

ASYMPTOTIC FORMULA FOR THE SECOND MOMENT:

$$
V_{2,\infty} = \frac{a_2{}^2 + B_1 V_{1,\infty}}{1 - B_2},
$$
(5.47)

where $V_{1,\infty}$ is given by equation 5.30. Since B_1 and B_2 are simple algebraic functions of a_1, a_2, and α, we have expressed $V_{2,\infty}$ in terms of those parameters. The asymptotic variance is obtained from the relation $\sigma_\infty^2 = V_{2,\infty} - V_{1,\infty}^2$ and so we have from equation 5.47

$$(5.48) \qquad \sigma_\infty^2 = \frac{a_2^2 + B_1 V_{1,\infty}}{1 - B_2} - V_{1,\infty}^2.$$

In Fig. 5.1 we illustrate how σ_n varies with n.

5.6 SUBJECT-CONTROLLED EVENTS WITH $\lambda_1 = 1$ AND $\lambda_2 = 0$

In many experimental applications we would be willing to assume that the repeated performance of a response would increase the likelihood of that response occurring until it was almost certain to occur. In mathematical language, this means that the limit point λ_1 is unity if the response considered is identified with alternative A_1. In other words a repeated application of Q_1 to p would tend to make the probability of A_1 unity. We might make the same assumption about response A_2; if A_2 occurred repeatedly its probability $q = 1 - p$ would tend to unity and so p would tend to zero, implying that $\lambda_2 = 0$. In Chapter 7 we consider some of the mathematical properties of the case of $\lambda_1 = 1$ and $\lambda_2 = 0$, but in this section we consider these conditions when we also have $\alpha_1 = \alpha_2$. The two operators are then

$$(5.49) \qquad \begin{aligned} Q_1 p &= \alpha p + (1 - \alpha), \\ Q_2 p &= \alpha p. \end{aligned}$$

We see that only the single parameter α is involved in these operators and so the equations should be especially simple.

A possible application of the operators of equations 5.49 is the Brunswik T-maze with 100 percent reinforcement on both sides of the maze. (In the Introduction and in Section 3.9 we have discussed the Brunswik experiment.) If the rat is rewarded on the right side, p (the probability of turning right) should increase towards unity. If the rat is rewarded for going left, this should decrease p towards zero. Moreover, since the situation is symmetric we would expect to have $\alpha_1 = \alpha_2$. Thus the foregoing operators might be quite adequate for describing the data.

From the relation $a_i = (1 - \alpha)\lambda_i$ we see that $\lambda_1 = 1$ implies that $a_1 = 1 - \alpha$ and that $\lambda_2 = 0$ implies that $a_2 = 0$. Hence the recurrence formula for the means, equation 5.22, becomes

$$(5.50) \qquad V_{1,n+1} = V_{1,n},$$

and so we conclude that

$$(5.51) \qquad V_{1,n} = V_{1,0} \qquad (n = 0, 1, 2, \cdots).$$

In other words, the mean is constant from trial to trial and is equal to the initial mean. This result can be obtained also from the explicit formula 5.31, since $a_1 = 1 - \alpha$ and $a_2 = 0$ imply that $\beta = 1$.

The formula for the second raw moment can be obtained directly from equation 5.46. We have $a_2 = 0$ and $\beta = 1$, and from definitions 5.42 we get $B_1 = (1 - \alpha)^2$ and $B_2 = \alpha(2 - \alpha)$. Thus

$$(5.52) \qquad V_{2,n} = (1 - B_2{}^n)V_{1,0} + B_2{}^n V_{2,0}.$$

For $0 \leq \alpha < 1$ we see that $B_2 = \alpha(2 - \alpha)$ is in the range $0 \leq B_2 < 1$ and so with this condition $B_2{}^n$ tends to zero as $n \to \infty$. Therefore

$$(5.53) \qquad V_{2,\infty} = V_{1,0} \qquad (0 \leq \alpha < 1).$$

The variance is of course

$$(5.54) \qquad \sigma_n{}^2 = V_{2,n} - V_{1,n}^2,$$

and so we have

$$(5.55) \qquad \begin{aligned} \sigma_\infty{}^2 &= V_{2,\infty} - V_{1,\infty}^2 \\ &= V_{1,0} - V_{1,0}^2 \\ &= V_{1,0}(1 - V_{1,0}). \end{aligned}$$

The fact that the asymptotic variance has this form suggests that the asymptotic distribution is binomial with all its density at zero and unity. We see in Chapter 7 that this is indeed true whenever $\lambda_1 = 1$ and $\lambda_2 = 0$.

5.7 SUBJECT-CONTROLLED EVENTS WITH $\lambda_1 = 0$ AND $\lambda_2 = 1$

The preceding discussion of the conditions $\lambda_1 = 1$ and $\lambda_2 = 0$ suggests the opposite case of $\lambda_1 = 0$ and $\lambda_2 = 1$. Whereas the previous case might correspond to equal rewards of two similar responses, the present case might correspond to extinction of two similar responses. If a rat in a Brunswik T-maze were to find no food on the right side, p might tend to zero; finding no food on the left side might tend to make p go to unity. This situation would require that $\lambda_1 = 0$ and $\lambda_2 = 1$. If in addition $\alpha_1 = \alpha_2 = \alpha$, we have the operators

$$(5.56) \qquad \begin{aligned} Q_1 p &= \alpha p, \\ Q_2 p &= \alpha p + (1 - \alpha). \end{aligned}$$

We now develop expressions for the asymptotic mean and variance for this pair of operators.

From the relation $a_i = (1 - \alpha)\lambda_i$ we see that $\lambda_1 = 0$ and $\lambda_2 = 1$ imply that $a_1 = 0$ and $a_2 = 1 - \alpha$. From definition 5.28 it follows that

$$(5.57) \qquad \beta = 2\alpha - 1.$$

The explicit formula for the means is equation 5.31 above, with the preceding value of β, but from equation 5.30 we see that the asymptotic mean is

$$(5.58) \qquad V_{1,\infty} = \frac{a_2}{1-\beta} = \frac{1-\alpha}{1-(2\alpha-1)} = \frac{1}{2}.$$

The asymptotic mean is 1/2 no matter what value of α or $V_{1,0}$ is involved. We next study other properties of the asymptotic distribution by looking at higher moments.

The asymptotic second raw moment is obtained directly from equation 5.47. As before, $\lambda_2 = 1$ implies that $a_2 = 1 - \alpha$, and from definitions 5.42 we get, since $a_1 = 0$,

$$(5.59) \qquad \begin{aligned} B_1 &= -(1-\alpha)^2 + 2(1-\alpha)\alpha = (1-\alpha)(3\alpha-1), \\ B_2 &= \alpha[-2(1-\alpha)+\alpha] = \alpha(3\alpha-2). \end{aligned}$$

With these equations we get from equations 5.47 and 5.30,

$$(5.60) \qquad V_{2,\infty} = \frac{1}{1-\alpha(3\alpha-2)}\left\{(1-\alpha)^2 + \frac{(1-\alpha)^2(3\alpha-1)}{1-(2\alpha-1)}\right\}.$$

Algebraic simplifications lead to the result

$$(5.61) \qquad V_{2,\infty} = \frac{1+\alpha}{2(1+3\alpha)}.$$

The asymptotic variance is then

$$(5.62) \qquad \begin{aligned} \sigma_\infty^2 &= V_{2,\infty} - V_{1,\infty}^2 \\ &= \frac{1+\alpha}{2(1+3\alpha)} - \left(\frac{1}{2}\right)^2. \end{aligned}$$

Minor simplifications then yield

$$(5.63) \qquad \sigma_\infty^2 = \frac{1-\alpha}{4(1+3\alpha)}.$$

For example, if $\alpha = 0$ we get the largest possible variance $\sigma_\infty^2 = 0.25$; if $\alpha = 0.5$ we get $\sigma_\infty^2 = 0.05$.

It may be instructive to consider one more moment of the asymptotic distribution, the third moment which has to do with skewness. It can be shown that the third moment *about the mean* is zero, which suggests that the asymptotic distribution is symmetric about the mean $V_{1,\infty} = 0.5$. The third raw moment of the asymptotic distribution turns out to be

$$(5.64) \qquad V_{3,\infty} = \frac{1}{2(1+3\alpha)}.$$

Further analysis shows that all the odd moments about the mean are zero, and so the asymptotic distribution is in fact symmetric. In Chapter 6 we illustrate this symmetry by displaying an approximation to the entire asymptotic distribution.

5.8 EXPERIMENTER-SUBJECT-CONTROLLED EVENTS

We now consider the problem discussed in Section 3.13 and 4.6 with the equal alpha restriction. We have two alternatives A_1 and A_2, with probabilities p_n and $1-p_n$ on trial n as in the preceding sections of this chapter. Each alternative may have outcome O_1 or O_2; the probability of O_1 following A_1 is π_1, and the probability of O_1 following A_2 is π_2. The operator applied to p_n when A_j and O_k occur on trial n is Q_{jk}. From equations 4.56 we have

(5.65) $$Q_{jk} p = \alpha_{jk} p + (1 - \alpha_{jk})\lambda_{jk}.$$

Since there are four parameters α_{jk} we have a choice between setting all four of these equal to one another, or imposing a less severe restriction. We first consider the consequences of the assumption that α_{jk} is independent of which response is made but does depend upon the outcome, that is, we let

(5.66) $$\alpha_{1k} = \alpha_{2k} = \alpha_k, \qquad k = 1, 2.$$

These conditions lead to the operators given by

(5.67)
$$Q_{11} p = \alpha_1 p + (1 - \alpha_1)\lambda_{11} \qquad [p\pi_1],$$
$$Q_{12} p = \alpha_2 p + (1 - \alpha_2)\lambda_{12} \qquad [p(1 - \pi_1)],$$
$$Q_{21} p = \alpha_1 p + (1 - \alpha_1)\lambda_{21} \qquad [(1 - p)\pi_2],$$
$$Q_{22} p = \alpha_2 p + (1 - \alpha_2)\lambda_{22} \qquad [(1 - p)(1 - \pi_2)].$$

The probability of applying each operator is shown in square brackets after each of the preceding equations. (For simplicity we have let $\pi_{j1} = \pi_j$ and $\pi_{j2} = 1 - \pi_j$ for $j = 1, 2$.) The mean parameters defined by equation 4.61 of Chapter 4 become

(5.68)
$$\bar{a}_1 = \pi_1(1 - \alpha_1)\lambda_{11} + (1 - \pi_1)(1 - \alpha_2)\lambda_{12},$$
$$\bar{a}_2 = \pi_2(1 - \alpha_1)\lambda_{21} + (1 - \pi_2)(1 - \alpha_2)\lambda_{22},$$
$$\bar{\alpha}_1 = \pi_1\alpha_1 + (1 - \pi_1)\alpha_2,$$
$$\bar{\alpha}_2 = \pi_2\alpha_1 + (1 - \pi_2)\alpha_2.$$

From equation 4.62 we then get the recurrence formula

(5.69) $$V_{1,n+1} = \bar{a}_2 + (\bar{a}_1 - \bar{a}_2 + \bar{\alpha}_2)V_{1,n} + (\pi_1 - \pi_2)(\alpha_1 - \alpha_2)V_{2,n}.$$

The annoying feature of equation 4.62—the presence of the term in $V_{2,n}$—has not been eliminated, unless either $\alpha_1 = \alpha_2$ or $\pi_1 = \pi_2$.

When we take $\alpha_1 = \alpha_2 = \alpha$, we then have the operators

$$(5.70) \qquad \begin{aligned} Q_{11}\,p &= \alpha p + (1 - \alpha)\lambda_{11} & [\,p\pi_1\,], \\ Q_{12}\,p &= \alpha p + (1 - \alpha)\lambda_{12} & [\,p(1 - \pi_1)], \\ Q_{21}\,p &= \alpha p + (1 - \alpha)\lambda_{21} & [(1 - p)\pi_2], \\ Q_{22}\,p &= \alpha p + (1 - \alpha)\lambda_{22} & [(1 - p)(1 - \pi_2)]. \end{aligned}$$

The probability of application of each of these operators is shown in square brackets on the right of each. In the last of equations 5.68 we see that, when $\alpha_1 = \alpha_2 = \alpha$,

$$(5.71) \qquad \bar{\alpha}_2 = \alpha,$$

and so equation 5.69 becomes

$$(5.72) \qquad V_{1,n+1} = \bar{a}_2 + (\bar{a}_1 - \bar{a}_2 + \alpha)V_{1,n}.$$

This result generalizes immediately to s possible outcomes O_k with conditional probabilities π_{jk}, provided that we re-define

$$(5.73) \qquad \begin{aligned} \bar{a}_1 &= \sum_{k=1}^{s} \pi_{1k}a_{1k}, \\ \bar{a}_2 &= \sum_{k=1}^{s} \pi_{2k}a_{2k}, \end{aligned}$$

and recall the condition that

$$(5.74) \qquad \sum_{k=1}^{s} \pi_{jk} = 1 \qquad (j = 1, 2).$$

Equation 5.72 can be solved immediately for $V_{1,n}$:

$$(5.75) \qquad V_{1,n} = V_{1,\infty} - (V_{1,\infty} - V_{1,0})\gamma^n,$$

where

$$(5.76) \qquad V_{1,\infty} = \frac{\bar{a}_2}{1 - \gamma},$$

and

$$(5.77) \qquad \gamma = (\bar{a}_1 - \bar{a}_2 + \alpha).$$

Higher moments of the distributions can be computed in a similar manner.

5.9 EXPERIMENTER-SUBJECT-CONTROLLED EVENTS WITH LIMITS ZERO AND UNITY

When we identify outcome O_1 with reward and outcome O_2 with non-reward, we are often willing to let $\lambda_{11} = \lambda_{22} = 1$ and $\lambda_{12} = \lambda_{21} = 0$.

The operators then are given by

$$Q_{11} p = \alpha_1 p + (1 - \alpha_1)$$

$$Q_{12} p = \alpha_2 p$$

(5.78)

$$Q_{21} p = \alpha_1 p$$

$$Q_{22} p = \alpha_2 p + (1 - \alpha_2).$$

The recurrence formula for the mean becomes

(5.79) $V_{1,n+1}$

$$= (1 - \pi_2)(1 - \alpha_2) + [1 - 2(1 - \pi_2)(1 - \alpha_2)$$

$$+ (\pi_1 - \pi_2)(1 - \alpha_1)]V_{1,n} + (\pi_1 - \pi_2)(\alpha_1 - \alpha_2)V_{2,n}.$$

This formula again suggests two special cases: $\pi_1 = \pi_2 = \pi$ and $\alpha_1 = \alpha_2 = \alpha$. For the equal π case, we have

(5.80) $V_{1,n+1} = (1 - \pi)(1 - \alpha_2) + [1 - 2(1 - \pi)(1 - \alpha_2)]V_{1,n}.$

This has the solution

(5.81) $V_{1,n} = V_{1,\infty} - (V_{1,\infty} - V_{1,0})[1 - 2(1 - \pi)(1 - \alpha_2)]^n,$

where

(5.82) $V_{1,\infty} = 1/2.$

For $\alpha_1 = \alpha_2 = \alpha$, recurrence formula 5.79 becomes

(5.83) $V_{1,n+1} = (1 - \pi_2)(1 - \alpha) + [1 - (2 - \pi_1 - \pi_2)(1 - \alpha)]V_{1,n},$

and its solution is

(5.84) $V_{1,n} = V_{1,\infty} - (V_{1,\infty} - V_{1,0})[1 - (2 - \pi_1 - \pi_2)(1 - \alpha)]^n$

where

(5.85) $V_{1,\infty} = \dfrac{1 - \pi_2}{2 - \pi_1 - \pi_2}.$

     The parameter α does not appear in the expression for the asymptotic mean, which can be computed from the reward probabilities π_1 and π_2 in advance of collecting data. For example, if we reward A_1 on 50 percent of the trials it occurs and reward A_2 on 10 percent of the trials it occurs, this result says that, for a group of subjects, A_1 should occur about 64 percent of the time after learning is complete. (It is worth noting that this is a good way from 83 percent, a result obtained by simple proportions.) In Chapter 13 we use this result in examining several sets of data from two-choice experiments.

*5.10 EXTENSION TO r RESPONSES AND s OUTCOMES

In this section we show how the analysis of the preceding sections can be generalized to any number r of response classes and any number s of outcomes. We use the matrix operators of Section 1.8; when response A_j and outcome O_k occur we apply the operator,

$$(5.86) \qquad \mathbf{T}_{jk} = \alpha_{jk}\mathbf{I} + (1 - \alpha_{jk})\mathbf{\Lambda}_{jk},$$

to the vector $\mathbf{p}_n$ to obtain the vector $\mathbf{p}_{n+1}$. We impose the equal alpha restriction

$$(5.87) \qquad \alpha_{jk} = \alpha \qquad (j = 1, 2, \cdots r, \quad k = 1, 2, \cdots s),$$

and so we have

$$(5.88) \qquad \mathbf{T}_{jk} = \alpha\mathbf{I} + (1 - \alpha)\mathbf{\Lambda}_{jk}.$$

The probability of applying this operator to the vector $\mathbf{p}_n$ on trial n is $p_{n,j}\pi_{jk}$, where $p_{n,j}$ is the jth component of $\mathbf{p}_n$ and π_{jk} is the conditional probability of outcome O_k, given alternative A_j occurred.

As in Section 4.4 we are involved in a multivariate distribution of p values on each trial, but we restrict our attention to the marginal means. We define a vector of marginal means

$$(5.89) \qquad \mathbf{V}_{1,n} = \sum_{\nu=1}^{(rs)^n} P_{\nu n}\mathbf{p}_{\nu n}.$$

Following the procedure of Section 4.4, we have for the next trial the vector

$$(5.90) \qquad \mathbf{V}_{1,n+1} = \sum_{\nu=1}^{(rs)^n} P_{\nu n}\Big\{\sum_{j=1}^{r} \sum_{k=1}^{s} p_{\nu n,j}\pi_{jk}\mathbf{T}_{jk}\mathbf{p}_{\nu n}\Big\}.$$

Using

$$(5.91) \qquad \mathbf{T}_{jk}\mathbf{p}_{\nu n} = \alpha\mathbf{p}_{\nu n} + (1 - \alpha)\mathbf{\lambda}_{jk},$$

and the relations

$$(5.92) \qquad \begin{aligned} &\sum_{k=1}^{s} \pi_{jk} = 1 \qquad (j = 1, 2, \cdots, r), \\ &\sum_{j=1}^{r} p_{\nu n,j} = 1, \end{aligned}$$

we have

$$(5.93) \qquad \mathbf{V}_{1,n+1} = \sum_{\nu=1}^{(rs)^n} P_{\nu n}\Big\{\alpha\mathbf{p}_{\nu n} + (1 - \alpha) \sum_{j=1}^{r} \sum_{k=1}^{s} p_{\nu n,j}\pi_{jk}\mathbf{\lambda}_{jk}\Big\}.$$

We then define an average vector $\bar{\mathbf{\lambda}}_j$ by

$$(5.94) \qquad \bar{\mathbf{\lambda}}_j = \sum_{k=1}^{s} \pi_{jk}\mathbf{\lambda}_{jk}.$$

Now consider the sum

$$(5.95) \qquad \sum_{j=1}^{r} \sum_{k=1}^{s} p_{\nu n,j} \pi_{jk} \lambda_{jk} = \sum_{j=1}^{r} p_{\nu n,j} \bar{\lambda}_j.$$

We show that this sum can be obtained by applying a new matrix $\bar{\Lambda}$ to the vector $\mathbf{p}_{\nu n}$. Let $\bar{\Lambda}$ be formed by the r vectors $\bar{\lambda}_j$ with $\bar{\lambda}_j$ in the jth column, that is, if

$$(5.96) \qquad \bar{\lambda}_j = \begin{bmatrix} \bar{\lambda}_{j,1} \\ \bar{\lambda}_{j,2} \\ \cdot \\ \cdot \\ \cdot \\ \bar{\lambda}_{j,r} \end{bmatrix},$$

then

$$(5.97) \qquad \bar{\Lambda} = \begin{bmatrix} \bar{\lambda}_{1,1} & \bar{\lambda}_{2,1} & \cdots & \bar{\lambda}_{r,1} \\ \bar{\lambda}_{1,2} & \bar{\lambda}_{2,2} & \cdots & \bar{\lambda}_{r,2} \\ \cdot & \cdot & & \cdot \\ \cdot & \cdot & & \cdot \\ \cdot & \cdot & & \cdot \\ \bar{\lambda}_{1,r} & \bar{\lambda}_{2,r} & \cdots & \bar{\lambda}_{r,r} \end{bmatrix}.$$

When we apply this matrix to $\mathbf{p}_{\nu n}$ we have for the uth element of $\bar{\Lambda}\,\mathbf{p}_{\nu n}$, the sum

$$(5.98) \qquad \sum_{j=1}^{r} \bar{\lambda}_{j,u}\, p_{\nu n,j}.$$

Hence we have

$$(5.99) \qquad \bar{\Lambda}\, \mathbf{p}_{\nu n} = \sum_{j=1}^{r} p_{\nu n,j} \bar{\lambda}_j.$$

Using this result in equation 5.93 gives

$$(5.100) \qquad \mathbf{V}_{1,n+1} = \sum_{\nu=1}^{(rs)^n} P_{\nu n}\{\alpha\mathbf{p}_{\nu n} + (1-\alpha)\,\bar{\Lambda}\,\mathbf{p}_{\nu n}\}.$$

Using the definition of $\mathbf{V}_{1,n}$ we then get

$$(5.101) \qquad \mathbf{V}_{1,n+1} = \alpha\mathbf{V}_{1,n} + (1-\alpha)\,\bar{\Lambda}\,\mathbf{V}_{1,n}.$$

Therefore we define an expected operator $\bar{\mathbf{T}}$ by

$$(5.102) \qquad \bar{\mathbf{T}} = \alpha\mathbf{I} + (1-\alpha)\,\bar{\Lambda},$$

so that

(5.103) $$\mathbf{V}_{1,n+1} = \overline{\mathbf{T}}\mathbf{V}_{1,n}.$$

The fixed point of the operator $\overline{\mathbf{T}}$ will be the vector of asymptotic marginal means $\mathbf{V}_{1,\infty}$. We solve for this by setting

(5.104) $$\mathbf{V}_{1,\infty} = \overline{\mathbf{T}}\mathbf{V}_{1,\infty} = \alpha\mathbf{V}_{1,\infty} + (1 - \alpha)\,\overline{\mathbf{\Lambda}}\,\mathbf{V}_{1,\infty}.$$

Hence

(5.105) $$\overline{\mathbf{\Lambda}}\,\mathbf{V}_{1,\infty} = \mathbf{V}_{1,\infty}.$$

This vector equation can be solved to yield the *j*th component $V_{1,\infty,j}$ of the vector $\mathbf{V}_{1,\infty}$. We have the set of equations

(5.106) $$\sum_{j=1}^{r} \bar{\lambda}_{j,u} V_{1,\infty,j} = V_{1,\infty,u} \qquad (u = 1, 2, \cdots, r),$$

and the necessary condition

(5.107) $$\sum_{u=1}^{r} V_{1,\infty,u} = 1.$$

These equations can be solved for the asymptotic marginal means.

We illustrate the solution of the preceding equations for the simpler case of two outcomes O_1 and O_2. The example to be given would be appropriate if endlessly repeated reinforcement of a response made it certain to occur, whereas endlessly repeated extinction of a response (were this possible) would eliminate it, but make all other responses equally likely. Accordingly, we let

(5.108)
$$\lambda_{j1,u} = \begin{cases} 1 & \text{for } u = j \\ 0 & \text{for } u \neq j, \end{cases}$$

$$\lambda_{j2,u} = \begin{cases} 0 & \text{for } u = j \\ \dfrac{1}{r-1} & \text{for } u \neq j. \end{cases}$$

Then we get

(5.109) $$\bar{\lambda}_{j,u} = \begin{cases} \pi_{j1} & \text{for } u = j \\ \dfrac{1 - \pi_{j1}}{r-1} & \text{for } u \neq j. \end{cases}$$

Equation 5.106 then gives

(5.110) $$\pi_{u1} V_{1,\infty,u} + \frac{1}{r-1} \sum_{j \neq u} (1 - \pi_{j1}) V_{1,\infty,j} = V_{1,\infty,u}.$$

Re-arrangements then give

(5.111) $$(1 - \pi_{u1})V_{1,\infty,u} = \frac{1}{r}\sum_{j=1}^{r}(1 - \pi_{j1})V_{1,\infty,j}.$$

The right side of this equation is a constant which we may call C and so

(5.112) $$V_{1,\infty,u} = \frac{C}{1 - \pi_{u1}}.$$

When we impose condition 5.107 we then can evaluate C and get finally

(5.113) $$V_{1,\infty,u} = \frac{\dfrac{1}{1 - \pi_{u1}}}{\displaystyle\sum_{j=1}^{r}\dfrac{1}{1 - \pi_{j1}}}.$$

Hence we obtain the asymptotic marginal means in terms of the conditional probabilities π_{j1}.

*5.11 MARKOV SEQUENCES OF EXPERIMENTER-CONTROLLED EVENTS

In Section 3.11 it was pointed out that a Markov chain had been used for constructing a schedule of events in at least one experimental study. Hake and Hyman [1] had subjects predict on each of 240 trials which of two symbols would appear. The sequences of the symbols were prepared in advance and so were independent of the responses made by the subjects. In two of the four groups of subjects, the sequences of symbols were Markov chains.

Formulas for the moments of the distributions of response probabilities were not developed in Chapter 4 for a Markov sequence of operators. The analysis is quite complicated and the results unsightly except when the equal-alpha restriction is imposed. However, this restriction appears appropriate when we are describing experiments such as the one of Hake and Hyman; the events—the appearance of one symbol or the other—have an intrinsic symmetry. If the roles of the two symbols were reversed, the data should be the same except for relabeling of the two responses—the prediction of one symbol or the other. Therefore, in this section, we compute the trial by trial means of the distributions of response probabilities for this problem.

The two event operators are

(5.114) $$\begin{aligned} Q_1 p &= \alpha p + (1 - \alpha)\lambda_1, \\ Q_2 p &= \alpha p + (1 - \alpha)\lambda_2. \end{aligned}$$

These operators are applied according to the rules of a Markov chain. If Q_1 has just been applied, it will be applied again with probability $\pi(1|1)$, whereas if Q_2 has just been applied, Q_1 will be applied with probability $\pi(1|2)$. The two conditional probabilities of applying Q_2 are then

(5.115)
$$\pi(2|1) = 1 - \pi(1|1),$$
$$\pi(2|2) = 1 - \pi(1|2).$$

Hence the Markov chain is specified by these conditional probabilities and an initial probability $\pi_0(1)$ of applying Q_1 on trial 0. The probability of applying Q_2 on trial 0 is

(5.116)
$$\pi_0(2) = 1 - \pi_0(1).$$

With these rules of application of the operators we wish to compute the means $V_{1,n}$ of the resulting distributions of response probabilities.

On any trial beyond the initial one we have several possible sequences of events up through that trial; there are 2^n possible sequences up through trial n. Half of these sequences terminate in event E_1 and half in event E_2. The probabilities of events E_1 and E_2 on trial $n+1$ depend upon the event which actually occurred on trial n, and so we must distinguish between the two classes of sequences. Let the sequences terminating in E_1 on trial n be labeled $\nu = 1, 2, \cdots, 2^{n-1}$, and those ending in an E_2 on trial n be labeled $\mu = 1, 2, \cdots, 2^{n-1}$. Furthermore, denote the probability of the νth sequence ending in E_1 by $P_{\nu n}$ and the response probability associated with this sequence by $p_{\nu n}$. Similarly, let the probability of the μth sequence ending in E_2 be $P_{\mu n}$ and the associated response probability be $p_{\mu n}$.

The unconditional probability of event E_1 on trial n of a Markov chain was computed in Section 3.11, but in the notation just introduced it is

(5.117)
$$\pi_n(1) = \sum_{\nu=1}^{2^{n-1}} P_{\nu n}.$$

Similarly, the unconditional probability of E_2 on trial n is

(5.118)
$$\pi_n(2) = \sum_{\mu=1}^{2^{n-1}} P_{\mu n},$$

and we have of course

(5.119)
$$\pi_n(1) + \pi_n(2) = 1.$$

The mean response probability on trial n is given by

(5.120)
$$V_{1,n} = \sum_{\nu} P_{\nu n} p_{\nu n} + \sum_{\mu} P_{\mu n} p_{\mu n}.$$

On trial $n + 1$ the mean will be

$$V_{1,n+1} = \sum_{\nu} P_{\nu n}\{\pi(1|1)Q_1 p_{\nu n} + \pi(2|1)Q_2 p_{\nu n}\}$$

(5.121)

$$+ \sum_{\mu} P_{\mu n}\{\pi(1|2)Q_1 p_{\mu n} + \pi(2|2)Q_2 p_{\mu n}\}.$$

We then insert the expressions for $Q_1 p$ and $Q_2 p$ given above:

$$V_{1,n+1} = \sum_{\nu} P_{\nu n}\{\pi(1|1)[\alpha p_{\nu n} + (1 - \alpha)\lambda_1] + \pi(2|1)[\alpha p_{\nu n} + (1 - \alpha)\lambda_2]\}$$

(5.122) $+ \sum_{\mu} P_{\mu n}\{\pi(1|2)[\alpha p_{\mu n} + (1 - \alpha)\lambda_1] + \pi(2|2)[\alpha p_{\mu n} + (1 - \alpha)\lambda_2]\}.$

Using relations 5.115, we get after simplifications:

$$V_{1,n+1} = \sum_{\nu} P_{\nu n}\{\alpha p_{\nu n} + (1 - \alpha)[\pi(1|1)\lambda_1 + \pi(2|1)\lambda_2]\}$$

(5.123)

$$+ \sum_{\mu} P_{\mu n}\{\alpha p_{\mu n} + (1 - \alpha)[\pi(1|2)\lambda_1 + \pi(2|2)\lambda_2]\}.$$

We then use the definitions 5.117, 5.118, and 5.120 and get

$$V_{1,n+1} = \alpha V_{1,n} + (1 - \alpha)\{\pi_n(1)[\pi(1|1)\lambda_1 + \pi(2|1)\lambda_2]$$

(5.124)

$$+ \pi_n(2)[\pi(1|2)\lambda_1 + \pi(2|2)\lambda_2]\}.$$

This formula, which is a recurrence formula in the means, can be solved to give an explicit formula.

From the analysis in Section 3.11, we obtain the expressions

(5.125) $\pi_n(1) = \pi_\infty(1) - [\pi_\infty(1) - \pi_0(1)] [1 - \pi(2|1) - \pi(1|2)]^n,$

(5.126) $\pi_\infty(1) = \dfrac{\pi(1|2)}{\pi(1|2) + \pi(2|1)} .$

These relations can be inserted in the foregoing recurrence formula for $V_{1,n}$ and the resulting difference equation solved. We do not carry this out here, but impose an additional restriction, namely, we take $\pi_0(1) = \pi_\infty(1)$. This simply means that the initial probability of event E_1 is chosen to match the asymptotic proportion; this was done in the Hake-Hyman experiment, for example. When this is true, $\pi_n(1) = \pi_\infty(1)$ and equation 5.124 simplifies to

(5.127) $V_{1,n+1} = \alpha V_{1,n} + (1 - \alpha)\{\pi_\infty(1)\lambda_1 + [1 - \pi_\infty(1)]\lambda_2\}.$

In the limit when $n \to \infty$, we have $V_{1,n+1} = V_{1,n} = V_{1,\infty}$ and

(5.128) $V_{1,\infty} = \pi_\infty(1)\lambda_1 + [1 - \pi_\infty(1)]\lambda_2.$

Finally, if we further assume that $\lambda_1 = 1$ and $\lambda_2 = 0$, assumptions which seem appropriate for the Hake-Hyman experiment, we get

(5.129) $$V_{1,\infty} = \pi_\infty(1).$$

This result agrees with the experimental finding of Hake and Hyman: The asymptotic frequency of predictions of one symbol is equal to the frequency of occurrence of that symbol in the sequence.

5.12 SUMMARY

The equal alpha condition examined in this chapter is of particular interest in analyzing data from choice experiments that involve similar alternative responses. In Chapter 13 several such experiments are discussed, and the machinery of this chapter applied. When the events are subject controlled or experimenter-subject controlled, the equal alpha condition leads to some major mathematical simplifications. The recurrence formulas for the moments can be solved explicitly.

The three types of event sequences considered in the previous two chapters—experimenter-controlled, subject-controlled, and experimenter-subject-controlled—are re-examined in this chapter with the equal alpha condition. Formulas for the means and variances are derived. The analysis is extended to arbitrary numbers of response classes and outcomes in Section 5.10. A Markov sequence of experimenter-controlled events is considered in Section 5.11, and formulas for the means are developed.

REFERENCE

1. Hake, H. W., and Hyman, R. Perception of the statistical structure of a random series of binary symbols. *J. exp. Psychol.*, 1953, **45**, 64–74.

CHAPTER 6

Approximate Methods

6.1 INTRODUCTION

In Chapter 4 we studied the distributions of p values when event occurrences were uncertain. Our main interest was in the moments of those distributions, especially the mean and variance. When events were experimenter controlled, that is, when the event probabilities were fixed, there was no difficulty in computing the mean and variance of the distribution from trial to trial; but when the events were subject controlled or experimenter-subject controlled, difficulties arose. Only with equal alphas, discussed in the last chapter, were we able to compute the moments exactly without expending a great amount of labor. As a result, we now turn to some approximate methods for computing trial-to-trial means for subject-controlled and experimenter-subject-controlled events without equal alphas.

Two main types of approximate methods are developed. The first general method involves a random number scheme for studying how *individual* model organisms behave. This method can be applied to many organisms, thereby giving estimates of means and variances. Furthermore, it may provide some insight into the kind of sequential behavior which is implied by the model. In Section 6.2 we introduce this method, and in Section 6.3 we apply it to a study of the asymptotic distributions. The second general method provides upper and lower *bounds* on the distribution means. This method leads to a statement that a mean is between two numbers that are readily computed from the formulas developed. When the upper and lower bounds are close together, we may know all that we need to know about the mean. An example discussed in Section 6.8 gives bounds of 0.676 and 0.655. We are ordinarily quite satisfied with bounds as "tight" as this, but we are not always so fortunate. Table 6.3 shows several sets of bounds; some sets are so far apart that they are of little value, whereas others are quite close together.

6.2 STAT-RATS

One method of obtaining approximate estimates of means, moments, and percentage points of the probability value distribution uses random numbers. If we are given the constants for the operators and a starting probability in numerical form, we can mechanically carry a hypothetical animal through a sequence of trials—noting which operator is to be applied on each trial, applying it, and keeping track of the p value at each stage. We have christened such hypothetical animals "stat-rats." When enough stat-rats are run we can obtain very good estimates of any constant of the p-value distribution that we wish for any trial number. The idea of the method is old. Before gamblers became so educated that they could compute probabilities or so wealthy they could hire mathematicians to do it for them, the standard method of estimating probabilities was to keep track of the outcomes of a large number of trials. In scientific work Student (William S. Gossett) used this method plus some intuition to obtain the sampling distribution of the correlation coefficient for small samples when the true correlation is zero [1]. Some time later R. A. Fisher demonstrated mathematically that Student's answer was correct. More recently such random number techniques have been called "Monte Carlo methods," and they are used in solving quite advanced problems in mathematical physics.

We illustrate the method by applying it for 25 trials to two subject-controlled events with $p_0 = 0.2$, $a_1 = 0.3$, $\alpha_1 = 0.6$, $a_2 = 0.01$, $\alpha_2 = 0.9$. In Table 6.1 the second column gives the p values for each trial. The third column is a set of 2-digit random numbers,* and the fourth column gives the operator to be applied on each trial. The 2-digit random numbers are of the forms 00, 01, 02, $\cdots$, 99. The number 00 stands for all decimal numbers from $0.00000\cdots$, to $0.009999\cdots$. We choose the numbers at the low end of the scale to go with applying Q_1, and the high numbers with Q_2. Thus if the probability of Q_1 is 0.344 on a particular trial the random numbers 00, 01, 02, $\cdots$, 33 all call for application of Q_1; and the numbers 35, 36, $\cdots$, 99 call for application of Q_2. The number 34 is indeterminate because it may mean any number from $0.34000\cdots$ to $0.34999\cdots$. For numbers from $0.3400\cdots$ to $0.343999\cdots$ we wish to apply Q_1, and for numbers from $0.34400\cdots$ to $0.34999\cdots$ we apply Q_2. The ambiguity is resolved, *not* by throwing the number 34 away and getting another, but by adding additional random digits to the end of the number until a decision is reached. This is usually easy because it is convenient to lay out in advance the random numbers for several stat-rats at once in parallel columns. When a tie occurs we join

---

* Numerous tables of random numbers are available [2, 3, 4, 5, 6].

the first digit of the random number of the next stat-rat on that trial to the end of the number we already have, leaving the number for the next stat-rat unchanged. For example, if in the present tie the next digit is 0, 1, 2, or 3 we must have a number less than 0.344 and we apply Q_1; but, if the digits 4, 5, 6, 7, 8, or 9 appear, we apply Q_2. It is not worth carrying more than two digits in the random numbers in the original layout because ambiguities seldom occur.

Inspection of Table 6.1 clarifies the procedure used. A p value of 0.2000 was chosen for trial 0; the first random number chosen was 84,

TABLE 6.1

Illustration of a computation sheet for 25 trials of one stat-rat. The operations are $Q_1p = 0.3 + 0.6p$ and $Q_2p = 0.01 + 0.9p$, and $p_0 = 0.2$.

Trial Number	p value	Random Number	Operator
0	0.2000	84	Q_2
1	0.1900	29	Q_2
2	0.1810	35	Q_2
3	0.1729	69	Q_2
4	0.1656	53	Q_2
5	0.1590	37	Q_2
6	0.1531	05	Q_1
7	0.3919	50	Q_2
8	0.3627	57	Q_2
9	0.3364	60	Q_2
10	0.3128	55	Q_2
11	0.2915	58	Q_2
12	0.2724	79	Q_2
13	0.2552	50	Q_2
14	0.2397	56	Q_2
15	0.2257	01	Q_1
16	0.4354	51	Q_2
17	0.4019	65	Q_2
18	0.3717	92	Q_2
19	0.3445	32	Q_1
20	0.5067	21	Q_1
21	0.6040	66	Q_2
22	0.5536	35	Q_1
23	0.6322	18	Q_1
24	0.6793	65	Q_1
25	0.7076		

which is above 20, and so Q_2 was applied to the initial p value and this gave a new p value of 0.1900. Then another random number, 29, was selected, and this was above 19, indicating again that Q_2 should be applied.

This procedure was repeated and on trial number 6 the random number was 05, which was less than 15, the number corresponding to the p value of 0.1531. Hence Q_1 was applied to 0.1531, giving a new p value of 0.3919. In Fig. 6.1 the small circles represent the successive means for 84 stat-rats, each run through 25 trials, with the parameter values given above.

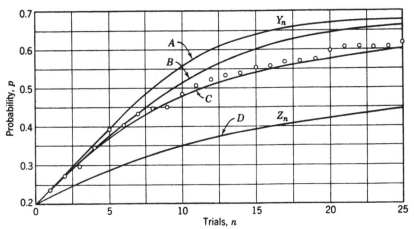

Fig. 6.1. Mean p values obtained from 84 stat-rats for 25 trials. The small circles denote these means. Also shown are upper and lower bounds, discussed in Section 6.7, for the true distribution means. Curve A is the expected operator bound computed from equation 6.69. Curve D is a lower bound computed from equation 6.68. Curves B and C are the bounds computed from equations 6.71, 6.72, and 6.73. All computations used the values $a_1 = 0.3$, $\alpha_1 = 0.6$, $a_2 = 0.01$, $\alpha_2 = 0.9$, and $p_0 = 0.2$.

In a sense, the procedure just described can provide us with all the available implications of our basic postulates. With enough patience, we could run hundreds of stat-rats for each set of parameter values that interested us. Only because we are not so patient as that do we bother with the mathematical analysis contained in most of Part I. On the other hand, stat-rat computations serve a slightly different purpose. As we have already suggested, a stat-rat is a sort of theoretical organism, a mathematical "robot." It will generate sequences of "responses" similar to the sequences of real organisms. The adequacy of the model can be judged in part by comparing stat-rat sequences with experimentally observed sequences. In Part II we shall make such comparisons.

6.3 THE ASYMPTOTIC DISTRIBUTIONS

The stat-rat procedure of making computations is now used to obtain the approximate form of the asymptotic distribution for a number of special cases. Two somewhat different methods are available. The first involves making a large number of stat-rat runs—say a thousand—for a number of trials. When the number of trials is sufficiently large, the final p values obtained from each run form a distribution which approximates the asymptotic distribution as closely as we like. The initial p value of these runs is arbitrary since the final p value is nearly independent of the initial p value for the cases we consider.

For a given numerical example we would like to estimate how many trials are needed to approximate the asymptotic distribution with a specified accuracy. Suppose that we have decided to use a class interval of γ in computing the approximate distribution. For example, we might be quite satisfied with percentage points of the cumulative distribution a distance of 0.01 apart. Suppose we have two sequences beginning at p_0 and p_0', but using the same sequence of random numbers. We would want to choose the trial number n so that p_n and p_n' for these two sequences would lie within the same class interval or at least in neighboring classes of p values. Thus we want to choose n so that

$$(6.1) \qquad |p_n - p_n'| \leq \gamma.$$

For experimenter-controlled events, consider two initial p values p_0 and p_0'. We can easily show that

$$Q_i p_0 - Q_i p_0' = \alpha_i(p_0 - p_0').$$

Thus the difference between the two initial values is multiplied by some number $\alpha_i < 1$, depending upon which operator is applied on the zeroth trial. On the next trial, the difference $p_1 - p_1'$ is multiplied again by one of the α_i. Let β stand for the largest α_i $(i = 1, 2, \cdots, t)$. Then on each trial the difference of the p values of the two sequences is multiplied by a number at least as small as β. Hence, for $p_0 > p_0'$,

$$(6.2) \qquad p_n - p_n' \leq \beta^n(p_0 - p_0').$$

Since all the asymptotic p values are between λ_{max} and λ_{min}, by the trapping theorem of Section 4.7, we would certainly choose p_0 and p_0' within these limits. Hence $p_0 - p_0'$ is at most $(\lambda_{max} - \lambda_{min})$, and so we would want n to satisfy

$$(6.3) \qquad \beta^n(\lambda_{max} - \lambda_{min}) \leq \gamma.$$

Consider the following example: There are two operators with $\alpha_1 = 0.6$, $\alpha_2 = 0.9$, $\lambda_1 = 0.75$, $\lambda_2 = 0.1$. Then $\beta = 0.9$, $\lambda_{max} = 0.75$, and $\lambda_{min} = 0.1$. If we select a class interval of $\gamma = 0.01$, our inequality is

$$(0.9)^n(0.75 - 0.10) \leq 0.01,$$

or

$$(0.9)^n \leq 0.0154.$$

The smallest integer which satisfies this is $n = 40$ since $(0.9)^{40} = 0.0148$

and $(0.9)^{39} = 0.0164$.   Therefore we need 40 trials of stat-rat computations to be certain that the most extreme values of p_0 lead to final p values in the same or adjacent classes in the desired distribution.   Ordinarily, a smaller number of trials would be adequate since random number sequences which would keep the p values as far apart as $\beta^n(p_0 - p_0')$ are rare.

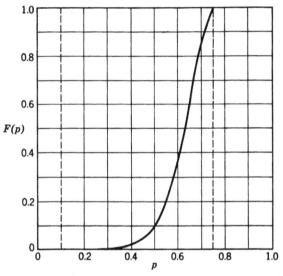

Fig. 6.2.   The approximate asymptotic cumulative distribution of p values obtained from a 1000-trial stat-rat for the example of experimenter-controlled events. $Q_1 p = 0.3 + 0.6p$, $Q_2 p = 0.01 + 0.9p$, $\pi_1 = \pi_2 = 0.5$.   The vertical dotted lines show the trapping limits $\lambda_2 = 0.10$ and $\lambda_1 = 0.75$.

For two subject-controlled events, it is rather difficult to obtain a good estimate of the number of trials necessary to obtain the desired approximation to the asymptotic distribution, so we do not attempt it here.

The method discussed above is rather wasteful because only the p value on the last trial of each stat-rat is used to approximate the asymptotic distribution.   However, the ergodic theorem for single sequences, stated in Section 4.7, provides us with a much less wasteful procedure.   According to that theorem, a single sequence (that is, a single stat-rat) will generate the asymptotic distribution when that sequence becomes infinitely long.   The only practical problem then is to estimate how long the sequence must be to give a reasonable approximation.   The main criterion is to let the total number of trials be large compared to the number of trials necessary for a single stat-rat.   In the previous numerical example we needed at most 40 trials per stat-rat, and so a single stat-rat of say 1000 trials might be adequate.

We have used the single stat-rat procedure to obtain the asymptotic distribution for a number of numerical examples. In Table 6.2 and Fig. 6.2 we give the results for a case of two experimenter-controlled events

TABLE 6.2

Frequencies of p values which occurred in a 1000 trial stat-rat with $Q_1 p = 0.3 + 0.6p$, $Q_2 p = 0.01 + 0.9p$, $\pi_1 = \pi_2 = 0.5$. All p values were rounded to two decimals.

p	f	p	f	p	f
≤ 0.26	0	0.43	6	0.60	33
0.27	1	0.44	5	0.61	53
0.28	0	0.45	7	0.62	58
0.29	1	0.46	7	0.63	23
0.30	0	0.47	13	0.64	58
0.31	1	0.48	10	0.65	44
0.32	1	0.49	11	0.66	36
0.33	1	0.50	8	0.67	65
0.34	1	0.51	28	0.68	57
0.35	0	0.52	17	0.69	28
0.36	1	0.53	20	0.70	55
0.37	2	0.54	28	0.71	44
0.38	2	0.55	18	0.72	30
0.39	6	0.56	32	0.73	32
0.40	3	0.57	36	0.74	46
0.41	3	0.58	22	0.75	0
0.42	9	0.59	38		1000

and $Q_1 p = 0.3 + 0.6p$, $Q_2 p = 0.01 + 0.9p$, $\pi_1 = \pi_2 = 0.5$. In Section 4.3 we computed the asymptotic mean and standard deviation from the exact moment formulas for this numerical example. We obtained $V_{1,\infty} = 0.62$ and $\sigma_\infty = 0.08$. From the 1000-trial stat-rat used in obtaining Table 6.2 and Fig. 6.2 we obtained a mean of 0.6185 and a standard deviation of 0.086. This close agreement provides a further check on the usefulness of the method.

For two subject-controlled events we computed an example where $Q_1 p = 0.3 + 0.6p$ and $Q_2 p = 0.01 + 0.9p$. The results are shown in Fig. 6.3. From these computations we obtain an asymptotic mean of $V_{1,\infty} = 0.6731$ and an asymptotic standard deviation of $\sigma_\infty = 0.0784$. In Section 6.8 we compare this result for the mean with some computed bounds on $V_{1,\infty}$ for the same numerical example.

In Fig. 6.4 we provide another example of the asymptotic distribution for two subject-controlled events. This example is for the operators $Q_1 p = 0.3 + 0.6p$ and $Q_2 p = 0.06 + 0.7p$ ($\lambda_1 = 0.75$ and $\lambda_2 = 0.20$). From the computations we obtained a mean $V_{1,\infty} = 0.513$ and a standard deviation $\sigma_\infty = 0.173$.

The equal alpha condition discussed in Chapter 5 led to exact formulas for the moments, but it is still a great deal of labor to obtain the form of the asymptotic distribution from those moment formulas. Therefore,

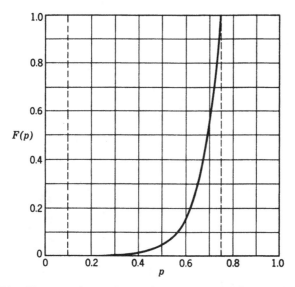

Fig. 6.3. The approximate asymptotic cumulative distribution of p values obtained from a 4000-trial stat-rat for an example of subject-controlled events. $Q_1 p = 0.3 + 0.6p$, $Q_2 p = 0.01 + 0.9p$. The vertical dotted lines show the trapping limits $\lambda_2 = 0.10$ and $\lambda_1 = 0.75$.

we provide two examples of distributions obtained by the single stat-rat procedure. The first, for $Q_1 p = 0.4 + 0.5p$ and $Q_2 p = 0.1 + 0.5p$ ($\lambda_1 = 0.8$ and $\lambda_2 = 0.2$), is presented in Fig. 6.5. The observed mean was 0.4826 as compared with the true mean, 0.500, computed from asymptotic formula 5.30. The observed standard deviation was 0.2464 as compared with the true value, 0.2236, obtained from asymptotic formula 5.48. Finally, in Fig. 6.6 we give an illustration for $Q_1 p = 0.5p$ and $Q_2 p = 0.5 + 0.5p$ ($\lambda_1 = 0$ and $\lambda_2 = 1$), an example of the special case discussed in Section 5.7. The true distribution parameters obtained from equations 5.58 and 5.63 are $V_{1,\infty} = 0.5000$ and $\sigma_\infty = 0.2236$. The values obtained from the single stat-rat computation were $V_{1,\infty} = 0.4958$ and $\sigma_\infty = 0.2227$.

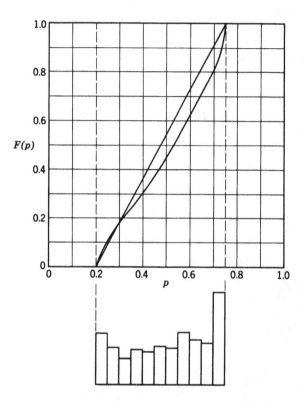

Fig. 6.4. The approximate asymptotic cumulative distribution of p values obtained from a 1000-trial stat-rat for another example of subject-controlled events. $Q_1p = 0.3 + 0.6p$, $Q_2p = 0.06 + 0.7p$. The vertical dotted lines show the trapping limits $\lambda_2 = 0.20$ and $\lambda_1 = 0.75$. Note that this distribution is much more symmetric about the center than the distribution in Fig. 6.3. For comparison, a straight line, corresponding to a uniform distribution, is also shown. The histogram below the cumulative diagram gives a rough picture of the density in class intervals.

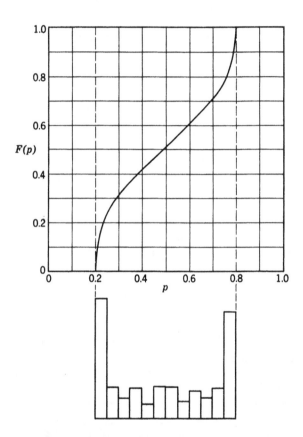

Fig. 6.5. The approximate asymptotic cumulative distribution of p values, obtained from a 4000-trial stat-rat for the equal alpha case of two subject-controlled events, with $\lambda_1 = 0.8$, $\lambda_2 = 0.2$, and $\alpha = 0.5$. The histogram below the cumulative diagram gives a rough picture of the density in class intervals.

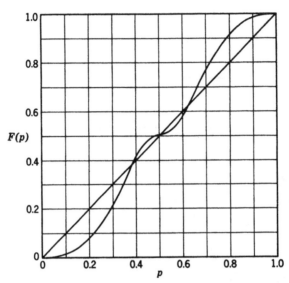

Fig. 6.6. The approximate asymptotic cumulative distribution of p values, obtained from a 2000-trial stat-rat, for the equal alpha case of two-subject-controlled events, with $\lambda_1 = 0$ and $\lambda_2 = 1$. This case is discussed in Section 5.7. For the computations, $Q_1 p = 0.5 p$, $Q_2 p = 0.5 + 0.5 p$. The straight line corresponds to a uniform distribution on the interval from zero to unity.

6.4 THE EXPECTED OPERATOR

In Section 3.12 we considered a rather obvious device for computing means from trial to trial for two subject-controlled events, and we consider it again here. This procedure involved computing the mathematical expectation of the p value on trial 1, using this mean p value for the probability of application of Q_1 on trial 1 to obtain the mean p value on trial 2, etc. The analogous procedure did yield the correct means for experimenter-controlled events, but, as we saw in Section 3.12, it did not work for subject-controlled events. The one exception is the equal alpha case described in Chapter 5. Nevertheless, we consider the procedure in this section without the equal alpha condition to provide a possible approximation scheme.

As in Section 5.4 we define an expected operator $\bar{Q}$, for two subject-controlled events, by the equation

(6.4) $$\bar{Q} V_{1,n} = V_{1,n} Q_1 V_{1,n} + (1 - V_{1,n}) Q_2 V_{1,n}.$$

Using the slope-intercept form for the operators, $Q_i p = a_i + \alpha_i p$, we have

(6.5)
$$\bar{Q} V_{1,n} = V_{1,n}(a_1 + \alpha_1 V_{1,n}) + (1 - V_{1,n})(a_2 + \alpha_2 V_{1,n})$$
$$= a_2 + (a_1 - a_2 + \alpha_2) V_{1,n} + (\alpha_1 - \alpha_2) V_{1,n}^2.$$

We then assume that this expression gives an approximation to $V_{1,n+1}$, that is,

(6.6) $$V_{1,n+1} \cong \bar{Q}V_{1,n}.$$

We thus have a quadratic difference equation which cannot be solved by elementary means, but we can study some of its properties.

First consider the asymptotic solution for which $\bar{Q}V_{1,\infty} = V_{1,\infty}$. We have

(6.7) $$V_{1,\infty} \cong a_2 + (a_1 - a_2 + \alpha_2)V_{1,\infty} + (\alpha_1 - \alpha_2)V_{1,\infty}^2.$$

The solution of this quadratic is

(6.8) $$V_{1,\infty} \cong \frac{(1 - a_1 + a_2 - \alpha_2) \pm \sqrt{(1 - a_1 + a_2 - \alpha_2)^2 - 4a_2(\alpha_1 - \alpha_2)}}{2(\alpha_1 - \alpha_2)}.$$

Usually only one of the roots falls within the possible range, and so it is easy to decide which sign before the radical is appropriate. This approximate formula for the asymptotic mean is sometimes useful. As we see in the next section, it provides either an upper or a lower bound on the true value.

Equation 6.6 may be used for trial by trial numerical computations, but it is often more convenient to consider the corresponding differential equation which is obtained by using the further approximation

(6.9) $$\frac{dV_{1,n}}{dn} \cong \bar{Q}V_{1,n} - V_{1,n}.$$

(Cf. Section 3.4.) We then have

(6.10) $$\frac{dV_{1,n}}{dn} \cong a_2 + (a_1 - a_2 + \alpha_2 - 1)V_{1,n} + (\alpha_1 - \alpha_2)V_{1,n}^2.$$

This equation is integrated directly [7] to yield

(6.11) $$V_{1,n} \cong \frac{1 - \beta}{2(\alpha_1 - \alpha_2)} + \frac{\rho}{2(\alpha_1 - \alpha_2)} \frac{1 + Ce^{\rho n}}{1 - Ce^{\rho n}},$$

where

$$\beta = a_1 - a_2 + \alpha_2,$$

(6.12) $$\rho = \sqrt{(1 - \beta)^2 - 4a_2(\alpha_1 - \alpha_2)},$$

$$C = \frac{2(\alpha_1 - \alpha_2)V_{1,0} - (1 - \beta) - \rho}{2(\alpha_1 - \alpha_2)V_{1,0} - (1 - \beta) + \rho}.$$

In the limit as $n \to \infty$, the ratio $(1 + Ce^{\rho n})/(1 - Ce^{\rho n})$ approaches -1 and so

(6.13) $$V_{1,\infty} \cong \frac{1 - \beta - \rho}{2(\alpha_1 - \alpha_2)},$$

in agreement with equation 6.8 above.

Consider our numerical example for which $a_1 = 0.3$, $a_2 = 0.01$, $\alpha_1 = 0.6$, $\alpha_2 = 0.9$, and $V_{1,0} = p_0 = 0.2$. We then get from the above definitions $\beta = 1.19$, $\rho = 0.2193$, and $C = -0.5161$, and so $V_{1,\infty} \cong 0.6822$. For the mean on trial n we have

(6.14) $$V_{1,n} \cong 0.3167 - 0.3655 \frac{1 - 0.5161e^{0.2193n}}{1 + 0.5161e^{0.2193n}}.$$

For experimenter-subject-controlled events with two response classes and s possible outcomes, we may define an expected operator $\bar{Q}$ by

(6.15) $$\bar{Q}V_{1,n} = V_{1,n} \sum_{k=1}^{s} \pi_{1k}Q_{1k}V_{1,n} + (1 - V_{1,n}) \sum_{k=1}^{s} \pi_{2k}Q_{2k}V_{1,n},$$

where the conditional probabilities satisfy the conditions

(6.16) $$\sum_{k=1}^{s} \pi_{jk} = 1 \qquad (j = 1, 2).$$

The operators are defined by the expressions

(6.17) $$Q_{jk}V_{1,n} = a_{jk} + \alpha_{jk}V_{1,n}.$$

When these expressions are inserted in the equation defining $\bar{Q}$ and the following abbreviations used,

(6.18) $$\bar{a}_j = \sum_{k=1}^{s} \pi_{jk}a_{jk},$$
$$\bar{\alpha}_j = \sum_{k=1}^{s} \pi_{jk}\alpha_{jk},$$

we obtain the equation

(6.19) $$\bar{Q}V_{1,n} = \bar{a}_2 + (\bar{a}_1 - \bar{a}_2 + \bar{\alpha}_2)V_{1,n} + (\bar{\alpha}_1 - \bar{\alpha}_2)V_{1,n}^2.$$

This equation is similar in form to equation 6.5 for subject-controlled events. Therefore, the approximate solution of that equation generalizes at once to the solution of equation 6.19, provided that we make the obvious parameter substitutions. Equations 6.11, 6.12, and 6.13 become

$$(6.20) \qquad V_{1,n} \cong \frac{1 - \bar{\beta}}{2(\bar{\alpha}_1 - \bar{\alpha}_2)} + \frac{\bar{\rho}}{2(\bar{\alpha}_1 - \bar{\alpha}_2)} \frac{1 + \bar{C}e^{\bar{\rho}n}}{1 - \bar{C}e^{\bar{\rho}n}},$$

$$(6.21) \qquad \bar{\beta} = \bar{a}_1 - \bar{a}_2 + \bar{\alpha}_2,$$

$$(6.22) \qquad \bar{\rho} = \sqrt{(1 - \bar{\beta})^2 - 4\bar{a}_2(\bar{\alpha}_1 - \bar{\alpha}_2)},$$

$$(6.23) \qquad \bar{C} = \frac{2(\bar{\alpha}_1 - \bar{\alpha}_2)V_{1,0} - (1 - \bar{\beta}) - \bar{\rho}}{2(\bar{\alpha}_1 - \bar{\alpha}_2)V_{1,0} - (1 - \bar{\beta}) + \bar{\rho}},$$

$$(6.24) \qquad V_{1,\infty} \cong \frac{1 - \bar{\beta} - \bar{\rho}}{2(\bar{\alpha}_1 - \bar{\alpha}_2)}.$$

*6.5 BOUNDS ON THE ASYMPTOTIC MEAN

We have just indicated that the expected operator usually gives only an approximation to the distribution mean. But we have no way to know how close the approximation is without computing the exact means. Therefore it is useful to obtain some bounds on the mean, that is, a quantity which is always greater than the mean and another quantity which is always less than the mean. When these bounds are close together we have a good estimate of the true mean. In this and the following section we restrict our attention to the asymptotic distributions for two subject-controlled events.

The recurrence relation, equation 4.52, gives for the mean, $V_{1,n+1}$, on the $(n + 1)$st trial

$$(6.25) \qquad V_{1,n+1} = C_{10} + C_{11}V_{1,n} + C_{12}V_{2,n}.$$

The coefficients are defined as before by

$$(6.26) \qquad \begin{aligned} C_{10} &= a_2, \\ C_{11} &= a_1 - a_2 + \alpha_2, \\ C_{12} &= \alpha_1 - \alpha_2. \end{aligned}$$

For very large n, we may take $V_{1,n} = V_{1,n+1} = V_{1,\infty} = V_1$; also $V_{2,n} = V_{2,n+1} = V_{2,\infty} = V_2$. Thus we get

$$(6.27) \qquad C_{10} + (C_{11} - 1)V_1 + C_{12}V_2 = 0.$$

Now if we knew the second raw moment V_2 there would be no problem; we could obtain V_1 immediately. However, V_2 is not known, but we might expect to obtain upper and lower bounds on V_1 by replacing V_2 with its upper and lower bounds. The task then reduces to finding bounds, V_2' and V_2'', on V_2. Of course we can say at once that

$$0 \leq V_2 \leq 1, \tag{6.28}$$

since our asymptotic distribution lies on the interval from zero to unity. If x is the variable along that interval and if $f(x)$ is the probability density function,* the second raw moment is by definition

$$V_2 = \int_0^1 x^2 f(x)\, dx. \tag{6.29}$$

The variable x is at most unity, and, moreover,

$$\int_0^1 f(x)\, dx = 1. \tag{6.30}$$

Hence we obtain inequalities 6.28. But we can do better than this, for we know from the trapping theorem of Section 4.7 that the asymptotic distribution must lie between λ_2 and λ_1. For present purposes we assume that $\lambda_2 < \lambda_1$, but we show later that interchanging λ_2 and λ_1 does not alter our final results. We therefore require that $0 \leq \lambda_2 \leq x \leq \lambda_1 \leq 1$. From equation 6.29 we then see that

$$\lambda_2^2 \leq V_2 \leq \lambda_1^2. \tag{6.31}$$

These limits on V_2 could be used in the relation 6.27, but ordinarily they will provide no better bounds on V_1 than the ones we already have, namely,

$$\lambda_2 \leq V_1 \leq \lambda_1. \tag{6.32}$$

From the foregoing discussion it should be clear that we are not satisfied with just *any* bounds on V_2—we are looking for bounds that are close together. In particular, we should like to find bounds on V_2 for a given value of V_1: How large and how small can V_2 become for any distribution between λ_2 and λ_1 and with mean V_1? The lower limit on V_2 is easy for we know that (see Section 4.2)

$$V_2 = V_1^2 + \sigma^2, \tag{6.33}$$

where σ^2 is the variance about the mean V_1. This variance is never negative and so we have

$$V_2 \geq V_1^2. \tag{6.34}$$

In order to ascertain whether this is the best possible lower bound on V_2 we ask whether there exists a distribution on the interval λ_2 to λ_1 with

---

* For simplicity of exposition we are using the density function instead of the cumulative. We point this out because the asymptotic density function may not exist though the cumulative does.

mean V_1 and second raw moment $V_2 = V_1^2$. If such a distribution does exist, we know that the bound is the best available unless further restrictions on the distribution are specified. The problem, then, is to find a distribution, if possible, with $V_2 = V_1^2$. This is accomplished immediately, for a distribution having all its density at V_1, where $\lambda_2 \leq V_1 \leq \lambda_1$, has a second moment equal to V_1^2. Hence, we have obtained the best possible lower bound on V_2 for given V_1.

The best possible upper bound on V_2 for given V_1 is a little more difficult to obtain. We begin by considering a distribution $g(z)$ where $0 \leq z \leq 1$. The mean, U_1, is

(6.35)
$$U_1 = \int_0^1 z g(z)\, dz,$$

and the second raw moment is

(6.36)
$$U_2 = \int_0^1 z^2 g(z)\, dz.$$

And, since z is between zero and unity, we have

(6.37)
$$\int_0^1 z^2 g(z)\, dz \leq \int_0^1 z g(z)\, dz,$$

and so

(6.38)
$$U_2 \leq U_1.$$

We now transform this distribution to the interval from λ_2 to λ_1 by letting

(6.39)
$$z = \frac{x - \lambda_2}{\lambda_1 - \lambda_2}.$$

We call the transformed distribution $f(x)$, and a necessary condition for making the transformation is that $f(x)\, dx = g(z)\, dz$. We need the transformation equations for the moments. The mean is

(6.40)
$$U_1 = \int_0^1 z g(z)\, dz = \int_{\lambda_2}^{\lambda_1} \frac{x - \lambda_2}{\lambda_1 - \lambda_2} f(x)\, dx$$
$$= \frac{1}{\lambda_1 - \lambda_2} \int_{\lambda_2}^{\lambda_1} x f(x)\, dx - \frac{\lambda_2}{\lambda_1 - \lambda_2} \int_{\lambda_2}^{\lambda_1} f(x)\, dx.$$

The first integral is the mean V_1 for our distribution on the interval from λ_2 to λ_1; the last integral is the total density and so is unity. Hence, we have

(6.41)
$$U_1 = \frac{V_1 - \lambda_2}{\lambda_1 - \lambda_2}.$$

By the same procedure we obtain

(6.42)
$$U_2 = \frac{V_2 - 2\lambda_2 V_1 + \lambda_2^2}{(\lambda_1 - \lambda_2)^2}.$$

We now insert these last two relations in the inequality 6.38 and obtain, after multiplying through by $(\lambda_1 - \lambda_2)^2$,

(6.43)
$$V_2 - 2\lambda_2 V_1 + \lambda_2{}^2 \le (V_1 - \lambda_2)(\lambda_1 - \lambda_2).$$

After re-arranging, we have

(6.44)
$$V_2 \le (\lambda_1 + \lambda_2)V_1 - \lambda_1\lambda_2.$$

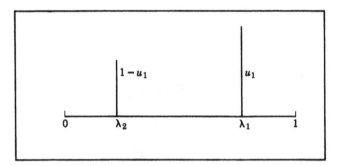

Fig. 6.7.   A distribution having all its density at the limits λ_2 and λ_1.

This is the upper bound on V_2 we are looking for.   We can establish that it is the best possible upper bound by finding a distribution on the interval between λ_2 and λ_1 having mean V_1 and second moment,

(6.45)
$$V_2 = (\lambda_1 + \lambda_2)V_1 - \lambda_1\lambda_2.$$

Such a distribution is the one shown in Fig. 6.7.   If the density at λ_1 is u_1 and the density at λ_2 is $1 - u_1$, we have for the mean and second raw moment

(6.46)
$$V_1 = u_1\lambda_1 + (1 - u_1)\lambda_2,$$

(6.47)
$$V_2 = u_1\lambda_1{}^2 + (1 - u_1)\lambda_2{}^2.$$

Solving equation 6.46 for u_1 we obtain

(6.48)
$$u_1 = \frac{V_1 - \lambda_2}{\lambda_1 - \lambda_2}.$$

If this result is substituted into equation 6.47 and simplifications are made, we at once obtain the upper bound for V_2 given earlier in equation 6.45.   Therefore, we have found a distribution on the interval from λ_2 to λ_1 with mean V_1 and second moment as large as the upper bound permitted by the inequality 6.44. Therefore, we have found the best possible upper bound on V_2 for given V_1. Combining the upper and lower bounds, we have finally the

BOUNDS ON V_2:

(6.49) $$V_1^2 \le V_2 \le (\lambda_1 + \lambda_2)V_1 - \lambda_1\lambda_2.$$

We observe that if λ_2 and λ_1 are interchanged we do not change the bounds on V_2, and so the results apply when $\lambda_1 < \lambda_2$ as well as when $\lambda_2 < \lambda_1$.

We now use these bounds on V_2 in equation 6.27 to obtain bounds on the mean V_1. We first consider the lower bound. In equation 6.27 we replace V_2 with V_1^2 and see that, when C_{12} is positive,

(6.50) $$C_{10} + (C_{11} - 1)V_1 + C_{12}V_1^2 \le 0 \qquad (0 < C_{12}),$$

and, when C_{12} is negative,

(6.51) $$C_{10} + (C_{11} - 1)V_1 + C_{12}V_1^2 \ge 0 \qquad (C_{12} < 0).$$

These inequalities place some restrictions on V_1. We denote the quadratic expression appearing in these inequalities by $J(V_1)$, that is,

(6.52) $$J(V_1) = C_{10} + (C_{11} - 1)V_1 + C_{12}V_1^2.$$

We see at once that when $V_1 = 0$ we have $J(0) = C_{10} = a_2$, which is positive. Furthermore, when $V_1 = 1$ we have $J(1) = C_{10} + (C_{11} - 1) + C_{12}$, and from the definitions of the C's we get $J(1) = (a_1 - a_2 + \alpha_2 - 1) + (\alpha_1 - \alpha_2) + a_2 = \alpha_1 + a_1 - 1$. Using the additional fact that $a_1 = \lambda_1(1 - \alpha_1)$, we have $J(1) = -(1 - \lambda_1)(1 - \alpha_1)$, and this is a negative quantity. Thus we have established that the quadratic function J is positive at zero and negative at unity, and so it follows that J is zero for only one value of V_1 between zero and unity. We denote this root by Y, that is, $J(Y) = 0$. In Fig. 6.8 we show an example of $J(V_1)$ for $a_1 = 0.3$, $a_2 = 0.01$, $\alpha_1 = 0.6$, and $\alpha_2 = 0.9$, giving the function

(6.53) $$J(V_1) = -0.30V_1^2 + 0.19V_1 + 0.01.$$

We observe from this figure that J is zero at about $Y = 0.68$.

We may solve the quadratic equation $J(Y) = 0$ for Y to obtain

(6.54) $$Y = \frac{(1 - C_{11}) - \sqrt{(1 - C_{11})^2 - 4C_{10}C_{12}}}{2C_{12}}.$$

It may be observed that this expression is identical to the approximate formula 6.8 obtained from the expected operator approximation. We see from Fig. 6.8 that when $J(V_1)$ is positive we have $V_1 < Y$, and when $J(V_1)$ is negative we have $Y < V_1$. But we already know from inequalities

6.50 and 6.51 that $J(V_1)$ is positive when $C_{12} < 0$, that is, when $\alpha_1 < \alpha_2$, and that $J(V_1)$ is negative when $0 < C_{12}$, that is, when $\alpha_2 < \alpha_1$. Thus we have

(6.55)
$$V_1 \leq Y \quad \text{for} \quad \alpha_1 < \alpha_2,$$
$$V_1 \geq Y \quad \text{for} \quad \alpha_2 < \alpha_1.$$

This is one of the desired bounds on V_1.

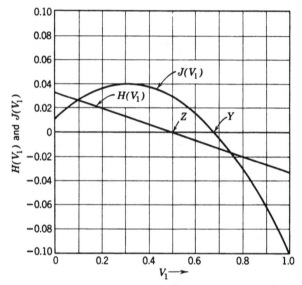

Fig. 6.8. Illustration of the functions $J(V_1)$ and $H(V_1)$ plotted from equations 6.53 and 6.57 with the parameters $a_1 = 0.3$, $\alpha_1 = 0.6$, $a_2 = 0.01$, $\alpha_2 = 0.9$. The two bounds, Y and Z of equations 6.61 and 6.62, are also indicated.

We next investigate the consequences of the upper bound on V_2 given by inequality 6.49. We replace V_2 by the expression $(\lambda_1 + \lambda_2)V_1 - \lambda_1\lambda_2$ in the left side of equation 6.27 and obtain a function

(6.56)
$$H(V_1) = C_{10} + (C_{11} - 1)V_1 + C_{12}[(\lambda_1 + \lambda_2)V_1 - \lambda_1\lambda_2]$$
$$= (C_{10} - \lambda_1\lambda_2 C_{12}) + [(C_{11} - 1) + C_{12}(\lambda_1 + \lambda_2)]V_1.$$

Using the definitions of the C's and the relation $a_i = \lambda_i(1 - \alpha_i)$, we can write this function in the form

(6.57)
$$H(V_1) = \lambda_2[(1 - \lambda_1)(1 - \alpha_2) + \lambda_1(1 - \alpha_1)]$$
$$- [\lambda_2(1 - \alpha_1) + (1 - \lambda_1)(1 - \alpha_2)]V_1.$$

When written in this form we see that the slope of the line $H(V_1)$ versus V_1 will be negative and $H(0)$ will be positive. We call Z the point at which H is zero, that is, $H(Z) = 0$. An example of the function $H(V_1)$ is given in Fig. 6.8.

Since V_2 is less than or equal to $(\lambda_1 + \lambda_2)V_1 - \lambda_1\lambda_2$, we have $H(V_1) \geq 0$ for $0 < C_{12}$, that is, for $\alpha_2 < \alpha_1$. But $H(V_1) \geq 0$ implies that $V_1 \leq Z$, and so

$$(6.58) \qquad\qquad V_1 \leq Z \quad \text{for} \quad \alpha_2 < \alpha_1.$$

Similarly, $H(V_1) \leq 0$ for $C_{12} < 0$ or $\alpha_1 < \alpha_2$, but $H(V_1) \leq 0$ implies that $V_1 \geq Z$, and so

$$(6.59) \qquad\qquad V_1 \geq Z \quad \text{for} \quad \alpha_1 < \alpha_2.$$

The last two inequalities provide the other bound on V_1. We obtain the value of Z by solving the equation $H(Z) = 0$.

We can summarize the preceding results to obtain

BOUNDS ON V_1 FROM BOUNDS ON V_2:

$$(6.60) \qquad \begin{aligned} Z \leq V_1 \leq Y \quad \text{for} \quad \alpha_1 < \alpha_2, \\ Y \leq V_1 \leq Z \quad \text{for} \quad \alpha_2 < \alpha_1, \end{aligned}$$

where

$$(6.61) \qquad Y = \frac{(1 - C_{11}) - \sqrt{(1 - C_{11})^2 - 4C_{10}C_{12}}}{2C_{12}},$$

$$(6.62) \qquad Z = \frac{\lambda_1\lambda_2 C_{12} - C_{10}}{(C_{11} - 1) + C_{12}(\lambda_1 + \lambda_2)},$$

and the C's are defined by equation 6.26:

$$C_{10} = a_2,$$
$$C_{11} = a_1 - a_2 + \alpha_2,$$
$$C_{12} = \alpha_1 - \alpha_2.$$

For the numerical example of $a_1 = 0.3$, $\alpha_1 = 0.6$, $a_2 = 0.01$, and $\alpha_2 = 0.9$, we have $\lambda_1 = 0.75$, $\lambda_2 = 0.10$, $C_{11} = 1.19$, and $C_{12} = -0.30$. Thus we get for the bounds

$$0.500 \leq V_1 \leq 0.682.$$

These bounds may be compared with the approximate value of 0.673 obtained from the single stat-rat distribution shown in Fig. 6.3. In Table 6.3 we show several other numerical examples of these bounds.

TABLE 6.3

Bounds on the asymptotic mean, $V_{1,\infty}$, for seven numerical examples. The limits λ_1 and λ_2 are the asymptotes obtained by applying one operator only. The bounds on $V_{1,\infty}$ were obtained by maximizing and minimizing the second raw moment, V_2, and the third raw moment, V_3.

Parameter Values				Limits		Bounds on $V_{1,\infty}$	
a_1	a_2	α_1	α_2	λ_2	λ_1	From V_2	From V_3
0.300	0.010	0.6	0.9	0.10	0.75	0.500–0.682	0.655–0.676
0.300	0.001	0.6	0.9	0.01	0.75	0.112–0.668	0.418–0.658
0.300	0.0001	0.6	0.9	0.001	0.75	0.013–0.667	0.093–0.654
0.396	0.001	0.6	0.9	0.01	0.99	0.394–0.987	0.967–0.986
0.396	0.003	0.6	0.7	0.01	0.99	0.184–0.961	0.723–0.878
0.360	0.03	0.6	0.7	0.10	0.90	0.557–0.718	0.637–0.657
0.300	0	0.6	0.9	0	0.75	0 –0.667	0 –0.656

*6.6   IMPROVED BOUNDS ON THE ASYMPTOTIC MEAN

In the preceding section we found the best possible bounds on V_2 for a distribution on the interval from λ_2 to λ_1 with mean V_1. These bounds on V_2 led to the bounds Y and Z on V_1. Better bounds on V_1 can be obtained from the best possible bounds on the third raw moment V_3 for given V_1 and V_2.* By arguments similar to those already used, it has been shown [8] that for a distribution on the interval $0 \leq z \leq 1$, the third raw moment U_3 satisfies the conditions

$$(6.63) \qquad \frac{U_2{}^2}{U_1} \leq U_3 \leq U_2 + \frac{(U_2 - U_1)^2}{U_1 - 1},$$

where U_2 is the corresponding second raw moment and U_1 is the mean. When this distribution is transformed into one on the interval $\lambda_2 \leq x \leq \lambda_1$, the conditions become

$$(6.64) \qquad \lambda_2 V_2 + \frac{(V_2 - \lambda_2 V_1)^2}{V_1 - \lambda_2} \leq V_3 \leq \lambda_1 V_2 + \frac{(V_2 - \lambda_1 V_1)^2}{V_1 - \lambda_1}.$$

These bounds on V_3 can be inserted in the recurrence formulas developed in Chapter 4 to eliminate V_3 and thereby make it possible to solve for bounds on V_1. In Section 6.8 we give formulas in full and illustrate the computations; in Table 6.3 we show the bounds for several numerical examples.

---

* We are indebted to Lotte Bailyn for suggesting this procedure.

*6.7 BOUNDS ON THE PRE-ASYMPTOTIC MEANS

The last two sections were restricted to a discussion of the asymptotic distributions for two subject-controlled events. The restriction to asymptotic distributions simplifies the exposition, but the development is readily extended to give bounds on the mean for any trial n. In this section we give the results for any trial, but the derivations are omitted because they are tedious though straightforward.

The distribution means $V_{1,n}$ lie between two bounds Z_n and Y_n. Analogous to inequalities 6.60, we have

(6.65)
$$Z_n \leq V_{1,n} \leq Y_n \quad \text{for} \quad \alpha_1 < \alpha_2,$$
$$Y_n \leq V_{1,n} \leq Z_n \quad \text{for} \quad \alpha_2 < \alpha_1.$$

The bound Z_n can be computed from the recurrence formula

(6.66)
$$Z_{n+1} = \alpha' Z_n + (1 - \alpha')Z,$$

provided that the parameter α', defined by

(6.67)
$$\alpha' = \alpha_2 + a_1 \frac{1 - \alpha_2}{1 - \alpha_1} - a_2 \frac{1 - \alpha_1}{1 - \alpha_2},$$

is nonnegative. This recurrence formula is readily solved to give

(6.68)
$$Z_n = \alpha'^n Z_0 + (1 - \alpha'^n)Z.$$

The limit point Z in this equation is the expression given by equation 6.62, and it may be seen that $Z_\infty = Z$. When α' is negative, these equations may not yield a bound on $V_{1,n}$ for all n. The initial value Z_0 is of course set equal to $V_{1,0}$.

The other bound Y_n is obtained from the expected operator discussed in Section 6.4. We have

(6.69)
$$Y_{n+1} = a_2 + (a_1 - a_2 + \alpha_2)Y_n + (\alpha_1 - \alpha_2)Y_n^2$$

provided that

(6.70)
$$(a_1 - a_2 + \alpha_2) + 2(\alpha_1 - \alpha_2)Y_n \geq 0$$

for all values of Y_n in the range of interest. When this condition is not satisfied, the recurrence formula for Y_n may not yield a bound on $V_{1,n}$. We take $Y_0 = V_{1,0}$ and it is readily shown that $Y_\infty = Y$, where Y is given by equation 6.61. The recurrence formula for Y_n cannot be solved by elementary means, but the differential equation solution, equation 6.11, can be used as an approximation. In Fig. 6.1 we show the bounds Z_n and Y_n for a numerical example.

The bounds obtained from the upper and lower bounds on V_3 can also be extended to any trial n. We use the bounds

(6.71)
$$V'_{3,n} = \lambda_2 V_{2,n} + \frac{(V_{2,n} - \lambda_2 V_{1,n})^2}{V_{1,n} - \lambda_2},$$

and

(6.72)
$$V''_{3,n} = \lambda_1 V_{2,n} + \frac{(V_{2,n} - \lambda_1 V_{1,n})^2}{V_{1,n} - \lambda_1},$$

in conjunction with the recurrence formulas

(6.73)
$$V_{1,n+1} = C_{10} + C_{11}V_{1,n} + C_{12}V_{2,n},$$
$$V_{2,n+1} = C_{20} + C_{21}V_{1,n} + C_{22}V_{2,n} + C_{23}V_{3,n},$$

to obtain upper and lower bounds on $V_{1,n}$. These equations can be solved to give recurrence formulas in the means alone, but we will spare the reader the sight of the result since a straightforward computation procedure requires only the preceding four equations. We begin with values of $V_{1,0}$ and $V_{2,0}$, and compute $V_{1,1}$ from equation 6.73 and $V'_{3,0}$ from equation 6.71. The latter result is used in equation 6.73 to obtain a bound on $V_{2,1}$. The value of $V_{1,1}$ and the bound on $V_{2,1}$ are then used in equation 6.71 to obtain a value of $V'_{3,1}$, and the procedure is repeated. Such computations are very laborious indeed but can be mechanized readily if there is serious interest in obtaining the bounds on the means $V_{1,n}$. In Fig. 6.1 we give an example of these bounds.

The entire analysis of bounds in this chapter was restricted to two subject-controlled events. The derivations can be generalized to experimenter-subject-controlled events in a straightforward way, but the formulas become lengthy and the computations tedious. Therefore we do not display them.

6.8 SUMMARY

In this chapter we present some approximate methods for determining properties of the distributions of p values. Most of the discussion is restricted to two subject-controlled events, but no special restrictions are placed on the parameters of the two operators Q_1 and Q_2.

A random number scheme for making approximate computations is discussed in Section 6.2. These "stat-rat" or "Monte Carlo" runs lead to estimates of the moments of the distribution of p values and are useful when exact computations are very laborious. In Section 6.3 we use the method to obtain the approximate form of the asymptotic distributions for several numerical examples.

The expected operator $\bar{Q}$, which was used for the equal alpha case in

Chapter 5, is considered for the more general case of $\alpha_1 \neq \alpha_2$ in Section 6.4. Approximate explicit formulas for the distribution means are developed from the expected operator for subject-controlled and experimenter-subject-controlled events.

Sections 6.5, 6.6, and 6.7 discuss some upper and lower bounds on the means. In Section 6.5 we establish that the asymptotic mean V_1 satisfies inequalities 6.60.

In Section 6.6 we mention an improved pair of bounds V_1' and V_1'' on the asymptotic mean V_1. These bounds are solutions of the two quadratic equations

$$(6.74) \qquad \begin{aligned} D_{20} + D_{21}V_1' + D_{22}V_1'^2 = 0, \\ D_{10} + D_{11}V_1'' + D_{12}V_1''^2 = 0, \end{aligned}$$

where the coefficients are for $i = 1, 2$,

$$(6.75) \qquad \begin{aligned} D_{i0} &= C_{23}C_{10}^2 - C_{20}C_{12}^2\lambda_i + C_{10}C_{12}(C_{22} - 1 + C_{23}\lambda_i)\lambda_i, \\ D_{i1} &= C_{12}^2(C_{20} - \lambda_i C_{21}) - C_{12}C_{10}(C_{22} - 1 - C_{23}\lambda_i) \\ &\quad + (C_{11} - 1)[C_{12}\lambda_i(C_{22} - 1 + C_{23}\lambda_i) + 2C_{23}C_{10}], \\ D_{i2} &= C_{12}^2(C_{21} + C_{23}\lambda_i^2) - C_{12}(C_{22} - 1 - C_{23}\lambda_i)(C_{11} - 1) \\ &\quad + C_{23}(C_{11} - 1)^2, \end{aligned}$$

and where

$$(6.76) \qquad \begin{aligned} C_{20} &= a_2^2, \qquad C_{21} = a_1^2 - a_2^2 + 2a_2\alpha_2, \\ C_{22} &= 2(a_1\alpha_1 - a_2\alpha_2) + \alpha_2^2, \qquad C_{23} = \alpha_1^2 - \alpha_2^2. \end{aligned}$$

To illustrate the procedure for obtaining the two foregoing sets of bounds on the asymptotic mean, we consider the parameters $a_1 = 0.3$, $a_2 = 0.01$, $\alpha_1 = 0.6$, and $\alpha_2 = 0.9$. From the relations $\lambda_i = a_i/(1 - \alpha_i)$ we see that $\lambda_1 = 0.75$ and $\lambda_2 = 0.10$. We then get the coefficients $C_{10} = 0.01$, $C_{11} = 1.19$, and $C_{12} = -0.30$. Using these numerical results in the equations for Y and Z, we have the bounds expressed by

$$0.500 \leq V_1 \leq 0.682.$$

For computing the improved pair of bounds, the necessary coefficients are $C_{20} = 0.0001$, $C_{21} = 0.1079$, $C_{22} = 1.152$, and $C_{23} = -0.45$. These values then give $D_{20} = -0.78 \times 10^{-4}$, $D_{21} = -26.91 \times 10^{-4}$, and $D_{22} = 42.90 \times 10^{-4}$. Thus the quadratic equation for V_1' is, after multiplication by 10^4,

$$-0.78 - 26.91V_1' + 42.90V_1'^2 = 0.$$

The positive root of this quadratic is 0.6550. In a similar way, we obtain for the quadratic in V_1'',

$$3.65625 + 4.14375 V_1'' - 14.1375 V_1''^2 = 0.$$

The positive root of this quadratic is 0.6758, and so the final result is

$$0.655 \le V_1 \le 0.676.$$

The corresponding stat-rat mean, obtained in Section 6.3, is 0.6731.

In Section 6.7 we describe a computation procedure for obtaining the bounds on the pre-asymptotic means. It is stated that under certain restrictions two operators, when applied repeatedly to $V_{1,0}$ generate upper and lower bounds on $V_{1,n}$. The reader is referred to that section for the details.

REFERENCES

1. Student. The probable error of a correlation coefficient. *Biometrika*, 1908, VI, 302.
2. Tippett, L. H. C. *Random sampling numbers, tracts for computors, No. XV.* London: Cambridge University Press, 1927.
3. Kendall, M. G., and Smith, M. B. *Random sampling numbers, tracts for computors, No. XXIV.* London: Cambridge University Press, 1939.
4. Fisher, R. A., and Yates, F. *Statistical tables.* Edinburgh: Oliver and Boyd, 1939.
5. Snedecor, G. W. *Statistical methods.* Ames, Iowa: Iowa State College Press, 1946, pp. 10–13.
6. Arkin, H., and Colton, R. R. *Tables for statisticians.* New York: Barnes and Noble, 1950, pp. 142–145.
7. Peirce, B. O. *A short table of integrals.* Boston: Ginn and Co., 1929, third revised edition, p. 10, integral 68.
8. Bush, R. R., and Mosteller, F. A stochastic model with applications to learning. *Annals of math. Stat.*, 1953, **24**, 559–585.

CHAPTER 7

Operators with Limits
Zero and Unity

7.1 INTRODUCTION

In many experimental situations "perfect" learning is possible and is nearly achieved by many subjects. If a particular response occurs repeatedly and is rewarded each time, that response may tend to occur with certainty. In a choice situation, one alternative may be rewarded while the others are not; we might expect that the organisms would tend to choose the rewarded alternative with probability one. For such problems, the model would use event operators that carry a response probability p towards an asymptote of one, that is, operators with limit points $\lambda = 1$. From the fixed-point form of the operators, we can write

$$(7.1) \qquad Q_i p = \alpha_i p + (1 - \alpha_i) \qquad (\lambda_i = 1).$$

Likewise, we are interested in operators which tend to make the response probability zero—complete extinction may be possible—and so we consider operators for which $\lambda_i = 0$, that is, operators of the form

$$(7.2) \qquad Q_i p = \alpha_i p \qquad (\lambda_i = 0).$$

The discussion in this chapter is restricted to two subject-controlled events.

From a mathematical point of view there are four main cases with limits zero and unity for two events:

$$(\text{I}) \quad \lambda_1 = 1, \quad \lambda_2 = 0,$$

$$(\text{II}) \quad \lambda_1 = 0, \quad \lambda_2 = 1,$$

$$(\text{III}) \quad \lambda_1 = \lambda_2 = 1,$$

$$(\text{IV}) \quad \lambda_1 = \lambda_2 = 0.$$

The latter two cases are discussed in the next chapter under the more general case of $\lambda_1 = \lambda_2 = \lambda$. In this chapter we discuss only the first two cases listed above.   In both, operator Q_1 is applied with probability p, and operator Q_2 is applied with probability $1 - p$, since the events are subject controlled.

Case I above involves the operators

(7.3)
$$Q_1 p = \alpha_1 p + (1 - \alpha_1),$$
$$Q_2 p = \alpha_2 p.$$

The first operator moves p toward $p = 1$, and this operator is applied with probability p.   Therefore, if we should ever achieve $p = 1$, the operator Q_1 is applied with certainty, and so $p = 1$ is a stable point in the process.   We refer to $p = 1$ as an "absorbing barrier."*   Similarly, Q_2 moves p toward $p = 0$ and is applied with probability $1 - p$.   Thus, if $p = 0$, Q_2 is certain to be applied and p remains at zero.   The point $p = 0$ is another stable point of the process, and we call it an absorbing barrier also.   With operators of this form we might expect that all p values would be "absorbed" at $p = 1$ or $p = 0$ in the limit as $n \to \infty$. We shall find that this is exactly what happens.

Case II listed above leads to the operators

(7.4)
$$Q_1 p = \alpha_1 p,$$
$$Q_2 p = \alpha_2 p + (1 - \alpha_2).$$

This is the reverse of Case I just discussed.   The first operator moves p toward zero, and the second moves p toward unity.   But Q_1 is applied with probability p; therefore, if we ever had $p = 0$, Q_2 would be applied with certainty, and so p would not remain at zero.   In the same way we see that if $p = 1$, then Q_1 is certain to be applied, taking p below unity. Hence, in Case II, $p = 1$ and $p = 0$ are not absorbing barriers but instead are "reflecting barriers."

The Brunswik T-maze experiment again supplies examples for which the two cases above may be appropriate.   If both right and left turns are rewarded on every trial they occur, the operators of equations 7.3 could be used.   Turning right (response A_1) and finding food increases p, the probability of turning right, whereas turning left (response A_2) and finding food increases the probability of turning left, and so decreases p. If, on the other hand, both right and left turns fail to lead to reinforcement, the operators of equations 7.4 seem more appropriate.   Extinction of

---

* In most physical problems involving an absorbing barrier, absorption is possible in a finite time.   This is not the case here; "absorption" occurrs only in the limit as $n \to \infty$ except when $\alpha_1 = 0$ or $\alpha_2 = 0$.

turning right decreases p, and extinction of turning left decreases $q = 1 - p$ and so increases p. In the following sections we consider some of the mathematical properties of Cases I and II. In Chapter 5 we handled these cases when $\alpha_1 = \alpha_2$; since there are many problems for which we would not care to make the equal alpha assumption, we treat unequal alphas here.

7.2 THE ASYMPTOTIC DISTRIBUTION FOR CASE I

We saw in the preceding section that Case I involves two absorbing barriers. The points $p = 0$ and $p = 1$ are stable points in the sense that, once one of these points is reached, a sequence of p values forever remains there. As a result we anticipated that the asymptotic distribution would have all its density at $p = 0$ and $p = 1$. We now prove that this is true by showing that the first and second asymptotic raw moments are equal, that is,

(7.5) $V_{1,\infty} = V_{2,\infty}.$

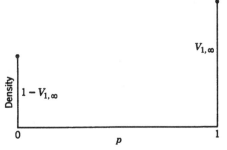

Once we have demonstrated that this is correct it is easy to show that the only distribution with such moments is a binomial one with density only at the points zero and unity. Moreover, from the definition of the mean

Fig. 7.1. The asymptotic distribution for Case I. All the density is at $p = 0$ and $p = 1$; the amount at $p = 1$ is $V_{1,\infty}$.

it must be that the amount of density at $p = 1$ is $V_{1,\infty}$, as shown in Fig. 7.1.

We now show that the second raw moment $V_{2,\infty}$ equals the mean $V_{1,\infty}$. From equation 4.52 and the asymptotic conditions, $V_{1,n+1} = V_{1,n} = V_{1,\infty}$ and $V_{2,n} = V_{2,\infty}$, we have

(7.6) $C_{10} + (C_{11} - 1)V_{1,\infty} + C_{12}V_{2,\infty} = 0.$

The coefficients are obtained from equations 4.53 with $a_2 = 0$ and $a_1 = 1 - \alpha_1$:

$$C_{10} = 0,$$

(7.7) $$C_{11} = 1 - \alpha_1 + \alpha_2,$$

$$C_{12} = \alpha_1 - \alpha_2.$$

Substituting these in equation 7.6 gives

(7.8) $(\alpha_1 - \alpha_2)V_{1,\infty} = (\alpha_1 - \alpha_2)V_{2,\infty},$

and, since we are considering only the cases for which $\alpha_1 \neq \alpha_2$, we conclude

(7.9) $$V_{2,\infty} = V_{1,\infty}.$$

It is an easy matter to prove by induction that all the asymptotic raw moments are equal (see footnote, Section 3.3), but this follows immediately once we have established that the distribution is a binomial as shown in Fig. 7.1. The definition of the raw moments, equation 4.1, gives for $m = 1$ and $m = 2$, for *discrete* distributions,

(7.10)
$$V_1 = \sum_\nu p_\nu P_\nu,$$

$$V_2 = \sum_\nu p_\nu^2 P_\nu.$$

We then equate these and get

$$\sum_\nu p_\nu P_\nu = \sum_\nu p_\nu^2 P_\nu,$$

or

(7.11) $$\sum_\nu p_\nu(1 - p_\nu)P_\nu = 0.$$

All terms in this sum are nonnegative by definition since $0 \leq p_\nu \leq 1$ and $0 \leq P_\nu \leq 1$ for all ν. Thus each term must be zero. This requires that $P_\nu = 0$ except when $p_\nu = 0$ or $p_\nu = 1$, because in these special cases $p_\nu(1 - p_\nu)$ vanishes instead of P_ν. In other words, no density exists at points other than zero and unity when the first and second raw moments of a distribution on the unit interval are equal. This completes the proof that the asymptotic distribution for Case I is the simple binomial distribution in Fig. 7.1. (This proof for discrete distributions is only suggestive for more general ones.)

The only problem which remains is to determine the one parameter $V_{1,\infty}$ of the asymptotic distribution. Unfortunately it does not appear possible to determine this quantity as an elementary function of the operator parameters. In the next section we develop some bounds on $V_{1,\infty}$ and show that it depends upon the initial distribution of p values. In Section 7.5 we mention a more elaborate approach for determining $V_{1,\infty}$.

7.3 BOUNDS ON THE ASYMPTOTIC MEAN FOR CASE I

In the preceding section we proved that *all* the density finally reaches zero or unity. This includes the possibility that all the density reaches unity, that is, that $V_{1,\infty} = 1$, which in turn would imply that all moments about the mean were zero. Such an asymptotic result would be an especially simple state of affairs, and the interpretation would be that *all* organisms would ultimately achieve perfect learning. In this section we demonstrate that this is *not* the case except when the initial probability is already unity or when $\alpha_2 = 1$. We further prove that the asymptotic mean depends upon the initial probability p_0 except when $\alpha_1 = 1$ or $\alpha_2 = 1$. (The asymptotic distribution theorem given in Section 4.7 does not apply to the present problem since $\lambda_1 - \lambda_2 = 1$.)

To carry out the proofs we need some upper and lower bounds on $V_{1,\infty}$. First we derive an expression for the probability $P_{2,\infty}$ of an infinite chain of A_2 responses; such sequences will terminate at $p = 0$. Since the asymptotic mean, $V_{1,\infty}$, is identical to the sum of the probabilities of all sequences which terminate at unity, $V_{1,\infty}$ cannot be greater than $1 - P_{2,\infty}$. This then provides an upper bound for $V_{1,\infty}$. A lower bound on $V_{1,\infty}$ is obtained in a similar way. The probability $P_{1,\infty}$, of an infinite chain of A_1 responses is computed; because such a sequence terminates at $p = 1$, $V_{1,\infty}$ is at least as large as $P_{1,\infty}$. We thus have the

BOUNDS ON THE ASYMPTOTIC MEAN:

$$(7.12) \qquad P_{1,\infty} \leq V_{1,\infty} \leq 1 - P_{2,\infty}.$$

We now proceed to compute these bounds.

The probability, $P_{2,n}$, of a chain of precisely n responses of type A_2 is

$$P_{2,n} = (1 - p_0)(1 - Q_2 p_0)(1 - Q_2^2 p_0) \cdots (1 - Q_2^{n-1} p_0)$$

$$(7.13) \qquad = \prod_{\nu=0}^{n-1} (1 - Q_2^\nu p_0).$$

From the second of equations 7.3, we have $Q_2 p = \alpha_2 p$, and so

$$(7.14) \qquad Q_2^\nu p_0 = \alpha_2^\nu p_0.$$

Thus

$$(7.15) \qquad P_{2,n} = \prod_{\nu=0}^{n-1} (1 - \alpha_2^\nu p_0).$$

In the limit when $n \to \infty$, we have

$$(7.16) \qquad P_{2,\infty} = \prod_{\nu=0}^{\infty} (1 - \alpha_2^\nu p_0).$$

Hence, $P_{2,\infty}$ is a function of two parameters α_2 and p_0. It is an example of the function

$$(7.17) \qquad P(\alpha, \beta) = \prod_{\nu=0}^{\infty} (1 - \alpha^\nu \beta)$$

which we have computed and present in Table 7.1.

For some purposes it is convenient to have an approximation to $P_{2,\infty}$, and this is developed below.* We take the logarithm of both sides of equation 7.15:

$$(7.18) \qquad \log P_{2,n} = \sum_{v=0}^{n-1} \log (1 - \alpha_2^v p_0).$$

Provided that $-1 < \alpha_2^v p_0 < +1$, we may expand each of the logarithm terms on the right and change signs (*note:* $\log (1 - x) = -x - x^2/2 - x^3/3 \cdots$):

$$(7.19) \qquad \begin{aligned} -\log P_{2,n} &= \sum_{v=0}^{n-1} \left\{ \alpha_2^v p_0 + \frac{(\alpha_2^v p_0)^2}{2} + \frac{(\alpha_2^v p_0)^3}{3} + \cdots \right\} \\ &= \sum_{v=0}^{n-1} \sum_{u=1}^{\infty} \frac{(\alpha_2^v p_0)^u}{u}. \end{aligned}$$

We then interchange the order of summation and perform the summation over v to obtain

$$(7.20) \qquad -\log P_{2,n} = \sum_{u=1}^{\infty} \frac{p_0^u}{u} \frac{1 - \alpha_2^{nu}}{1 - \alpha_2^u}.$$

When $\alpha_2 < 1$, we have in the limit as $n \to \infty$,

$$(7.21) \qquad -\log P_{2,\infty} = \sum_{u=1}^{\infty} \frac{p_0^u}{u} \frac{1}{1 - \alpha_2^u}.$$

This sum is difficult to evaluate, but if we replace α_2^u with α_2 we shall certainly cause each term beyond the first to increase, and so we have the inequality

$$(7.22) \qquad -\log P_{2,\infty} < \sum_{u=1}^{\infty} \frac{p_0^u}{u} \frac{1}{1 - \alpha_2}.$$

This sum involves the expansion of $-\log (1 - p_0)$, and we see that

$$(7.23) \qquad -\log P_{2,\infty} < \frac{-\log (1 - p_0)}{1 - \alpha_2}.$$

It then follows that

$$(7.24) \qquad P_{2,\infty} > (1 - p_0)^{1/(1-\alpha_2)}.$$

---

* We are grateful to William J. McGill for assistance in carrying out these computations.

The approximation used in going from equation 7.21 to inequality 7.22 is useful since it permitted us to complete the summation and thus obtain expression 7.24, which leads to an upper bound on the asymptotic mean:

$$(7.25) \qquad V_{1, \infty} < 1 - (1 - p_0)^{1/(1 - \alpha_2)}.$$

This bound already shows that the asymptotic mean cannot be unity unless either $p_0 = 1$ or $\alpha_2 = 1$. (When $\alpha_2 = 1$, we have a special case considered in the next chapter.)

The reader may be curious at this point why we did not obtain an upper bound, corresponding to equation 7.25, for the general case when

$$(7.26) \qquad Q_2 p = \alpha_2 p + (1 - \alpha_2) \lambda_2.$$

The answer is simply this. The probability $P_{2, \infty}$ of an infinite sequence of A_2 responses is zero unless $\lambda_2 = 0$. Let us refer to equation 3.5, which may be written as

$$(7.27) \qquad Q_2{}^r p_0 = \alpha_2{}^r p_0 + (1 - \alpha_2{}^r) \lambda_2,$$

where λ_2 is the asymptote achieved when Q_2 is applied repeatedly to p_0. First consider $p_0 < \lambda_2$. Then we observe that $Q_2{}^r p_0 > p_0$ and so $1 - Q_2{}^r p_0 < 1 - p_0$. From equation 7.13 we get

$$(7.28) \qquad P_{2, n} < (1 - p_0)^n.$$

Since $(1 - p_0)^n$ approaches zero as n becomes infinite, $P_{2, \infty}$ is zero unless $p_0 = 0$. Second, consider $p_0 > \lambda_2$. Note that $Q_2{}^r p_0 > \lambda_2$ and hence that $1 - Q_2{}^r p_0 < 1 - \lambda_2$. So again from equation 7.13 we have

$$(7.29) \qquad P_{2, n} < (1 - \lambda_2)^n,$$

and so in the limit we have $P_{2, \infty} = 0$ unless $\lambda_2 = 0$. Only in the special cases for which either $p_0 = 0$ or $\lambda_2 = 0$ do we obtain a $P_{2, \infty} > 0$. The condition $p_0 = 0$ needs no further discussion, and the case of $\lambda_2 = 0$ is being considered in this chapter.

We may also compute the probability $P_{1, n}$ of a chain of n responses of type A_1, to obtain another bound on $V_{1, \infty}$. We have

$$(7.30) \qquad P_{1, n} = \prod_{\nu = 0}^{n-1} Q_1{}^\nu p_0.$$

But from equation 3.5 we see that, for $\lambda_1 = 1$,

$$(7.31) \qquad Q_1{}^\nu p_0 = \alpha_1{}^\nu p_0 + (1 - \alpha_1{}^\nu) = 1 - \alpha_1{}^\nu q_0,$$

and so

$$(7.32) \qquad P_{1, n} = \prod_{\nu = 0}^{n-1} (1 - \alpha_1{}^\nu q_0).$$

In the limit as $n \to \infty$ we have

$$(7.33) \qquad P_{1,\infty} = \prod_{\nu=0}^{\infty} (1 - \alpha_1{}^\nu q_0) = P(\alpha_1, q_0),$$

where $P(\alpha_1, q_0)$ is the function defined by equation 7.17 and presented in Table 7.1. By analogy with our previous development of an approximation for $P_{2,\infty}$, we obtain

$$(7.34) \qquad P_{1,\infty} > (1 - q_0)^{1/(1-\alpha_1)}.$$

TABLE 7.1

The function

$$P(\alpha, \beta) = \prod_{\nu=0}^{\infty} (1 - \alpha^\nu \beta).$$

β	α										
	0	0.1	0.2	0.3	0.4	0.5	0.6	0.7	0.8	0.9	1.0
0	1.000	1.000	1.000	1.000	1.000	1.000	1.000	1.000	1.000	1.000	1
0.1	0.900	0.890	0.878	0.862	0.841	0.813	0.772	0.709	0.598	0.358	0
0.2	0.800	0.783	0.760	0.733	0.698	0.650	0.586	0.492	0.346	0.120	0
0.3	0.700	0.677	0.648	0.612	0.568	0.510	0.434	0.331	0.193	0.038	0
0.4	0.600	0.574	0.541	0.501	0.452	0.390	0.313	0.216	0.102	0.011	0
0.5	0.500	0.473	0.439	0.398	0.349	0.289	0.217	0.134	0.051	0.003	0
0.6	0.400	0.373	0.342	0.303	0.258	0.204	0.143	0.079	0.024	0.001	0
0.7	0.300	0.277	0.249	0.216	0.178	0.134	0.087	0.042	0.010	0.000	0
0.8	0.200	0.182	0.161	0.137	0.109	0.078	0.047	0.020	0.003	0.000	0
0.9	0.100	0.090	0.078	0.065	0.050	0.034	0.019	0.007	0.001	0.000	0
1.0	0	0	0	0	0	0	0	0	0	0	0

This result yields a lower bound on the asymptotic mean:

$$(7.35) \qquad V_{1,\infty} > p_0{}^{1/(1-\alpha_1)}.$$

Combining this with inequality 7.25 we have a new pair of

BOUNDS ON THE ASYMPTOTIC MEAN:

$$(7.36) \qquad p_0{}^{1/(1-\alpha_1)} < V_{1,\infty} < 1 - (1 - p_0)^{1/(1-\alpha_2)}.$$

These bounds on the asymptotic mean can now be used to demonstrate that $V_{1,\infty}$ depends in general on the initial probability p_0. A numerical example will suffice. Let $\alpha_1 = 2/3$, and $\alpha_2 = 3/4$, so that the bounds are

$$(7.37) \qquad p_0{}^3 < V_{1,\infty} < 1 - (1 - p_0)^4.$$

For $p_0 = 0.1$ we have

(7.38) $0.001 < V_{1,\infty} < 0.344,$

and for $p_0 = 0.9$ we have

(7.39) $0.729 < V_{1,\infty} < 0.9999.$

Clearly since $V_{1,\infty}$ must be below 0.344 when $p_0 = 0.1$ but must be above 0.729 when $p_0 = 0.9$, *it follows that $V_{1,\infty}$ must indeed depend upon p_0.* In other words, the proportion of organisms reaching $p = 1$ depends upon the initial response probability.

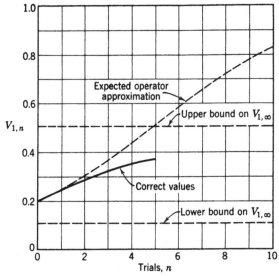

Fig. 7.2. The correct means of the p-value distributions for the first five trials, the expected operator approximation for ten trials, and the bounds (0.109, 0.508) on the asymptotic mean, $V_{1,\infty}$. The values $\lambda_1 = 1$, $\lambda_2 = 0$, $\alpha_1 = 0.4$, $\alpha_2 = 0.7$, and $p_0 = 0.2$ were used for the illustration.

The dependence upon p_0 of $V_{1,\infty}$ is further substantiated by what happens in two trivial cases. We know that if $p_0 = 0$, the operator Q_2 is always applied and so the probability forever remains at zero, that is, $V_{1,\infty} = 0$. We also know that when $p_0 = 1$ the operator Q_1 is always applied and the probability remains at unity, that is, $V_{1,\infty} = 1$. We conclude, therefore, that as p_0 goes from zero to unity, $V_{1,\infty}$ increases from zero to unity.

The reader may have detected that the upper and lower bounds of inequalities 7.36 become very near unity and zero, respectively, when α_1 and α_2 are near unity, provided that p_0 is not too near zero or unity.

It should not be inferred from this that the asymptotic mean becomes less and less sensitive to p_0 as α_1 and α_2 approach unity, however, because the bounds of inequalities 7.36 may be far from the correct mean. Those bounds were obtained from a consideration of only the sequences of all A_1 responses and all A_2 responses. As the parameters α_1 and α_2 approach unity, other sequences become more important, and hence the bounds we derived become less sensitive.

To illustrate how the upper and lower bounds surround the true means, we have computed the means of all possible sequences for five trials. These are shown in Fig. 7.2 along with the limits computed from inequalities 7.12. Also shown in that figure are means obtained from the expected operator approximation discussed in Section 6.4.

7.4 FURTHER RESTRICTIONS ON CASE I

We now consider a very special case which is easy to compute. Let $\alpha_1 = 0$ so that the two operators are

$$(7.40) \qquad \begin{aligned} Q_1 p &= 1, \\ Q_2 p &= \alpha_2 p. \end{aligned}$$

The first asserts that whenever a response A_1 occurs the p value jumps immediately to unity, where it remains. Thus the only sequence that terminates at zero p value is the one containing an infinite chain of A_2 responses. It follows then that

$$(7.41) \qquad V_{1,\infty} = 1 - P_{2,\infty} = 1 - P(\alpha_2, p_0),$$

where $P(\alpha_2, p_0)$ is the function given in Table 7.1.

We have another very special case when we take $\alpha_2 = 0$ so that the operators are

$$(7.42) \qquad \begin{aligned} Q_1 p &= \alpha_1 p + (1 - \alpha_1), \\ Q_2 p &= 0. \end{aligned}$$

This case is complementary to the preceding one; whenever an A_2 response occurs the probability goes immediately to zero and remains there. Hence the asymptotic mean $V_{1,\infty}$ must equal the probability $P_{1,\infty}$ of an infinite chain of A_1 responses, that is,

$$(7.43) \qquad V_{1,\infty} = P_{1,\infty} = P(\alpha_1, q_0).$$

We have still another special case when we set $\alpha_1 = \alpha_2 = \alpha$. From the analysis in Chapter 5 we know that for $\lambda_1 = 1$ and $\lambda_2 = 0$ we have $V_{1,\infty} = p_0$.

*7.5 THE FUNCTIONAL EQUATION FOR CASE I

In Case I the asymptotic distribution has all its density at $p = 0$ and $p = 1$ and is therefore described by a single parameter, $V_{1,\infty}$, which gives the amount of density at $p = 1$. In the preceding sections we derived some bounds on $V_{1,\infty}$ and considered the special cases when $\alpha_1 = 0$, when $\alpha_2 = 0$, and when $\alpha_1 = \alpha_2 = \alpha$. Moreover, we demonstrated that $V_{1,\infty}$ depends on p_0, the initial probability. The asymptotic mean, then, is a function of α_1, α_2, and p_0, so that we write

$$(7.44) \qquad V_{1,\infty} = \eta(\alpha_1, \alpha_2, p_0).$$

This function η may be considered to be the probability that a "particle" beginning at p_0 will end up at unity when the operator parameters are α_1 and α_2. On the first trial either Q_1 or Q_2 is applied to p_0. Suppose that Q_1 is applied. The particle is then at $Q_1 p_0$, and the probability that it will be at unity asymptotically is $\eta(\alpha_1, \alpha_2, Q_1 p_0)$. Similarly, if Q_2 is applied, the probability that the particle will then go to unity asymptotically is $\eta(\alpha_1, \alpha_2, Q_2 p_0)$. We know, of course, that Q_1 is applied with probability p_0 and Q_2 with probability $1 - p_0$. Therefore, we may write

$$(7.45) \quad \eta(\alpha_1, \alpha_2, p_0) = p_0 \eta(\alpha_1, \alpha_2, Q_1 p_0) + (1 - p_0)\, \eta(\alpha_1, \alpha_2, Q_2 p_0),$$

or, using the expressions for the operators, $Q_1 p_0$ and $Q_2 p_0$, given by equations 7.3, we write

$$(7.46) \quad \eta(\alpha_1, \alpha_2, p_0) = p_0 \eta(\alpha_1, \alpha_2, 1 - \alpha_1 + \alpha_1 p_0) + (1 - p_0)\, \eta(\alpha_1, \alpha_2, \alpha_2 p_0).$$

This equation is a particular type of *functional equation*. Its properties and solution have been investigated by Shapiro and Bellman [1]. They have shown that no simple solution exists in closed form, but they have developed a numerical procedure for obtaining values of η.

Shapiro and Bellman have also shown that the functional equation has some special symmetry properties which result from the symmetry in the basic operators Q_1 and Q_2. We may define a complementary function $\tilde{\eta}(\alpha_2, \alpha_1, q_0)$ which represents the probability that a particle beginning at $q_0 = 1 - p_0$ will terminate at $q = 0$ when the operator parameters are α_2 and α_1. By the same arguments as those given above, this function $\tilde{\eta}$ must satisfy

$$(7.47) \quad \tilde{\eta}(\alpha_2, \alpha_1, q_0) = q_0 \tilde{\eta}(\alpha_2, \alpha_1, 1 - \alpha_2 + \alpha_2 q_0) + (1 - q_0)\, \tilde{\eta}(\alpha_2, \alpha_1, \alpha_1 q_0).$$

We see, then, that $\tilde{\eta}$ must satisfy the same functional equation as η when α_1 and α_2 are interchanged and when p_0 and $q_0 = 1 - p_0$ are interchanged. This property is of considerable importance in the computation scheme developed by Shapiro and Bellman.

We have already obtained solutions of the functional equation for some special cases. We see from Section 7.3 that, when $\alpha_1 = 1$, then $V_{1,\infty} = 0$, and when $\alpha_2 = 1$, $V_{1,\infty} = 1$. Hence

(7.48)
$$\eta(1, \alpha_2, p_0) = 0,$$
$$\eta(\alpha_1, 1, p_0) = 1.$$

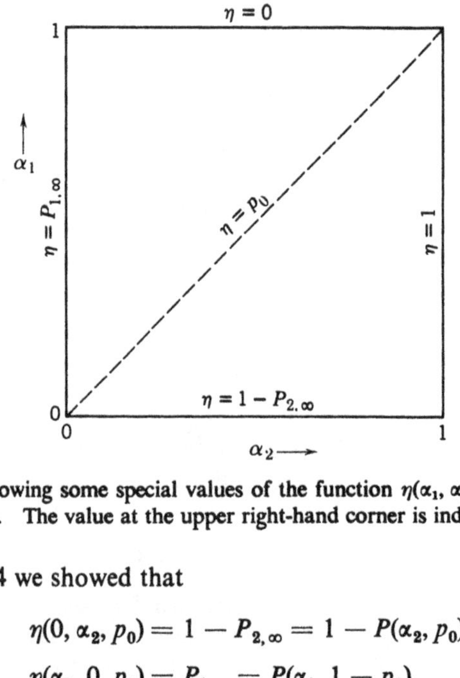

Fig. 7.3. Showing some special values of the function $\eta(\alpha_1, \alpha_2, p_0)$ in the α_1, α_2 plane. The value at the upper right-hand corner is indeterminate.

In Section 7.4 we showed that

(7.49)
$$\eta(0, \alpha_2, p_0) = 1 - P_{2,\infty} = 1 - P(\alpha_2, p_0),$$
$$\eta(\alpha_1, 0, p_0) = P_{1,\infty} = P(\alpha_1, 1 - p_0),$$

and

(7.50)
$$\eta(\alpha, \alpha, p_0) = p_0.$$

These five special solutions may be summarized if we consider η to be a function in the α_1, α_2 plane with a parameter p_0. In Fig. 7.3 we show such a plot.

7.6 THE ASYMPTOTIC DISTRIBUTION FOR CASE II

Let us return to Case II defined in Section 7.1, by equations 7.4. Neither $p = 0$ nor $p = 1$ is an absorbing barrier, and so the asymptotic distribution is independent of the initial p values. The asymptotic distribution depends upon α_1 and α_2, of course, and so we investigate this dependence.

First consider the asymptotic mean $V_{1,\infty}$. From equation 4.52 we have as $n \to \infty$,

$$(7.51) \qquad C_{10} + (C_{11} - 1)V_{1,\infty} + C_{12}V_{2,\infty} = 0.$$

The coefficients of equations 4.53 are for $a_1 = 0$ and $a_2 = (1 - \alpha_2)$

$$C_{10} = 1 - \alpha_2,$$

$$(7.52) \qquad C_{11} = 2\alpha_2 - 1,$$

$$C_{12} = \alpha_1 - \alpha_2.$$

When we insert these coefficients in equation 7.51 and solve for $V_{1,\infty}$ we have

$$(7.53) \qquad V_{1,\infty} = \frac{1}{2} + \frac{\alpha_1 - \alpha_2}{2(1 - \alpha_2)} V_{2,\infty}.$$

The second raw moment $V_{2,\infty}$ is never negative, and so when $\alpha_2 < 1$ we have

$$(7.54) \qquad V_{1,\infty} \lessgtr 1/2 \quad \text{for} \quad \alpha_1 \lessgtr \alpha_2.$$

We can obtain better bounds than these on $V_{1,\infty}$ by the methods used in Chapter 6. The first bound is readily obtained from inequality 6.44, which can be written

$$(7.55) \qquad V_{2,\infty} \leq V_{1,\infty}.$$

Using this in equation 7.53 gives

$$(7.56) \qquad V_{1,\infty} \gtrless \frac{1}{2} + \frac{\alpha_1 - \alpha_2}{2(1 - \alpha_2)} V_{1,\infty} \quad \text{for} \quad \alpha_1 \lessgtr \alpha_2.$$

Solving for $V_{1,\infty}$ in these last two inequalities we have

$$(7.57) \qquad V_{1,\infty} \gtrless \frac{1 - \alpha_2}{2 - \alpha_2 - \alpha_1} \quad \text{for} \quad \alpha_1 \lessgtr \alpha_2.$$

The expected operator bound described in Section 6.5 leads to another set of bounds, namely,

$$(7.58) \qquad V_{1,\infty} \lessgtr \frac{(1 - \alpha_2) - \sqrt{(1 - \alpha_2)(1 - \alpha_1)}}{\alpha_1 - \alpha_2} \quad \text{for} \quad \alpha_1 \lessgtr \alpha_2.$$

As a numerical example, let $\alpha_1 = 0.9$ and $\alpha_2 = 0.6$. Relations 7.57 and 7.58 then give

$$(7.59) \qquad\qquad 0.667 \leq V_{1,\infty} \leq 0.800.$$

Improved bounds may be obtained by the procedures given in Section 6.6.

The form of the asymptotic distribution may be approximated by Monte Carlo computations as described in Section 6.3. We provide an example in Fig. 7.4.

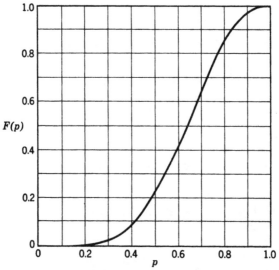

Fig. 7.4. The approximate asymptotic cumulative distribution for Case II, discussed in Section 7.6, obtained from a 1000-trial stat-rat with $\alpha_1 = 0.8$ and $\alpha_2 = 0.5$. The mean and standard deviation of the stat-rat p values are $V_{1,\infty} = 0.634$ and $\sigma_\infty = 0.158$, respectively. The bounds given by relations 7.57 and 7.58 in the text give $0.613 \leq V_{1,\infty} \leq 0.714$.

7.7 CASES WITH ONLY ONE ABSORBING BARRIER

In previous sections of this chapter we considered two cases: (I) $\lambda_2 = 0$ and $\lambda_1 = 1$, and (II) $\lambda_1 = 0$ and $\lambda_2 = 1$. For Case I we said that there were two absorbing barriers, at $p = 0$ and $p = 1$, because once a p value reached one of those points it forever remained there. This case of two absorbing barriers suggests the possibility of only one such barrier. In applications to experimental problems we might like to use an event operator Q_1 which permitted perfect learning ($\lambda_1 = 1$) but an operator Q_2 which did not lead to complete extinction ($\lambda_2 = 0$). Or we might want to allow complete extinction but rule out perfect learning.

When $\lambda_1 = 1$ absorption at $p = 1$ may occur, but when $0 < \lambda_2$ absorption at $p = 0$ is impossible. We might expect that nearly all sequences of p values would eventually reach $p = 1$. In the next section we prove that this is correct. Similarly, when $\lambda_1 < 1$ and $\lambda_2 = 0$, eventual absorption at $p = 0$ will occur. (From stat-rat computations, as described in Section 6.2, we have found that the absorption can be extremely slow even for values of α_1 and α_2 as small as 0.8.)

The case of one absorbing barrier could be used in an attempt to apply association theory to our mathematical system [2]. One of the basic principles of association theory (or contiguity theory) is that the response which occurs last in a stimulus situation is more likely to occur upon the next presentation of that situation. Suppose that we consider response A_1 to be "success" and response A_2 to be "failure." On each trial success or failure terminates the stimulus situation, and so whichever occurred should tend to have greater probability of occurrence on the next trial, according to the principles of association theory. Hence Q_1 should increase p and Q_2 should increase $q = 1 - p$, that is, Q_2 should decrease p. But suppose that we demanded that "success" eventually occur on all trials. This could be approximated by letting $\lambda_1 = 1$ but requiring that $\lambda_2 \neq 0$. The latter requirement would mean that an occurrence of failure would usually increase q, the probability of failure, but that q could never reach unity; q would be at most $1 - \lambda_2$. The "rate of learning" would then depend upon the values of α_1, α_2, and λ_2.

*7.8   THE ASYMPTOTIC DISTRIBUTION FOR
ONE ABSORBING BARRIER

When there is only one absorbing barrier, all density is ultimately absorbed at that barrier. We consider the case of $\lambda_2 = 0$ and $\lambda_1 < 1$ in detail because the proof for $0 < \lambda_2$ and $\lambda_1 = 1$ follows by a simple symmetry argument. We have the operators

(7.60)
$$Q_1 p = \alpha_1 p + (1 - \alpha_1)\lambda_1 \qquad (\lambda_1 < 1),$$
$$Q_2 p = \alpha_2 p \qquad\qquad (\alpha_2 < 1),$$

and Q_1 is applied with probability p, and Q_2 is applied with probability $1 - p$. This case is included under the theorem stated in Section 4.7, and so we know that the asymptotic distribution is independent of the initial p values. But we now prove that the foregoing operators lead to an asymptotic distribution that has all its density at $p = 0$.*

From the trapping theorem of Section 4.7 we know that absorption

---

* A similar proof was developed by A. Birnbaum [3].

cannot occur at $p = 1$, that is, all the density of the asymptotic distribution is in the range from zero to λ_1. Thus, if absorption occurs it is certain to occur at $p = 0$; if $p = 0$, then $Q_2 p = 0$ and Q_2 is certain to be applied. To prove that absorption at $p = 0$ will occur with probability one as the trial number n becomes infinite, we show that the conditional probability, γ_n, of an infinite sequence of A_2 responses beginning on trial n, given an A_1 response on trial $n - 1$, is greater than zero. When an A_2 response occurs on trial n and on all future trials we say that the sequence of p values is in a *state of pre-absorption*. The probability that the sequence is in this state on trial n we denote by ψ_n, whereas the conditional probability of the sequence being in a state of pre-absorption on trial n, given that it was not in that state on trial $n - 1$, is γ_n. We then see that

(7.61) $$\psi_{n+1} = \psi_n + (1 - \psi_n)\gamma_{n+1}.$$

This recurrence formula has the formal solution,

(7.62) $$\psi_n = 1 - (1 - \psi_0)\prod_{n'=1}^{n}(1 - \gamma_{n'}).$$

Clearly, $\psi_n \to 1$ as $n \to \infty$, provided that

(7.63) $$\lim_{n \to \infty} \prod_{n'=1}^{n}(1 - \gamma_{n'}) = 0.$$

This in turn will be true if $\gamma_{n'}$ is greater than some positive quantity ϵ for all n', because one condition that the product tends to zero is that the sum of the $\gamma_{n'}$ diverges with n [4]; if $\gamma_{n'} \geq \epsilon$, then $\sum_{n'=1}^{n} \gamma_{n'} \geq n\epsilon$, and of course $n\epsilon$ diverges. We see at once that

(7.64)
$$\gamma_n = (1 - p_n)(1 - \alpha_2 p_n)(1 - \alpha_2^2 p_n) \cdots$$
$$= \prod_{\nu=0}^{\infty}(1 - \alpha_2^\nu p_n).$$

In Section 7.3 we encountered such a product for $P_{2,\infty}$. Analogous to inequality 7.24 we have

(7.65) $$\gamma_n > (1 - p_n)^{1/(1-\alpha_2)}.$$

Provided that $p_n \neq 1$ for any n, we shall have $\gamma_n > \epsilon$ for all n. But since the limit point λ_1 is less than unity and since $\lambda_2 = 0$ for the case being considered in this proof, we can never have $p_n = 1$ except possibly for $n = 0$. If, however, $p_0 = 1$, we need only consider the process starting on trial $n = 1$. Therefore γ_n is greater than some positive quantity ϵ for all n, and, as we previously saw, this implies that $\psi_n \to 1$ as $n \to \infty$. This means that a p-value sequence will get in a state of pre-absorption with probability one, and as a result absorption at $p = 0$ must eventually occur.

The proof that absorption at $p = 1$ eventually occurs when we have the operators

$$(7.66) \quad \begin{aligned} Q_1 p &= \alpha_1 p + (1 - \alpha_1) & (\alpha_1 < 1), \\ Q_2 p &= \alpha_2 p + (1 - \alpha_2)\lambda_2 & (0 < \lambda_2), \end{aligned}$$

follows immediately by symmetry. We merely interchange p and $q = 1 - p$, and the above proof is applicable; eventual absorption occurs at $q = 0$.

7.9 SUMMARY

In this chapter we discuss cases of the general model for which the limits λ_1 and λ_2 are zero or unity and the events are subject controlled. These cases are of particular interest for applications in which "perfect" learning or "complete" extinction or both are possible.

Case I, discussed in Sections 7.2, 7.3, 7.4, and 7.5, involves the operators

$$(7.3) \quad \begin{aligned} Q_1 p &= \alpha_1 p + (1 - \alpha_1), \\ Q_2 p &= \alpha_2 p. \end{aligned}$$

Since all sequences tend toward $p = 0$ or $p = 1$, we call these points "absorbing barriers." The probability that a sequence gets arbitrarily close to $p = 1$ depends upon α_1, α_2, and p_0, as is shown in Section 7.3. Bounds on the asymptotic mean are provided.

Case II, discussed in Section 7.6, involves the operators

$$(7.4) \quad \begin{aligned} Q_1 p &= \alpha_1 p, \\ Q_2 p &= \alpha_2 p + (1 - \alpha_2). \end{aligned}$$

The points $p = 0$ and $p = 1$ are "reflecting barriers" because if a sequence ever reached one of these points it would be certain to move away from it on the next trial. Some bounds on the asymptotic mean for this case are provided.

Cases with only one absorbing barrier are discussed in Sections 7.7 and 7.8. One of two such cases employs the operators

$$(7.60) \quad \begin{aligned} Q_1 p &= \alpha_1 p + (1 - \alpha_1)\lambda_1 & (\lambda_1 < 1), \\ Q_2 p &= \alpha_2 p & (\alpha_2 < 1). \end{aligned}$$

For this case it is proved that sequences tend toward $p = 0$ with probability one. The other such case arises when the operators are

$$(7.66) \qquad \begin{aligned} Q_1 p &= \alpha_1 p + (1 - \alpha_1) & (\alpha_1 < 1), \\ Q_2 p &= \alpha_2 p + (1 - \alpha_2)\lambda_2 & (0 < \lambda_2). \end{aligned}$$

In this case, sequences tend toward $p = 1$ with probability one.

REFERENCES

1. Harris, T. E., Bellman, R., Shapiro, H. N. *Studies in functional equations occurring in decision processes.* Research Memorandum RM-878, RAND Corporation, Santa Monica, Calif., July 1, 1952.
2. Guthrie, E. R. *The psychology of learning.* New York: Harper, 1935.
3. Birnbaum, A., personal communication.
4. Bromwich, T. J. I'a. *An introduction to the theory of infinite series.* London: Macmillan, 1942, second edition, pp. 104–106.

CHAPTER 8

Commuting Operators

8.1 THE COMMUTATIVITY CONDITIONS

If two operators yield the same result when they are applied in either order we say that those operators commute. In this chapter we consider cases of the general mathematical system which arise when two operators commute with each other. The mathematical system described in the first four chapters has already been specialized to yield a number of cases which lead to mathematical simplifications and which are of interest in various kinds of applications to learning. Experimenter-controlled events led to few problems but subject-controlled events led us into serious mathematical difficulties. Because we believe that such events are important in experimental problems, most of the last three chapters were devoted to their analysis and discussion. This chapter also is devoted mainly to subject-controlled events for which the probabilities of the two responses on trial n are p_n and $1 - p_n$.

The notion of commutativity of two operators was first introduced in Section 1.4, but in Section 3.6 we gave a more explicit discussion of the commutativity of the operators Q_1 and Q_2. Those operators commute provided that

$$(8.1) \qquad Q_1 Q_2 p = Q_2 Q_1 p,$$

for all values of p. If this condition holds, it can be inferred that success followed by failure gives the same net result as failure followed by success, if Q_1 and Q_2 corresponded to success and failure in an experiment. Thus, with commutativity there is no "recency" effect—the more recent event does not have a greater effect. The commutativity condition, however, has even more far-reaching implications. Suppose that we have a sequence of n events, E_1 and E_2, and that some number k of them are E_1's and $n - k$ of them are E_2's. Corresponding to this event sequence will be a sequence of operators which leads to a particular final p value. Now if the event operators commute, any re-arrangement of the order of the events will not change the final p value, provided the total number of E_1's and the number of E_2's are kept the same. If the p value on trial zero is p_0, the p value on trial n is

(8.2) $$p_{n,k} = Q_1^k Q_2^{n-k} p_0.$$

This fact simplifies matters considerably.

We saw in Section 3.6 that Q_1 and Q_2 commute if and only if one or more of the following conditions holds:

$$
\begin{align*}
&(a) \quad \alpha_1 = 1, \\
(8.3) \quad &(b) \quad \alpha_2 = 1, \\
&(c) \quad \lambda_1 = \lambda_2 = \lambda.
\end{align*}
$$

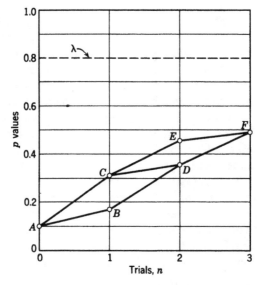

Fig. 8.1. The three possible sequences of p values for three trials when the operators commute. The paths $ACEF$, $ACDF$, and $ABDF$ show the sequences $E_1E_1E_2$, $E_1E_2E_1$, and $E_2E_1E_1$, respectively. The figure was drawn using equations 8.4 with $\lambda = 0.8$, $p_0 = 0.1$, $\alpha_1 = 0.7$, and $\alpha_2 = 0.9$.

Condition a implies that $Q_1 p = p$, that is, that Q_1 is the identity operator which does not change p. Similarly, condition b implies that $Q_2 p = p$. Condition c says that Q_1 and Q_2 have the same limit point λ, that is, $Q_1^n p$ and $Q_2^n p$ approach the same asymptote as n goes to infinity. These three conditions, taken together, are equivalent to requiring that the operators have the forms

$$
\begin{align*}
&Q_1 p = \alpha_1 p + (1 - \alpha_1)\lambda, \\
(8.4) \\
&Q_2 p = \alpha_2 p + (1 - \alpha_2)\lambda.
\end{align*}
$$

We see that $\alpha_1 = 1$ and $\alpha_2 = 1$ are special restrictions that yield operators which already satisfy condition c above. It is easily demonstrated that these operators commute; by direct computation we see that

$$(8.5) \quad \begin{aligned} Q_1 Q_2 p &= \alpha_1 [\alpha_2 p + (1 - \alpha_2)\lambda] + (1 - \alpha_1)\lambda \\ &= \alpha_1 \alpha_2 p + (1 - \alpha_1 \alpha_2)\lambda. \end{aligned}$$

The reader may easily verify that $Q_2 Q_1 p$ gives the same result. Moreover, it is a simple matter to generalize this result to k applications of Q_1 and $n - k$ applications of Q_2 to obtain

$$(8.6) \quad p_{n,k} = Q_1{}^k Q_2{}^{n-k} p_0 = \alpha_1{}^k \alpha_2{}^{n-k} p_0 + (1 - \alpha_1{}^k \alpha_2{}^{n-k})\lambda.$$

In Fig. 8.1 we illustrate the three possible sequences for $n = 3$ and $k = 2$. These sequences are $E_1 E_1 E_2$, $E_1 E_2 E_1$, and $E_2 E_1 E_1$. Each sequence leads to the same value of $p_{3,2}$ but the paths are different. In the following sections we discuss some of the properties of such sequences of p values.

8.2 EQUAL LIMIT POINTS OF THE OPERATORS

When neither operator, Q_1 nor Q_2, is an identity operator they have limit points λ_1 and λ_2, respectively. As we have just seen, Q_1 and Q_2 commute when $\lambda_1 = \lambda_2 = \lambda$. We discuss this condition in this section; we require that α_1 and α_2 be strictly less than unity. Such a set of restrictions seems plausible for a number of experimental problems. For example, in the Brunswik T-maze experiment first described in the Introduction, a rat may always be rewarded on the right side and never be rewarded on the left side. Both types of events might increase p, the probability of turning right, towards a limit point of unity, as assumed in the Introduction; but, contrary to an assumption made there, the magnitude of the effects of the two events may be different. Thus, for this problem, we might have $\lambda_1 = \lambda_2 = \lambda$, but $\alpha_1 \neq \alpha_2$. Moreover, we may not wish to assume that $\lambda = 1$.

Another example for which equal limit points would seem reasonable is extinction of an operant response, such as bar pressing. When the operant response occurs and is not reinforced, "inhibition" might result, and so the probability p of that response should decrease. Furthermore, when other responses occur, we might expect them to increase in probability even if they are not rewarded by the experimenter. Therefore, the occurrence of both the unreinforced operant response and of other responses might tend to decrease the probability of the operant response towards an asymptote of zero. In such a situation we would have $\lambda_1 = \lambda_2 = 0$. We would not want to assume that the two types of events had effects that were of equal magnitude and so we would not set α_1 equal to α_2.

The asymptotic distribution of p values is especially simple when we have equal limit points. All the density is at the point $\lambda = \lambda_1 = \lambda_2$. Intuitively this is clear since each operator moves p towards λ. Moreover, it may be seen that the expression in equation 8.6 tends towards λ as n tends to infinity because both α_1 and α_2 are less than unity, by assumption. Our chief interest, then, is in the distributions of p values during the early

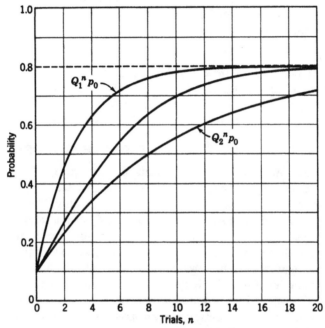

Fig. 8.2. The limiting p values $Q_1^n p_0$ and $Q_2^n p_0$ for the case of equal limit points. The computations were made with $\lambda = 0.8$, $p_0 = 0.1$, $\alpha_1 = 0.7$, and $\alpha_2 = 0.9$. The middle curve shows the approximate means $V_{1,n}$ computed from equation 8.8.

part of the learning process. First we consider the problem of computing the distribution means from trial to trial and then we provide some approximate formulas for computing the expected cumulative number of occurrences of each response.

The recurrence formula for the means is obtained from equations 4.52 and 4.53 with the relations $a_i = (1 - \alpha_i)\lambda$ for $i = 1, 2$. We have

$$(8.7) \quad V_{1,n+1} = (1 - \alpha_2)\lambda + [\alpha_2 - (\alpha_1 - \alpha_2)\lambda]V_{1,n} + (\alpha_1 - \alpha_2)V_{2,n}.$$

This formula has not been materially simplified by the equal-λ assumption because $\alpha_1 \neq \alpha_2$, so we use the approximate methods given in Chapter 6. To begin, we note that the distribution of p values on trial n is contained entirely between $Q_1^n p_0$ and $Q_2^n p_0$. In Fig. 8.2 we illustrate these extreme sequence bounds computed with the aid of equation 3.5. The upper and lower bounds developed in Section 6.7 can be used to obtain limits on the means $V_{1,n}$, but we use the expected operator approximation described in Section 6.4. When the difference between α_1 and α_2 is not large, the limits $Q_1{}^n p_0$ and $Q_2{}^n p_0$ are fairly close together. The means computed from the expected operator lie between those limits, and so we can expect the error to be small. Using the conditions $a_i = (1 - \alpha_i)\lambda$, we obtain from equations 6.11 and 6.12 the

APPROXIMATE EXPLICIT FORMULA FOR THE MEANS:

$$(8.8) \quad V_{1,n} \cong \frac{1}{2}\left\{(\lambda + \mu) + (\lambda - \mu)\frac{(p_0 - \mu)e^{\rho n} + (p_0 - \lambda)}{(p_0 - \mu)e^{\rho n} - (p_0 - \lambda)}\right\},$$

where

$$(8.9) \quad \begin{aligned} \mu &= \frac{1 - \alpha_2}{\alpha_1 - \alpha_2}, \\ \rho &= (1 - \alpha_2) - (\alpha_1 - \alpha_2)\lambda, \end{aligned}$$

and where we have taken $V_{1,0} = p_0$. In Fig. 8.2 we give an illustration of the use of this approximate formula.

If $\lambda = 0$ we have two "extinction" operators. With this condition the preceding equations give

$$(8.10) \quad V_{1,n} \cong \frac{\mu p_0}{p_0 + (\mu - p_0)e^{\rho n}},$$

with

$$(8.11) \quad \rho = 1 - \alpha_2, \qquad \mu = \frac{1 - \alpha_2}{\alpha_1 - \alpha_2}.$$

The expected total number of A_1 occurrences as $n \to \infty$, for example, the total number of responses in extinction, can be estimated by the area under the curve of $V_{1,n}$ versus n. We know that $V_{1,n}$ must lie between $Q_1{}^n p_0$ and $Q_2{}^n p_0$, and so limits on the area under the curve of $V_{1,n}$ are

$$(8.12) \quad \sum_{n=0}^{\infty} Q_1{}^n p_0 = \sum_{n=0}^{\infty} \alpha_1{}^n p_0 = \frac{p_0}{1 - \alpha_1},$$

and

$$(8.13) \quad \sum_{n=0}^{\infty} Q_2{}^n p_0 = \frac{p_0}{1 - \alpha_2}.$$

Furthermore, an approximation to the area under the curve is obtained from equation 8.10:

$$(8.14) \qquad \int_0^\infty V_{1,n}\, dn \cong \int_0^\infty \frac{\mu p_0\, dn}{p_0 + (\mu - p_0)e^{\rho n}}.$$

This is readily integrated to yield (logarithms are to the base e)

AN ESTIMATE OF THE EXPECTED TOTAL NUMBER OF A_1 RESPONSES ($\lambda = 0$):

$$(8.15) \qquad \begin{aligned} E(T_1) &\cong \int_0^\infty V_{1,n}\, dn \cong -\frac{\mu}{\rho}\log\left(1 - \frac{p_0}{\mu}\right) \\ &= \frac{-\log\left(1 - \dfrac{\alpha_1 - \alpha_2}{1 - \alpha_2}p_0\right)}{\alpha_1 - \alpha_2}. \end{aligned}$$

As a numerical example, consider the values $\alpha_1 = 0.95$, $\alpha_2 = 0.75$, and $p_0 = 1.00$. We then obtain for the integral

$$(8.16) \qquad E(T_1) \cong 5\log 5 = 8.05.$$

The limits given by equations 8.12 and 8.13 are

$$(8.17) \qquad \frac{p_0}{1 - \alpha_1} = 20, \qquad \frac{p_0}{1 - \alpha_2} = 4.$$

We made one hundred stat-rat runs of 25 trials each, as described in Section 6.2, and obtained a mean number of 8.99 A_1 responses with a standard deviation of 3.1. Hence the standard deviation of the estimate of the mean, 8.99, is about 0.31. Whereas the integral underestimates the mean number of A_1 responses in this problem, the result may be close enough for some purposes.

The preceding approximations were derived for the case of $\lambda = 0$. When we have the complementary case of $\lambda = 1$, similar results may be obtained. We need merely to replace p_0 with q_0, $V_{1,n}$ with $(1 - V_{1,n})$, and interchange α_1 and α_2 to obtain as

AN ESTIMATE OF THE EXPECTED TOTAL NUMBER OF A_2 RESPONSES ($\lambda = 1$):

$$(8.18) \qquad E(T_2) \cong \int_0^\infty (1 - V_{1,n})\, dn \cong \frac{-\log\left(1 - \dfrac{\alpha_2 - \alpha_1}{1 - \alpha_1}q_0\right)}{\alpha_2 - \alpha_1}.$$

We find this equation useful in Chapter 11. It will be interpreted as the total number of errors prior to "perfect" learning.

When $p_0 = 1$, as well as $\lambda = 0$, equation 8.15 reduces to

$$(8.19) \qquad E(T_1) \cong \frac{-\log\left[(1 - \alpha_1)/(1 - \alpha_2)\right]}{\alpha_1 - \alpha_2}.$$

Similarly, when $q_0 = 1$, equation 8.18 for $\lambda = 1$ reduces to

$$(8.20) \qquad E(T_2) \cong \frac{-\log\left[(1 - \alpha_2)/(1 - \alpha_1)\right]}{\alpha_2 - \alpha_1}.$$

The right side of each of the preceding two equations is an example of a function,

$$(8.21) \qquad T(\alpha, \beta) = \frac{-\log\left[(1 - \alpha)/(1 - \beta)\right]}{\alpha - \beta},$$

which is symmetric in α and β, that is,

$$(8.22) \qquad T(\alpha, \beta) = T(\beta, \alpha.)$$

This function arises in Part II. We present its values for various values of α and β in Table B at the end of the book.

8.3 THE FIRST OCCURRENCE OF RESPONSE A_1

In the preceding section we considered the expected total number of occurrences of one response or the other in the special cases when $\lambda = 0$ and $\lambda = 1$. In this section we consider another property of the sequences of p values—the mean number of trials prior to the first occurrence of one response or the other. Suppose that "failure" in an experiment increases the probability of "success" and that we wish to know how many trials, on the average, are required before the first success occurs. In the Solomon-Wynne experiment on avoidance training (first discussed in the Introduction) we might like to know about the mean trial of the first avoidance.

The mathematical problem is a rather simple one in the theory of runs. In Section 4.8 we developed a framework for discussing such questions, and so we rely upon those results. We consider only the case of $\lambda_2 = 1$, giving the operator

$$(8.23) \qquad Q_2 p = \alpha_2 p + (1 - \alpha_2),$$

and compute the mean number of trials before the first A_1 occurrence. (We make no assumptions about the value of λ_1 in this section.) If an A_2 response occurs on trial 0, then there will be A_2 runs of length $1, 2, 3, \cdots$. The mean length of this run of A_2's is obtained from equation 4.90:

$$(8.24) \qquad E(R_{2,0}) = \sum_{\nu=0}^{\infty} \tilde{\tau}_{0,\nu},$$

where we have used the abbreviation of equation 4.92:

(8.25)
$$\tilde{\tau}_{0,\nu} = \prod_{K=1}^{\nu} q_K, \qquad \tilde{\tau}_{0,0} = 1.$$

The q_K are the probabilities of response A_2 for trials $K = 1, 2, \cdots$ prior to the first A_1 occurrence. For the operator of equation 8.23,

(8.26)
$$q_K = \alpha_2^{K} q_0.$$

Using this expression for q_K in the previous equation gives

(8.27)
$$\tilde{\tau}_{0,\nu} = \prod_{K=1}^{\nu} \alpha_2^{K} q_0 = (\alpha_2 q_0)(\alpha_2^{2} q_0) \cdots (\alpha_2^{\nu} q_0)$$
$$= \alpha_2^{\nu(\nu+1)/2} q_0^{\nu}.$$

Hence the mean number of A_2's before the first A_1, given an A_2 on trial 0, is

(8.28)
$$E(R_{2,0}) = \sum_{\nu=0}^{\infty} \alpha_2^{\nu(\nu+1)/2} q_0^{\nu}.$$

We encountered this sum in Section 4.8 and considered it a special case of a function $\Phi(\alpha, \beta)$ defined by

(8.29)
$$\Phi(\alpha, \beta) = \sum_{\nu=0}^{\infty} \alpha^{\nu(\nu+1)/2} \beta^{\nu}.$$

This function is shown in Table A at the end of the book. In these terms we have

(8.30)
$$E(R_{2,0}) = \Phi(\alpha_2, q_0).$$

This equation gives the mean number of A_2's under the condition that an A_2 occurs on trial 0. The probability that an A_2 occurs on trial 0 is q_0, and so the unconditional mean number of trials, $E(F_1)$, before the first A_1 occurrence is

(8.31)
$$E(F_1) = q_0 \Phi(\alpha_2, q_0).$$

For given values of α_2 and q_0 we can compute the value of $E(F_1)$ from Table A.

In Chapter 11 we are also interested in the variance of the number of trials before the first A_1 occurrence. Equation 4.91 leads to the result

(8.32)
$$\sigma^2(F_1) = 2q_0\Psi(\alpha_2, q_0) + q_0\Phi(\alpha_2, q_0) - [q_0\Phi(\alpha_2, q_0)]^2,$$

where the function $\Psi(\alpha_2, q_0)$ is an example of the function

(8.33)
$$\Psi(\alpha, \beta) = \sum_{\nu=1}^{\infty} \nu\alpha^{\nu(\nu+1)/2}\beta^{\nu},$$

which is also given in Table A.

As a numerical example of the use of the preceding equations, let $\alpha_2 = 0.94$ and $q_0 = 1.00$. From Table A we get $\Phi(0.94, 1.00) = 5.0776$ and $\Psi(0.94, 1.00) = 13.7904$. These values then give $E(F_1) = 5.0776$ and $\sigma^2(F_1) = 6.8764$.

8.4 THE SECOND OCCURRENCE OF RESPONSE A_1

In the preceding section we considered the mean number of trials before the first occurrence of response A_1 when $\lambda_2 = 1$, that is, when Q_2 has the form

$$(8.34) \qquad Q_2 p = \alpha_2 p + (1 - \alpha_2).$$

We also presented a formula for the variance of the number of trials before the first A_1 occurrence. We now consider the distribution of the number of trials before the *second* A_1 occurrence when Q_2 is as given above and

$$(8.35) \qquad Q_1 p = \alpha_1 p + (1 - \alpha_1),$$

that is, when $\lambda_2 = \lambda_1 = 1$.

Let S_1 be a random variable denoting the number of trials before the second A_1 occurrence. The probability distribution of this random variable can be computed and its mean obtained; we omit the derivation but present the results. The distribution is given by

$$(8.36)$$
$$Pr\{S_1 = \nu\} = \alpha_2^{(\nu-1)(\nu-2)/2} q_0^{\nu-1}(1 - \alpha_1\alpha_2^{\nu-1}q_0) \left\{ \frac{1 - \alpha_1^\nu}{1 - \alpha_1} - \frac{\alpha_2^\nu - \alpha_1^\nu}{\alpha_2 - \alpha_1} q_0 \right\}.$$

When $\alpha_1 = \alpha_2 = \alpha$, this expression is indeterminate, but it can be shown that

$$(8.37) \quad Pr\{S_1 = \nu\} = \alpha^{(\nu-1)(\nu-2)/2} q_0^{\nu-1}(1 - \alpha^\nu q_0) \left\{ \frac{1 - \alpha^\nu}{1 - \alpha} - \nu\alpha^{\nu-1} q_0 \right\}.$$

The expected value of S_1 when $\alpha_1 \neq \alpha_2$ is given by

$$(8.38)$$
$$E(S_1) = 1 + \frac{\alpha_1 q_0}{\alpha_2 - \alpha_1} + \left[\frac{q_0}{1 - \alpha_1} - \frac{\alpha_1 q_0}{\alpha_2 - \alpha_1} \right] \Phi(\alpha_2, q_0)$$
$$+ \left[\frac{\alpha_1^2 q_0^2}{\alpha_2 - \alpha_1} - \frac{\alpha_1^2 q_0}{1 - \alpha_1} \right] \Phi(\alpha_2, \alpha_1 q_0),$$

where Φ is the function given in Table A. When $\alpha_1 = \alpha_2 = \alpha$, this expression is replaced by

$$(8.39) \qquad E(S_1) = \frac{1}{1 - \alpha} + \frac{q_0 - \alpha}{1 - \alpha} \Phi(\alpha, q_0) - q_0 \Psi(\alpha, q_0),$$

where Ψ is the other function given in Table A.

8.5 CASES WITH ONE IDENTITY OPERATOR

In Section 8.1 we pointed out that the two operators Q_1 and Q_2 commute either if they have the same limit point λ or if one of those operators is the identity operator. We now consider the latter case with Q_2 as the identity operator. We then have

$$(8.40) \qquad \begin{aligned} Q_1 p &= \alpha_1 p + (1 - \alpha_1)\lambda, \\ Q_2 p &= p. \end{aligned}$$

The operator Q_2 does not change p and so it does not have a unique limit point. The event E_2, associated with Q_2, has no influence on the response probabilities. It may be difficult to envisage such an event in a strict sense, but, if E_2 is believed to have a relatively small effect on subsequent behavior, it may be convenient to assume $Q_2 p = p$ as an approximation. In Part II we have occasion to make this assumption.

Obviously, nearly all sequences tend towards the limit point λ, and so the limiting distribution has all its density at the point λ. This may be shown by an appeal to the law of large numbers [1]. The only exception occurs when $p_0 = 0$; in this case operator Q_1 is never applied, and so the probability remains at zero forever.

The recurrence formula for the mean $V_{1,n+1}$ is

$$(8.41) \qquad V_{1,n+1} = [1 + \lambda(1 - \alpha_1)]V_{1,n} - (1 - \alpha_1)V_{2,n}.$$

We know that the distribution is contained between p_0 and λ on all trials, and so analogous to inequality 6.44 we have

$$(8.42) \qquad V_{2,n} \leq (\lambda + p_0)V_{1,n} - \lambda p_0.$$

Therefore, after using this inequality in equation 8.41, we get

$$(8.43) \quad V_{1,n+1} \geq [1 + \lambda(1 - \alpha_1)]V_{1,n} - (1 - \alpha_1)[(\lambda + p_0)V_{1,n} - \lambda p_0],$$

or

$$(8.44) \qquad V_{1,n+1} \geq (1 - \alpha_1)\lambda p_0 + [1 - (1 - \alpha_1)p_0]V_{1,n}.$$

We may obtain a lower bound on the mean $V_{1,n}$ by solving this expression with the equality sign. The result is

$$(8.45) \qquad V_{1,n} \geq \lambda - (\lambda - p_0)[1 - (1 - \alpha_1)p_0]^n.$$

An upper bound, Y_n, may be found from the expected operator procedure. Equation 6.69 with $a_2 = 0$, $\alpha_2 = 1$, $a_1 = (1 - \alpha_1)\lambda$ gives

$$(8.46) \qquad Y_{n+1} = [1 + \lambda(1 - \alpha_1)]Y_n - (1 - \alpha_1)Y_n{}^2.$$

Restriction 6.70 becomes

$$(8.47) \qquad 1 + \lambda(1 - \alpha_1) - 2(1 - \alpha_1)Y_n \geq 0.$$

The quadratic difference equation is troublesome, but the corresponding differential equation

(8.48)
$$\frac{dY_n}{dn} = Y_n(1 - \alpha_1)(\lambda - Y_n)$$

has the solution

(8.49)
$$Y_n = \frac{p_0\lambda}{p_0 + (\lambda - p_0)e^{-\lambda(1-\alpha_1)n}}.$$

Thus we have approximately

(8.50)
$$V_{1,n} \leq \frac{p_0\lambda}{p_0 + (\lambda - p_0)e^{-\lambda(1-\alpha_1)n}}.$$

A further special case of one identity operator obtains when the limit of the other operator is unity. This special case has been studied in detail by G. A. Miller and W. J. McGill [2]. From a different formulation, they obtain the very useful recurrence relation

(8.51)
$$V_{1,n} = p_0 + (1 - p_0)(1 - \alpha_1^n)V_{1,n-1} \qquad (\alpha_2 = 1, \lambda = 1).$$

This equation is exact and may be used for computing the mean on every trial. It does not involve higher moments than the mean, as did our previous recurrence formula, 8.41. In Fig. 8.3 we show an illustration

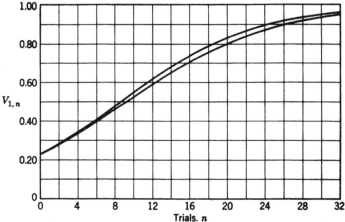

Fig. 8.3. Distribution means, $V_{1,n}$, versus trials for the case of one identity operator (Q_1). The lower curve shows the exact means computed from equation 8.51, with $p_0 = 0.230$ and $\alpha_1 = 0.860$. The upper curve shows the approximation given by equation 8.49 for $\lambda = 1$.

of how this equation can be used to compute an average learning curve. For comparison, we also show the result of using the differential equation approximation (8.49) with $\lambda = 1$.

From equations 8.41 and 8.51 we can obtain for the variance on the nth trial

$$(8.52) \qquad \sigma_n^2 = V_{1,n}(1 - V_{1,n}) + \frac{[1 - (1 - \alpha_1^{n+1})(1 - p_0)]V_{1,n} - p_0}{1 - \alpha_1}.$$

Hence the variance on each trial may be found in terms of the mean on that trial. For $0 \leq \alpha_1 < 1$, σ_n^2 approaches zero as n becomes large.

The distribution of the number of trials, F_1, before the first A_1 occurrence is especially simple when $Q_2 p = p$. It is

$$(8.53) \qquad Pr\{F_1 = \nu\} = q_0^\nu(1 - q_0).$$

The expected value is

$$(8.54) \qquad E(F_1) = \sum_{\nu=0}^{\infty} \nu Pr\{F_1 = \nu\} = q_0/p_0,$$

and the variance is

$$(8.55) \qquad \sigma^2(F_1) = E(F_1^2) - [E(F_1)]^2 = q_0/p_0^2.$$

For example, if $p_0 = 0.5$, then $E(F_1) = 1$ and $\sigma^2(F_1) = 2$. From one hundred stat-rat computations (see Section 6.2) we observed a mean value of 1.17 trials before the first A_1 occurrence. The variance of this estimate from one hundred stat-rats should be $\sigma^2(F_1)/100 = 0.02$, and so the standard deviation is $\sqrt{0.02} = 0.14$. Thus, the stat-rat mean of 1.17 is a little more than one σ above the true mean of 1.00.

The distribution of the number of trials S_1 before the second A_1 occurrence is also simplified when $Q_2 p = p$ and $\lambda = 1$. Equation 8.36 becomes, when $\alpha_2 = 1$,

$$(8.56) \qquad Pr\{S_1 = \nu\} = q_0^{\nu-1}(1 - q_0)(1 - \alpha_1 q_0) \frac{1 - \alpha_1^\nu}{1 - \alpha_1}.$$

The mean of this distribution may be obtained from equation 8.38 by noting that

$$(8.57) \qquad \Phi(1, \beta) = \sum_{\nu=0}^{\infty} \beta^\nu = \frac{1}{1 - \beta}.$$

Some algebra leads to the simplified result

$$(8.58) \qquad E(S_1) = \frac{1 - \alpha_1 q_0^2}{(1 - q_0)(1 - \alpha_1 q_0)}.$$

Furthermore, the variance can be computed without too much difficulty. The result is

$$(8.59) \qquad \sigma^2(S_1) = \frac{q_0}{(1 - q_0)^2} + \frac{\alpha_1 q_0}{(1 - \alpha_1 q_0)^2}.$$

When $q_0 = 0.5$ and $\alpha_1 = 0.9$ we have $E(S_1) = 2.818$ and $\sigma^2(S_1) = 3.488$. From the one hundred stat-rats mentioned above we obtain a mean $\bar{S}_1 = 3.27$; this is about 2.4σ above the true mean 2.818.

8.6   EXPERIMENTER-SUBJECT-CONTROLLED EVENTS WITH IDENTITY OPERATORS

In this section we consider a problem of interest in the applications described in Chapter 13. Assume that there are two responses, A_1 and A_2, and two outcomes, O_1 and O_2. For the events formed from O_2, we assume identity operators, and for the other two events we choose the limit points appropriate for perfect learning. The operators are then defined by

$$
\begin{aligned}
Q_{11}\, p &= \alpha_1\, p + (1 - \alpha_1) \\
Q_{12}\, p &= p \\
Q_{21}\, p &= \alpha_1\, p \\
Q_{22}\, p &= p,
\end{aligned}
$$

(8.60)

where the subscripts on the operators refer to the response and outcome, respectively. This is a specialization of the problem discussed in Section 5.9. We have taken the α's which correspond to O_1 equal to one another, and so only one parameter, α_1, remains. From equation 5.79 we see that the recurrence formula is

(8.61) $V_{1,n+1} = [1 + (\pi_1 - \pi_2)(1 - \alpha_1)]V_{1,n} - (\pi_1 - \pi_2)(1 - \alpha_1)V_{2,n}.$

The annoying second moment term remains, and so we consider the expected operator approximation; we replace $V_{2,n}$ by $V_{1,n}^2$ and get

(8.62) $V_{1,n+1} \cong V_{1,n} + (\pi_1 - \pi_2)(1 - \alpha_1)V_{1,n}(1 - V_{1,n}).$

This difference equation is awkward to solve but the corresponding differential equation,

(8.63) $\dfrac{dV_{1,n}}{dn} \cong (\pi_1 - \pi_2)(1 - \alpha_1)V_{1,n}(1 - V_{1,n}),$

has the solution

(8.64) $V_{1,n} \cong \dfrac{V_{1,0}}{V_{1,0} + (1 - V_{1,0})e^{-(\pi_1 - \pi_2)(1 - \alpha_1)n}}.$

From this result, we see that

(8.65) $V_{1,\infty} \cong \begin{cases} 1 & \text{when } \pi_1 > \pi_2 \\ 0 & \text{when } \pi_1 < \pi_2 \\ V_{1,0} & \text{when } \pi_1 = \pi_2. \end{cases}$

Only the equal π result is exact.

8.7 SUMMARY

The two operators Q_1 and Q_2 commute—yield the same result when applied to p in either order—if they have the same limit point λ or if one of the operators leaves p unchanged. When the events are represented by operators which commute, the order of the events in a sequence with a fixed number of each event does not affect the final p value. This fact simplifies computations somewhat and, as we see in Part II, considerably simplifies the task of estimating parameters from experimental data.

In Section 8.2 we present an approximate formula for the means $V_{1,n}$ as a function of n when the operators have the same limit point λ. We then obtain some estimates of the mean total number of A_1 responses when $\lambda = 0$ and the mean number of A_2 responses when $\lambda = 1$. In Sections 8.3 and 8.4 we develop expressions for the mean number of trials before the first and second A_1 responses when $\lambda = 1$. We use some of these properties in Chapter 11, where we analyze an experiment on avoidance training.

In Section 8.5 we discuss the special case of $\alpha_2 = 1$, that is, of $Q_2 p = p$. The previous results for equal limit points are applied to this case, and appreciable simplifications result. This analysis will be applied in Chapter 10 to an experiment on verbal learning. In Section 8.6 we extend the identity-operator condition to experimenter-subject-controlled events.

REFERENCES

1. Feller, W. *An introduction to probability theory and its applications.* New York: Wiley, 1950, p. 141.
2. Miller, G. A., and McGill, W. J. A statistical description of verbal learning. *Psychometrika*, 1952, **17**, 369–396.

PART II

APPLICATIONS

CHAPTER 9

Identification and Estimation

9.1 THE IDENTIFICATION PROBLEM

In the preceding eight chapters we have presented a mathematical system intended to be useful for some, but by no means all, learning problems. In addition, there are some computational schemes and discussions of special cases. Although we have had applications to psychological problems in mind throughout, the organization up to this point has been guided by the mathematical analysis. In a sense, all we have given the reader so far is a mathematical structure. Although the basic features of that mathematics were described in more or less psychological terms—responses, stimulus elements, trials, events, etc.—there exists no essential connection between the system presented so far and the experimental world.

The first step in applying the system to experimental data is to relate the basic elements to actual behavior and events. We must unambiguously identify system symbols with observables. For the most part we identify elements and quantities of the system with things which are defined experimentally. A response class in the system is identified with a class of behavior observed and recorded by an experimenter such as "turning left" in a maze or "pressing the bar" in a Skinner box. We do not identify system responses with more microscopic behavior such as a muscle flexion (Guthrie's "movements") unless those bits of behavior are being observed and recorded. We do not mean to imply that we consider studies of such behavior unimportant, or that a complete theory of learning can ignore such matters. We have a much more restricted goal; we wish to describe experimental data, and so we usually must accept the experimenters' definitions of responses. In this sense, the system is a descriptive theory rather than a new psychological theory of learning. As is pointed out in the next section, this mathematical system, in its most general form, can be made compatible, we believe, with several current learning theories.

A *trial* has been defined as an opportunity for choosing among a set of mutually exclusive and exhaustive alternatives or responses. In many problems the identifications of system responses with certain experimental behavior classes automatically identify system trials with experimental trials. How microscopic the system is depends on the choice of the unit which is to be called a trial. For example, each time a rat is placed at the starting end of a T-maze and is allowed to pass a choice point we have one experimental trial, and this will be identified with a trial in the system. But in some experiments the correspondence is not so obvious. In a runway, for example, an experimental trial is defined so that the rat *always* leaves the starting box. If we were to identify a system response with "leaving the starting box on an experimental trial," no choice is involved; there is but one response, and its probability of occurrence during an experimental trial is always unity (unless starting boxes are used as coffins). In runway experiments, we are concerned with changes in latency, that is, changes in the time which a rat spends in the starting box before going into the runway. A reasonable identification is between system response and "leaving the starting box during a small time interval." During each second (to pick a specific unit of time) the rat may either leave the box (response A_1) or not leave the box (response A_2). In this case a system trial is identified with a time increment of one second. In Chapter 14 we have more to say about these time problems. At this point we wish only to emphasize that the appropriate definition of a trial depends upon the identifications we make between system responses and experimental responses.

Learning is represented by orderly changes resulting from the occurrences of *events*. Again, the notion of an event is an abstract concept in the system, but it is necessary to identify these events with empirical events. As was suggested when the concept was introduced in Chapter 1, the events are identified with such things as stimulus changes, giving a reward or punishment, or actual response occurrences, depending upon the problem. In general, whenever an experimenter manipulates a subject's environment in a specified way, an event has occurred. Furthermore, the subject may change its own environment by making a particular response, and again this constitutes an event. In a somewhat degenerate sense, an event is sometimes associated with no environmental change; this will often be mathematically convenient, as is seen in later chapters. We have no simple formula for identifying system events with experimental events, but in each application we try to make the identifications as explicit as possible. Intuition is an important guide. If we do have a general principle, it is simply that any class of empirical events suspected of systematically changing the subject's behavior in an experiment should be identified with an event in the model.   The model, of course, helps us to cut down quickly the total number of variables involved, and helps categorize the possibilities.   (The reader may feel critical at this point because intuition is not ordinarily regarded as an important scientific principle.   The word "intuition" has been used because we are trying to make the difficulty overt, rather than conceal it beneath such phrases as "natural and obvious choices.")

The utility of the general model depends upon both the formal structure of the mathematical system and the appropriateness of the identifications made.   It could turn out that the general model might be quite adequate with one set of identifications but wholly inadequate with another set.   On the other hand, it might be that no set of identifications would lead to reasonable agreement between the model and data, in which case we would be forced to discard the basic mathematical framework as unsatisfactory for the problem.   For example, the basic assumption of linear operators may be untenable for many learning phenomena.   But before altering such a basic assumption, and thereby introducing major mathematical difficulties, we would search for a set of identifications which would lead to better agreement between the models and experimental data.

It is also possible that more than one set of identifications would lead to good agreement.   For example, it is conceivable that in a two-choice situation, say choosing "right" or "left," some people would behave as if the response were right or left, whereas for others "same" or "opposite" of previous trial would be the appropriate identifications of responses.

9.2   REINFORCEMENT THEORY VERSUS CONTIGUITY THEORY

Though psychologically trained readers are well acquainted with the material in this section, we include it for the benefit of others.

Two of the major psychological theories of learning differ in a fundamental respect.   On the one hand, reinforcement theory stems from Thorndike's law of effect [1], which assumes that living organisms seek to achieve or experience certain kinds of environmental changes and try to avoid certain other "noxious" stimuli.   All stimuli are assumed to possess intrinsic properties which make them rewarding or punishing to a given organism; certain basic needs or drives are postulated from extensive observations of animal and human behavior, and reduction of those drives is assumed to lead to learning.   Contiguity theory, on the other hand, postulates that the basic determinant of learning is association. Stimuli and responses become "connected" or associated merely by their being contiguous in time.   The connections are viewed as being continuously changing, but when an organism's environment is changed the connections between the stimuli perceived just prior to the change and the responses just made are preserved. When the organism next encounters a stimulus situation, the response last made in that situation will most likely occur. In contiguity theory, reinforcement is considered to be only a stimulus change which preserves the stimulus-response connections that existed just prior to the change. The distinction between reward and punishment, in the operational sense of the terms, is made by distinguishing between the kinds of responses which become associated to the stimuli present; punishment is viewed as preserving connections between the stimuli and withdrawal responses, that is, responses which cause the cessation of pain.

From a psychological point of view, the difference between reinforcement theory and contiguity theory is important. Hull's behavior system [2] along with Spence's refinements and extensions [3] are based upon the reinforcement concept, whereas Guthrie is the outstanding proponent of association theory [4, 5]. These learning theories have led to various lines of investigation, both theoretical and experimental. In some cases, differential predictions have been made and crucial experiments have resulted. Indeed this has been and continues to be a healthy state of affairs in the development of psychological thinking.

In presenting the mathematical system there is no need to take a definite position on the issue of reinforcement versus contiguity. In developing the mathematical framework little or no reference is made to either set of concepts. An event occurrence may be either a drive reduction, as Hull might have said, or a stimulus change, as Guthrie might prefer. Both schools of thought would agree that an event has occurred and that this event has a definite effect on future behavior in the stimulus situation in which it occurred. The major exception to this position of impartiality is contained in Chapter 2, where a set-theoretic model of conditioning, based primarily on association theory, is presented. This set-theoretic model, originally developed by Estes, is not an essential part of our mathematical system, however. As pointed out in Chapter 2, it was given only as an illustration of how the basic operators could be derived from more primitive assumptions in stimulus-response theory.

In principle, then, our mathematical model is not committed to either reinforcement theory or contiguity theory. In this sense it is more general than either of those theories! In a more important sense, however, it is much less general. The assumption of linear operators, the identifications we will make, and the assumed relations between the "true" probabilities and experimental measures of behavior all make the model much more restricted than the current learning theories. And it will be still more restricted when we make special assumptions about the parameters λ_i and α_i in applying the model to experimental problems. In making these special assumptions we shall have an opportunity to be influenced by the several learning theories. In some cases the special assumptions will be dictated by considerations such as symmetry (in experimental situations), the fact that people do learn short lists of words perfectly, the fact that pigeons do extinguish on a pecking response, etc. In other cases, however, we may have a choice. For example, if we were trying to describe extinction of an operant response A_1, we could assume that the operator Q_1, which is applied when A_1 occurs, decreases the probability p of A_1 to zero ($\lambda_1 = 0$, $\alpha_1 \neq 1$ so $Q_1 p = \alpha_1 p$) and that the operator Q_2, which is applied when other responses A_2 occur, does not change p ($\lambda_2 = 0$, $\alpha_2 = 1$ so $Q_2 p = p$). These assumptions would correspond most closely to the Hullian inhibition theory of extinction. On the other hand, we could argue that Q_1 does not appreciably change p ($\lambda_1 = 0$, $\alpha_1 = 1$ so $Q_1 p = p$) but that Q_2 decreases p to zero ($\lambda_2 = 0$, $\alpha_2 \neq 1$ so $Q_2 p = \alpha_2 p$). These assumptions would be consistent with the Guthrian concept that extinction occurs because other responses become associated with the stimuli. The mathematical model permits either extreme view of the extinction process or any combination.

9.3 EXPERIMENTAL VARIABLES

Experimenters studying animal and human learning have been concerned with a number of variables, including amount of reward, strength of drive, amount of work required in making the response, time interval between response and reward, and the intensity and duration of an electric shock. Our model does not explicitly involve variables of this kind. To be sure, a complete theory of learning would attempt to handle these variables in an explicit and unambiguous way. But our program is a more provisional one. The model uses linear operators which introduce parameters such as λ_i and α_i. These parameters must depend upon the experimental variables listed above and probably upon still others. However, in this work the parameters λ_i and α_i which remain after special assumptions are made in each problem must be estimated from the data. Hence, the relation of α_1, for example, to the amount of reward given is an empirical question. In an analogous way, Ohm's law relates current, voltage, and resistance. But it is still an engineering problem to determine the resistance for any particular object.

In a sense the model provides the experimental psychologist with statistics that summarize some kinds of data. We present a mathematical model and procedures for estimating parameters from data. Parametric studies can then determine how the parameters depend upon various experimental variables. One difficulty which has been evident in the interpretation of learning data from various experiments is the lack of obvious and useful summary statistics. How do we measure "speed of learning" from a sequence of right and left turns in a T-maze, for example? Or how does an experimenter summarize in one or two statistics how rapidly an operant response extinguishes? The measures which have been used, such as time or trials to a fixed criterion of conditioning or extinction, have serious shortcomings in view of the evident (to us) statistical and sequentially dependent nature of the processes. Extinction of bar pressing, for instance, is often defined as "complete" by an experimenter when no responses occur during a predetermined time interval. Yet he may find that one animal which has met the criterion will respond very rapidly a short time later whereas another animal may not. Time to a criterion may be quite sensitive to the time interval used in defining the criterion, and a comparison of data obtained under different conditions is difficult. While agreeing that such measures are useful, we feel that more stable and more psychologically meaningful statistics are needed and that we are not likely to discover them without a model. During the remainder of this book our chief concern is with estimating parameters from experimental data and with comparing these data with implications of the model.

9.4 THE ESTIMATION PROBLEM

We are now faced with the rather technical statistical question of how to estimate parameters of the model from actual data. The parameters are never strictly determined from any set of data since the model is intrinsically a statistical one. There are better and worse ways of estimating parameters. A group of subjects are considered to represent a sample from a population of all possible subjects, or a block of responses of a single subject are thought of as a sample of responses from a population. The probability of a response on a particular trial can never be strictly determined from the data and so the parameters which specify how that probability changes cannot be determined exactly. The problem is one of statistical inference. It is desired to obtain good estimates of parameters from data. A simple analogy is the estimation of the true probability of a head in flipping a coin. We flip the coin a necessarily finite number of times, and from the results we estimate the probability of a head. Or, if we have a large number of identical coins, we can flip all of them just once and from these results obtain an estimate of the probability. The procedure used in polling is another example. A population parameter, for example, proportion of people unemployed, is estimated from a sample of people.

The estimation problem in the learning models is usually more complicated than those involved in the foregoing examples. In fact, it might as well be admitted that the estimation process becomes quite forbidding in some situations. We have a double sampling process as we now indicate. For two subject-controlled events there are 2^n possible p values on trial n, since there are 2^n possible sequences of A_1's and A_2's. In general, these 2^n values of probability can all be different. In a group of k identical subjects the ith subject has a particular probability value p_{in} on trial n $(i = 1, 2, \cdots, k)$. (Identical subjects have the same values of λ_j, α_j, and initial probability p_0.) These k values p_{in} constitute a sample from the population of 2^n possible p values. Chapters 3 and 4 dealt with the properties of such populations, but we must now consider in detail various properties of samples drawn from these populations. Suppose that the sample of k probability values is a random sample. It will have a mean value denoted by $\bar{p}_n$. The population mean $V_{1,n}$ can be estimated by the sample mean $\bar{p}_n$. Similarly, higher moments of the sample, such as its variance, can be used to estimate the corresponding moments of the population of probability values. The estimation procedure would be as straightforward as this if we knew the k probabilities p_{in}. But, sad to state, they are not known. They are true probabilities and must in turn be estimated from the data just as we would estimate a single probability p of a head when a coin is flipped. The double estimation process is now evident—the k probability values p_{in} are estimated and then used to obtain estimates of properties of the p-value population. An analogy may help to make the point clearer. Consider a large population of coins, most of which are not "true," that is, they have a distribution of probabilities of coming up heads when flipped. (These probabilities correspond to p values.) It is desired to estimate the properties of this distribution; so first select a sample of k coins. Flip each coin a number of times to get an estimate of the probability for each coin in the sample. Then use these individual coin estimates to infer properties of the population of coins from which the sample was drawn.

The probability p_{in} for the ith subject on trial n may be regarded as the parameter of a binomial distribution of a random variable x_{in}. This random variable x_{in} has the value 1 if response A_1 occurs on trial n for the ith subject, and it has the value 0 if an A_2 occurs. The probability that $x_{in} = 1$ is p_{in} and the probability that $x_{in} = 0$ is $1 - p_{in} = q_{in}$. Given N trials with the same probability p_{in}, the probability of x occurrences of A_1 $(x = 0, 1, \cdots, N)$ is given by the familiar binomial distribution

$$(9.1) \qquad f(x) = \binom{N}{x} p_{in}{}^x q_{in}{}^{N-x}, \qquad \binom{N}{x} = \frac{N!}{x!(N-x)!}.$$

In applications of the general model, presented in the following chapters, we seldom have more than one trial for estimating a given probability p_{in}, because the probabilities change for each individual on each trial, and usually by a different amount. When we have but one trial, $N = 1$, and the above equation for the binomial reduces to the statement that $f(0) = q_{in}$ and $f(1) = p_{in}$. Then it may be proved that the unbiased estimate of p_{in} is 1 when $x_{in} = 1$ and is 0 when $x_{in} = 0$. With as little information as this to estimate the p_{in} the reader may feel that the whole estimation problem is a bit hopeless. But this is not so, at least in the special cases we present. When two operators Q_1 and Q_2 commute (see Chapter 8) the probability p_{in} on trial n is independent of the particular order of A_1's and A_2's up to trial n, and depends only upon the number k of A_1 occurrences and the number $n - k$ of A_2 occurrences. In other words, all sequences with k occurrences of A_1 and $n - k$ of A_2 terminate in the same probability value, which we denote by $p_{k,n}$. In experimental records of groups of subjects we may find several such equivalent sequences, and so have several subject-trials which can be used to estimate $p_{k,n}$. When this happens the estimation problem is simplified greatly. Of course, a price is paid for this. We are forced to assume that all subjects have approximately the same parameter values, or else we are estimating some sort of average parameter. Such simplifications are discussed in connection with several applications.

9.5 MONTE CARLO CHECKS ON ESTIMATES

Procedures for estimating the parameters are described in the following sections of this chapter and in later chapters. Often it is helpful to have a check on the estimation procedure. Suppose that data from an experiment did behave according to the model, that is, the true probabilities of the responses changed according to the postulated rules. We would still want to know how good a particular estimation procedure was in extracting from the data the true values of the parameters. Such a check is readily available. One need only fabricate a set of data using the Monte Carlo method or "stat-rat" procedure described in Section 6.2.

In making a set of Monte Carlo computations first set up the desired form of the operators Q_1 and Q_2. For example: $Q_1 p = \alpha_1 p + (1 - \alpha_1)$ and $Q_2 p = \alpha_2 p + (1 - \alpha_2)$. Next select a value of the initial probability, p_0, of response A_1 and choose numerical values for the remaining parameters, for instance, α_1 and α_2 for the example just given. Then make a number of Monte Carlo runs as described in Section 6.2. From these runs we obtain sequences of responses A_1 and A_2; these sequences are the data desired.

Having obtained a number of sequences of responses, proceed as with a set of experimental data by applying the particular estimation procedure under study to obtain numerical estimates of the parameters. Finally, these estimates can be compared with the known values used in making the Monte Carlo runs. Needless to say, the greater the number of sequences computed, the better are the estimates of the parameters if the estimation procedure is a good one. Some examples are given in the next section.

9.6 SIMPLE STATISTICS OF THE DATA

In most experimental applications considered in the following chapters, the data for a single subject can be characterized by a sequence of A_1's and A_2's. For example, a rat will generate a sequence of right and left turns in a T-maze, or a dog will produce a sequence of avoidances and non-avoidances in an avoidance training experiment. From such sequential data on a number of subjects, numerous simple statistics can be computed. Sometimes these statistics can be related to the parameters in the model to provide a direct method of estimation.

One simple statistic is the mean number of trials before the first A_1 occurrence. This is readily computed from the data, and we call it F_1 (F for "first" and the subscript 1 for A_1). We have been able to obtain a simple closed expression for the expected value of such a quantity only when $\lambda_2 = 1$, that is, when Q_2 is defined by

$$(9.2) \qquad\qquad Q_2 p = \alpha_2 p + (1 - \alpha_2).$$

Equation 8.31 gives the theoretical expression for the expected number of trials before the first A_1 occurrence in this case. Hence, we have the estimation equation*

$$(9.3) \qquad\qquad F_1 \triangleq q_0 \Phi(\alpha_2, q_0) \qquad (\lambda_2 = 1),$$

where Φ is a function given in Table A, and where q_0 is the initial probability of response A_2. We can readily compute a similar statistic, F_2, defined as the mean number of trials before the first A_2 occurrence. When $\lambda_1 = 0$, that is, when

$$(9.4) \qquad\qquad Q_1 p = \alpha_1 p,$$

we have the estimation equation

$$(9.5) \qquad\qquad F_2 \triangleq p_0 \Phi(\alpha_1, p_0) \qquad (\lambda_1 = 0).$$

---

* Here and elsewhere the symbol $\triangleq$ means "is estimated by" or "estimates." On one side of this symbol there is a function of the data, and on the other side a mathematical function of the parameters being estimated. From classical statistics, if m is the true mean of a distribution and $\bar{x}$ the mean of a random sample drawn from this distribution, the notation could read either $\bar{x} \triangleq m$ (the sample mean estimates the population mean) or $m \triangleq \bar{x}$ (the population mean is estimated by the sample mean).

When the model for a particular experiment involves $\lambda_2 = 1$, we can use the statistic F_1 to get a relationship between p_0 and α_2, and when it involves $\lambda_1 = 0$ we can use F_2 to get a relationship between p_0 and α_1. These relationships, along with others to be discussed next, can lead to estimates of the parameters. For the special case when $\alpha_2 = 1$, that is, when

$$(9.6) \qquad\qquad Q_2 p = p,$$

equation 9.3 simplifies to

$$(9.7) \qquad\qquad F_1 \triangleq q_0/p_0 \qquad (\alpha_2 = 1).$$

Similarly, when $\alpha_1 = 1$, that is, when

$$(9.8) \qquad\qquad Q_1 p = p,$$

equation 9.5 simplifies to

$$(9.9) \qquad\qquad F_2 \triangleq p_0/q_0 \qquad (\alpha_1 = 1).$$

Another simple statistic of the data, denoted by $\bar{S}_1$, is the mean number of trials before the second A_1 occurrence (S for "second"). We computed the expectation of this statistic when $\lambda_1 = \lambda_2 = 1$; from equation 8.38 we can get an estimation equation for this case. Similarly, when $\lambda_1 = \lambda_2 = 0$, we can get an estimation equation for $\bar{S}_2$, the mean number of trials before the second A_2 response. For the special case when $\lambda_1 = 1$ and $\alpha_2 = 1$, that is, when

$$(9.10) \qquad \begin{aligned} Q_1 p &= \alpha_1 p + (1 - \alpha_1) \\ Q_2 p &= p, \end{aligned}$$

we obtain, from equation 8.58,

$$(9.11) \qquad \bar{S}_1 \triangleq \frac{1 - \alpha_1 q_0^2}{(1 - q_0)(1 - \alpha_1 q_0)} \qquad (\lambda_1 = 1, \alpha_2 = 1).$$

Similarly, when $\lambda_2 = 0$ and $\alpha_1 = 1$, that is, when

$$(9.12) \qquad \begin{aligned} Q_1 p &= p \\ Q_2 p &= \alpha_2 p, \end{aligned}$$

we have the estimation equation

$$(9.13) \qquad \bar{S}_2 \triangleq \frac{1 - \alpha_2 p_0^2}{(1 - p_0)(1 - \alpha_2 p_0)} \qquad (\lambda_2 = 0, \alpha_1 = 1).$$

For the special cases mentioned, the statistic $\bar{S}_1$ or $\bar{S}_2$ yields another relation among p_0, α_1, and α_2. If one of these parameters is known or assumed, this relation along with the one obtained from F_1 or F_2 may be solved for the unknown parameters. We now give two examples.

From one hundred stat-rat runs, made with $\lambda_1 = 1$ and $\alpha_2 = 1$, mentioned in Section 8.5, we obtained 1.17 for the mean number of trials before the first A_1 and 3.27 for the mean number of trials before the second A_1. From equation 9.7 above we get

$$(9.14) \qquad\qquad F_1 = 1.17 \doteq \frac{q_0}{1 - q_0}.$$

We solve this equation and get as our estimate* of q_0, the value

$$(9.15) \qquad\qquad \hat{q}_0 = 0.54.$$

From equation 9.11 above we get

$$(9.16) \qquad\qquad \bar{S}_1 = 3.27 \doteq \frac{1 - \alpha_1 q_0^2}{(1 - q_0)(1 - \alpha_1 q_0)}.$$

When we use the estimate $\hat{q}_0 = 0.54$ in this equation, we obtain as our estimate of α_1,

$$(9.17) \qquad\qquad \hat{\alpha}_1 = 0.97.$$

The parameters actually used in making the stat-rat runs were $q_0 = 0.50$ and $\alpha_1 = 0.90$. Although the estimates are not as close to the true values as we would like, we were able to obtain these estimates very easily. Later we shall see how better estimates can be made from these stat-rat data.

As a second example, we consider thirty stat-rat computations made with $\lambda_1 = \lambda_2 = 1$ and $p_0 = 0$. The parameter values $\alpha_1 = 0.70$ and $\alpha_2 = 0.95$ were used, but α_1 and α_2 will be estimated from the data. From the thirty sequences we found that $F_1 = 5.67$. Equation 9.3 with $q_0 = 1$ gives

$$(9.18) \qquad\qquad F_1 = 5.67 \doteq \Phi(\alpha_2, 1).$$

From Table A, we find by interpolation that

$$(9.19) \qquad\qquad \hat{\alpha}_2 = 0.950.$$

This happens to be the precise value used in making the computations.

Another statistic of learning data which is often used when "perfect" learning is achieved is the total number of errors. For example, in the T-maze, with food on one side only, a rat will eventually stop making errors. Similarly, in the Solomon-Wynne experiment on avoidance training, a normal dog seems to have learned to avoid without fail after having been shocked on a certain number of trials. Denote the total

---

*Here, as elsewhere, the circumflex (^) over a parameter denotes an estimate of that parameter.

number of A_2 occurrences by T_2 (T for "total"). In Section 8.2 it was shown that, for $\lambda_1 = \lambda_2 = 1$, the expected value of T_2 is approximated by equation 8.18, and so we have the estimation equation

$$(9.20) \qquad T_2 \triangleq \frac{-\log\left[1 - \dfrac{(\alpha_2 - \alpha_1)q_0}{1 - \alpha_1}\right]}{\alpha_2 - \alpha_1} \qquad (\lambda_1 = \lambda_2 = 1).$$

When $\alpha_2 = 1$ this simplifies to

$$(9.21) \qquad T_2 \triangleq \frac{-\log p_0}{1 - \alpha_1} \qquad (\lambda_1 = 1, \alpha_2 = 1).$$

When all sequences approach zero asymptotically a useful statistic is T_1, the mean total number of A_1 responses. In experimental extinction, for example, T_1 is some finite number. In Section 8.2 we showed that for $\lambda_1 = \lambda_2 = 0$, the expected value of T_1 is approximated by equation 8.15. Hence we have the estimation equation

$$(9.22) \qquad T_1 \triangleq \frac{-\log\left[1 - \dfrac{(\alpha_1 - \alpha_2)p_0}{1 - \alpha_2}\right]}{\alpha_1 - \alpha_2} \qquad (\lambda_1 = \lambda_2 = 0),$$

and for $\alpha_1 = 1$ this becomes

$$(9.23) \qquad T_1 \triangleq \frac{-\log q_0}{1 - \alpha_2} \qquad (\lambda_2 = 0, \alpha_1 = 1).$$

We now give an example of the use of these estimation equations involving T_1 and T_2.

The thirty stat-rats, mentioned above, for which $p_0 = 0$ and $\lambda_1 = \lambda_2 = 1$, gave the value $T_2 = 8.4$. We may now use equation 9.20. We take $q_0 = 1$ and get

$$(9.24) \qquad T_2 = 8.4 \triangleq \frac{-\log \dfrac{1 - \alpha_2}{1 - \alpha_1}}{\alpha_2 - \alpha_1} = T(\alpha_2, \alpha_1),$$

where $T(\alpha_2, \alpha_1)$ is the function given in Table B. We then use the value $\hat{\alpha}_2 = 0.95$ obtained from the statistic F_1 above and, turning to Table B, we find

$$(9.25) \qquad \hat{\alpha}_1 = 0.77.$$

This is to be compared with the true value $\alpha_1 = 0.70$ used in making the stat-rat runs.

Various other statistics can be computed from sequential learning data.

For example, the number of alternations, or the mean number of trials before the *third* A_1 occurrence, might be of interest. Such statistics depend directly on the parameters in the model, but we have not investigated them. Instead we turn to some more efficient procedures for estimating the parameters.

9.7 BIAS AND VARIANCE OF ESTIMATORS

Various procedures for estimating parameters are known to statisticians. Some of these are better than others for particular purposes. For example, we might flip a coin 100 times to estimate the probability p of a head. It can be shown that the "best" estimate of p—best according to explicit criteria—is the proportion of the 100 trials on which a head appears. But there are other estimates of p available. We could use, say, the proportion of heads during the first thirty trials or the proportion of heads on every other trial. Or we might count the number of times we get two heads in a row, use this to estimate p^2, and take the square root of the result as an estimate of p. Many other estimators of p could be devised, but none seem better than the obvious one, the proportion of heads in the total 100 trials; it is an *unbiased* estimate.

It is desirable but not mandatory that an estimator be *unbiased*. To say that an estimate is unbiased is to say, in essence, that if we repeatedly draw samples and compute the estimate for each sample, the mean of these estimates will get nearer and nearer the true value of the parameter as the number of samples drawn gets large. More strictly, we may conceive of a distribution of all possible values of the estimate, and, if the mean of this distribution equals the true parameter value, the estimate is unbiased. If the estimate is biased, the mean will not equal the true parameter value. Although unbiasedness is desirable, statisticians will often sacrifice this property in favor of other important properties.

It is important to make clear what biased and unbiased estimates are, because the emotional tone of the word "bias" is so strong. We shall frequently work with biased estimates. One reason for this is that sometimes there are no unbiased estimates of the parameter we are interested in estimating. Sometimes there may be an unbiased estimate, and we are unaware of it. On the other hand, some of our biased estimates may have very little bias (difference between mean estimate and true parameter value) when the sample sizes are large.

Another desirable property of a good estimator is that it have a small variance. Again, we conceive of a distribution of all possible values of the estimate, obtained from a very large number of samples of the same size drawn from a population. This distribution of estimates, centered around the true value, will have a variance, and we want it to be as small as possible. We will not usually know whether our statistic has smallest variance, but we will usually have the property that as the sample sizes get large the variance of the estimator will tend to zero. If the bias tends to zero at the same time we shall have what is called a consistent estimator. Unbiasedness and minimum variance are two properties related to the more fundamental but vaguer notion that a good estimate has a distribution tightly clustered about the true value. For a searching discussion of criteria for estimates, see Savage [13].

Not for all the estimators we describe will it be possible to compute numerically the variance of the estimate. Whenever we can we shall do so. Though the variance of an estimate can be approximated by making Monte Carlo computations as described in Section 9.5, this technique ordinarily involves a great deal of labor; we need to make many sets of Monte Carlo runs to approximate the variance of the estimate.

9.8 MAXIMUM LIKELIHOOD ESTIMATORS

The principle of maximum likelihood, developed by R. A. Fisher in the early 1920's, yields one of the most important estimation procedures known to mathematical statistics. We shall use this method in some of the following chapters, and so we give a brief and elementary exposition here. The reader is referred to several standard texts for more complete discussions of maximum likelihood [6, 7, 8].

The basic idea is simple enough. We take, as the estimate of a parameter, that value which gives the greatest possible likelihood of obtaining the data actually observed. Moreover, the computational procedure is in principle straightforward. We write down the likelihood function P, which is the probability of obtaining the observed data, in terms of the parameter to be estimated. We then find the value of the parameter that makes P as large as possible. Consider a simple example. We have a coin with an unknown probability p of coming up heads. Suppose we flip the coin 10 times and obtain 7 heads and 3 tails. Common sense would tell us that a good estimate of p is 0.7 in this case, but let us see what the maximum likelihood estimate is. The probability P of getting precisely 7 heads and 3 tails in a particular order is

$$(9.26) \qquad P = p^7(1 - p)^3.$$

We now want to choose p so that P is a maximum. The standard procedure is to differentiate P with respect to p, set this derivative equal to zero, and solve for p:

$$(9.27) \qquad \frac{dP}{dp} = 7p^6(1 - p)^3 - 3p^7(1 - p)^2$$
$$= p^6(1 - p)^2\,[7(1 - p) - 3p] = 0.$$

The solution which makes P a maximum (rather than zero, the minimum) is $p = 0.7$, as we intuited! This is of course the value of p that makes the quantity in brackets vanish. Hence the procedure yields the obvious estimate. However, maximum likelihood estimates often do not turn out to be the intuitively obvious estimate, nor are they always, or even often, unbiased in small samples. They do have one property that endears them to us though; if there is an estimate which has in large samples the smallest variance, then the maximum likelihood procedure will usually find it.

As another example of the maximum likelihood procedure we choose a simple case of our learning model. Suppose that we have the operators

$$Q_1 p = \alpha p + (1 - \alpha)$$

(9.28)

$$Q_2 p = \alpha p.$$

Assume that we know that the initial probability p_0 has the value 0.2, and we wish to estimate the single parameter α. First suppose we observe the sequence $A_1 A_2$. The likelihood function is then

(9.29) $$P = p_0(1 - Q_1 p_0) = p_0 \alpha (1 - p_0).$$

It is obvious that P is a maximum when $\alpha = 1$ (the largest allowed value of α), and so this is the maximum likelihood estimate of α. Next, suppose we observed the sequence $A_1 A_1$. The likelihood function is then

(9.30) $$P = p_0 Q_1 p_0 = p_0[1 - \alpha(1 - p_0)],$$

and it is clearly largest when $\alpha = 0$. Therefore, this is the maximum likelihood estimate of α. Finally, consider the sequence $A_1 A_2 A_1$, which gives the likelihood function

$$P = p_0(1 - Q_1 p_0) Q_2 Q_1 p_0$$

(9.31)

$$= p_0 \alpha (1 - p_0) \alpha [1 - \alpha(1 - p_0)].$$

We now take the derivative:

(9.32) $$\frac{dP}{d\alpha} = p_0(1 - p_0)[2\alpha - 3\alpha^2(1 - p_0)] = 0.$$

The appropriate solution is

(9.33) $$\alpha = \frac{2}{3(1 - p_0)}.$$

Since we assumed p_0 was known to be 0.2 we find that $\alpha = 0.833$. We may note that for $p_0 > 1/3$, this solution gives a value of α greater than unity. In such a case we take $\alpha = 1$ to be the maximum likelihood estimate since α can never be greater than unity. In general, we must be cautious about setting the derivative of the likelihood function equal to zero, as the function may not have an analytic maximum (maximum with zero slope) in the allowed range of the parameter being estimated.

The method of maximum likelihood has very wide applicability. It usually leads to a solution, but unfortunately computational difficulties are often tremendous. This seems to be true when we try to use the method for obtaining simultaneous estimates of all our parameters in the general case. The likelihood function itself is embarrassingly lengthy, when we have, say, twenty-five trials for ten subjects. And then this function would have to be maximized with respect to the five variables λ_1, α_1, λ_2, α_2, and p_0, simultaneously. The procedure seems completely unfeasible except for high-speed machine computing of particular examples. But a program is available if we care to expend the necessary labor. In the following chapters we use the maximum likelihood method in a more restricted way. We use only a portion of the data and usually estimate only one parameter at a time, sometimes two or three. The results are quite simple in some cases. We give up some information to obtain this simplicity.

When we obtain a maximum likelihood estimate it is sometimes easy to compute the asymptotic variance of the estimate. It is well known from mathematical statistics that the asymptotic variance σ^2 is the negative reciprocal of the expected value of the quantity

$$(9.34) \qquad \frac{d^2}{d\theta^2} \log P,$$

where P is the likelihood function and θ is the parameter being estimated, that is,

$$(9.35) \qquad \sigma^2 = \frac{1}{-E\left(\dfrac{d^2 \log P}{d\theta^2}\right)}.$$

The second derivative ordinarily involves both the true parameter value and the observations. When we take the expected value we often are taking expected values of simple functions of the observations, and so it is sometimes easy to get the variance. For example, in the binomial case, if we observe x successes in n trials, and the probability of a success on a single trial is p, we have

$$P = p^x(1-p)^{n-x}$$

$$(9.36) \qquad \log P = x \log p + (n-x) \log (1-p)$$

$$\frac{d \log P}{dp} = \frac{x}{p} - \frac{n-x}{1-p}.$$

When this derivative is set equal to zero and solved for the maximum likelihood estimate, we get $\hat{p} = x/n$ as expected. Taking the second derivative,

$$(9.37) \qquad \frac{d^2 \log P}{dp^2} = -\frac{x}{p^2} - \frac{n-x}{(1-p)^2}.$$

But the expected value of x is np, and the expected value of $n - x$ is $n(1 - p)$, so

$$(9.38) \qquad -E\left(\frac{d^2 \log P}{dp^2}\right) = +\frac{n}{p} + \frac{n}{1-p} = \frac{n}{p(1-p)}.$$

The reciprocal of this is the well-known result

$$(9.39) \qquad \sigma^2(\hat{p}) = \frac{p(1-p)}{n}.$$

In this instance the variance and the asymptotic variance are identical.

We shall neither prove this theorem about the variance of maximum likelihood estimates nor state the conditions under which it holds, but we will make use of it in the following chapters. The description just given for estimating single parameters suggests that we would be in a position to get a good idea of the variance of our estimates when we use maximum likelihood and large samples. But we are cheating a little, and possibly a lot. If there are several parameters to estimate simultaneously, maximum likelihood methods can give a good idea of the variance and covariance of the estimates. Ordinarily, we estimate parameters singly, and in each case as if the values of all other parameters were known. Consequently our variances ought really to be adjusted for the fact that we do not know the values of all parameters except the one in question. Finally, of course, we will not really know the variances, but only have estimates of them on the basis of the estimated parameters. There are, therefore, a number of loose ends lying around.

9.9 A SPECIAL MAXIMUM LIKELIHOOD PROBLEM

We now develop some maximum likelihood equations which we use in the following chapters to estimate parameters from experimental data. The rest of this chapter is technical and specific to estimation of parameters in special cases of interest (see also [9]). The reader may care to skip forward to a treatment of an experiment, and return to this material when he needs it. This development is not applicable to all the problems we take up later, but it is used in Chapters 10 and 11. The procedure to be described is appropriate whenever we have a set of probabilities q_r related by the equation

$$(9.40) \qquad q_r = \alpha^r q_0.$$

The data provide information about the q_ν's, and the problem is to estimate α or q_0 or both.

Equations of this type occur, for example, when we apply an operator Q_2 of the form

$$(9.41) \qquad Q_2 p = \alpha_2 p + (1 - \alpha_2),$$

on every trial for several trials. Applying this operator is equivalent to applying an operator $\tilde{Q}_2$ to $q = 1 - p$:

$$(9.42) \qquad \tilde{Q}_2 q = \alpha_2 q.$$

As we have seen, when we apply such an operator on every trial we have for trial n, the probability

$$(9.43) \qquad q_n = \tilde{Q}_2{}^n q_0 = \alpha_2{}^n q_0.$$

In this application, ν in equation 9.40 is the trial number n. In other applications, however, ν is not the trial number but denotes the number of occurrences of one response. Specifically, if one operator is an identity operator and the other is of the form $Qp = \alpha p + (1 - \alpha)$ or $Qp = \alpha p$, then again we get equations of the type 9.40, but ν stands for number of occurrences of the response which increases or decreases p. In Chapter 10 we use the operators

$$(9.44) \qquad \begin{aligned} Q_1 p &= \alpha_1 p + (1 - \alpha_1) \\ Q_2 p &= p, \end{aligned}$$

or the equivalent operators

$$(9.45) \qquad \begin{aligned} \tilde{Q}_1 q &= \alpha_1 q \\ \tilde{Q}_2 q &= q. \end{aligned}$$

The probability q_ν of response A_2 after ν occurrences of response A_1 is

$$(9.46) \qquad q_\nu = \tilde{Q}_1{}^\nu q_0 = \alpha_1{}^\nu q_0.$$

The estimation problem is essentially the same, whether ν stands for trial number in the first example given above, or for number of occurrences of A_1 in the second example. For this reason we discuss this estimation problem here rather than in connection with the particular applications later.

The problem in estimation may now be summarized as follows. The probability of some alternative A_j is q_ν for a specified value of the index ν, and we have the relation

$$(9.47) \qquad q_\nu = \alpha^\nu q_0,$$

where we assume $0 < \alpha < 1$. For each value of ν, we have a number of observations concerning whether or not alternative A_j occurred. These observations may be on a single subject or from a number of subjects. We let the number of observations for a specified value of ν be N_ν, and we use the index μ to denote these N_ν observations, that is, $\mu = 1, 2, \cdots,$ N_ν. The μth observation for a particular value of ν is simply whether A_j occurred or not. We represent the data by a set of *random variables* $x_{\mu\nu}$. We let $x_{\mu\nu} = 0$ if A_j occurred and $x_{\mu\nu} = 1$ if it did not. The data are thereby reduced to a set of 0's and 1's. These could just as well have been checks and pluses or "yeses" and "noes," but the 0's and 1's have a distinct advantage, as we shall see. We can define a quantity x_ν by

$$(9.48) \qquad x_\nu = \sum_{\mu=1}^{N_\nu} x_{\mu\nu}.$$

This sum is simply the number of times A_j did not occur on N_ν observations, for we enter a zero in the sum when A_j occurs and the number one when A_j does not occur. The number of occurrences of A_j during the N_ν observations is of course $N_\nu - x_\nu$. The data give the values of all the $x_{\mu\nu}$'s, and from these we want to obtain the maximum likelihood estimates of α and q_0.

When we say that $x_{\mu\nu} = 0$ we are saying that A_j occurred on the μth observation for a specified value of ν. The probability that A_j occurred is q_ν, and so q_ν is the probability that $x_{\mu\nu} = 0$ and $1 - q_\nu$ is the probability that $x_{\mu\nu} = 1$. Therefore the expected value of $x_{\mu\nu}$ is (for fixed N_ν)

$$(9.49) \qquad E(x_{\mu\nu}) = 1 - q_\nu,$$

$$(9.50) \qquad E(x_\nu) = \sum_{\mu=1}^{N_\nu} E(x_{\mu\nu}) = N_\nu(1 - q_\nu).$$

Therefore, an unbiased estimate of $1 - q_\nu$ is the ratio x_ν/N_ν. Thus, an unbiased estimate of q_ν is

$$(9.51) \qquad \hat{q}_\nu = 1 - \frac{x_\nu}{N_\nu} = \frac{N_\nu - x_\nu}{N_\nu}.$$

This is the obvious estimate of q_ν, the proportion of occurrences of A_j in the N_ν observations. From our data we can thereby obtain estimates of all the q_ν. We could then combine these estimates in some way, for the various values of ν, and get estimates of α and q_0 through equation 9.47. But we wish to obtain the maximum likelihood estimates of α and q_0, and so we consider all values of ν simultaneously and avoid the necessity of estimating each of the q_ν and then combining them.

We want to write down the probability or likelihood of obtaining the entire set of $x_{\mu\nu}$'s. We begin by writing down the likelihood $P_{\mu\nu}$ of obtaining these numbers for single values of ν and μ. From the last paragraph we see that

$$(9.52) \qquad P_{\mu\nu} = \begin{cases} 1 - q_\nu, & \text{if } x_{\mu\nu} = 1 \\ q_\nu, & \text{if } x_{\mu\nu} = 0. \end{cases}$$

A convenient compact way of writing this is

$$(9.53) \qquad P_{\mu\nu} = (1 - q_\nu)^{x_{\mu\nu}}(q_\nu)^{1 - x_{\mu\nu}}.$$

This is equivalent to the preceding statement (equation 9.52) since, when $x_{\mu\nu} = 1$, the first factor is $1 - q_\nu$ and the second factor is unity; and, when $x_{\mu\nu} = 0$, the first factor is unity and the second factor is q_ν. Next we write down the likelihood P_ν of obtaining the set of $x_{\mu\nu}$ for $\mu = 1, 2, \cdots, N_\nu$ but for a single value of ν. It is just the product

$$(9.54) \qquad \begin{aligned} P_\nu &= \prod_{\mu=1}^{N_\nu} P_{\mu\nu} \\ &= \prod_{\mu=1}^{N_\nu} (1 - q_\nu)^{x_{\mu\nu}}(q_\nu)^{1 - x_{\mu\nu}}. \end{aligned}$$

This product is simply $(1 - q_\nu)$ to some power times q_ν to some other power. But in the whole product the exponent of $(1 - q_\nu)$ is just x_ν of equation 9.48, which is the number of non-occurrences of A_j. Similarly, the exponent of q_ν in the product is $N_\nu - x_\nu$, the number of occurrences of A_j. Hence,

$$(9.55) \qquad P_\nu = (1 - q_\nu)^{x_\nu}(q_\nu)^{N_\nu - x_\nu}.$$

Finally, we want the likelihood P of obtaining the whole set of data given by all the $x_{\mu\nu}$'s. We let ν range from 0 to some number Ω. The likelihood P is then the product

$$(9.56) \qquad \begin{aligned} P &= \prod_{\nu=0}^{\Omega} P_\nu \\ &= \prod_{\nu=0}^{\Omega} \{(1 - q_\nu)^{x_\nu}(q_\nu)^{N_\nu - x_\nu}\}. \end{aligned}$$

We now insert the expression for q_ν given by equation 9.47 and get

$$(9.57) \qquad P = \prod_{\nu=0}^{\Omega} \{(1 - \alpha^\nu q_0)^{x_\nu}(\alpha^\nu q_0)^{N_\nu - x_\nu}\}.$$

This is the likelihood expression we need. It gives the likelihood of getting the entire set of data in terms of the numbers x_ν and N_ν, obtained from the data, and the parameters α and q_0 to be estimated.

For computational purposes it is easier to work with the logarithm of P rather than with P itself. We want to maximize P with respect to α and q_0, that is, solve for the values of those parameters which make P as large as possible, but we can just as well maximize $\log P$. When P is at its maximum, $\log P$ will be at its maximum, and vice versa. From the last equation we then get

$$(9.58) \qquad \log P = \sum_{\nu=0}^{\Omega} \{x_\nu \log (1 - \alpha^\nu q_0) + (N_\nu - x_\nu) \log (\alpha^\nu q_0)\}.$$

We now use the standard procedure for maximizing this expression. First we take the partial derivative with respect to α and get

$$(9.59) \qquad \frac{\partial}{\partial \alpha} \log P = \sum_{\nu=0}^{\Omega} \left\{ x_\nu \frac{-\nu \alpha^{\nu-1} q_0}{1 - \alpha^\nu q_0} + (N_\nu - x_\nu) \frac{\nu}{\alpha} \right\}$$

$$= \sum_{\nu=0}^{\Omega} \frac{\nu}{\alpha} \left\{ (N_\nu - x_\nu) - x_\nu \frac{\alpha^\nu q_0}{1 - \alpha^\nu q_0} \right\}.$$

If we know the value of q_0, we need only set this derivative equal to zero and solve for α. When we do this, we replace α with $\hat{\alpha}$, which is the maximum likelihood estimate of α. We cancel out the common factor $1/\alpha$ and get

$$(9.60) \qquad \sum_{\nu=0}^{\Omega} \nu(N_\nu - x_\nu) = \sum_{\nu=0}^{\Omega} x_\nu \left\{ \frac{\nu \hat{\alpha}^\nu q_0}{1 - \hat{\alpha}^\nu q_0} \right\}.$$

This equation must be solved for $\hat{\alpha}$, but unfortunately this is not easy. In the following section we propose some procedures for doing so.

When we wish to estimate both α and q_0 from the data, the problem is a little more involved. We must return to the expression for $\log P$, equation 9.58 above, and take the partial derivative with respect to q_0:

$$(9.61) \qquad \frac{\partial}{\partial q_0} \log P = \sum_{\nu=0}^{\Omega} \left\{ x_\nu \frac{-\alpha^\nu}{1 - \alpha^\nu q_0} + (N_\nu - x_\nu) \frac{1}{q_0} \right\}$$

$$= \sum_{\nu=0}^{\Omega} \frac{1}{q_0} \left\{ (N_\nu - x_\nu) - x_\nu \frac{\alpha^\nu q_0}{1 - \alpha^\nu q_0} \right\}.$$

We then set equal to zero both of the partial derivatives of $\log P$. In so doing we replace α by its estimate $\hat{\alpha}$ and q_0 by its estimate $\hat{q}_0$ and get the pair of equations

$$\sum_{\nu=0}^{\Omega} \nu(N_\nu - x_\nu) = \sum_{\nu=0}^{\Omega} x_\nu \left\{ \frac{\nu \hat{\alpha}^\nu \hat{q}_0}{1 - \hat{\alpha}^\nu \hat{q}_0} \right\},$$

$$(9.62)$$

$$\sum_{\nu=0}^{\Omega} (N_\nu - x_\nu) = \sum_{\nu=0}^{\Omega} x_\nu \left\{ \frac{\hat{\alpha}^\nu \hat{q}_0}{1 - \hat{\alpha}^\nu \hat{q}_0} \right\}.$$

These two equations must be solved for the maximum likelihood estimates $\hat{\alpha}$ and $\hat{q}_0$. This creates some rather serious computational problems; we discuss them in Section 9.11.

9.10 PROCEDURES FOR COMPUTING $\hat{\alpha}$

We now describe some procedures for computing the maximum likelihood estimate $\hat{\alpha}$ when the value of q_0 is known. This involves solving equation 9.60 of the last section. The left-hand side causes no trouble for it is determined completely by the data; we call it D_1:

$$(9.63) \qquad D_1 = \sum_{\nu=0}^{\Omega} \nu(N_\nu - x_\nu).$$

From a particular set of data we can tabulate N_ν and x_ν for each value of ν and readily compute the value of the sum D_1. But the right-hand side of equation 9.60 involves the x_ν obtained from the data in addition to the known parameter q_0 and the unknown estimate $\hat{\alpha}$. The problem is to solve for $\hat{\alpha}$. We have found three procedures useful under different conditions.

The first procedure is appropriate for the special case when $q_0 = 1$. We then have from equations 9.63 and 9.60

$$(9.64) \qquad D_1 = \sum_{\nu=0}^{\Omega} x_\nu \frac{\nu \hat{\alpha}^\nu}{1 - \hat{\alpha}^\nu}.$$

Introducing the abbreviation

$$(9.65) \qquad g_\nu(\hat{\alpha}) = \frac{\nu \hat{\alpha}^\nu}{1 - \hat{\alpha}^\nu},$$

we can write

$$(9.66) \qquad D_1 = \sum_{\nu=0}^{\Omega} x_\nu g_\nu(\hat{\alpha}).$$

In Table C we give values of the function $g_\nu(\alpha)$ for a range of values of ν and α. These tables, along with the x_ν from a particular set of data, permit us to compute the sum on the right side of the last equation relatively easily for any value of $\hat{\alpha}$. This must be done for various values of $\hat{\alpha}$ until we get as close as possible to the correct value, D_1. This is a trial-and-error procedure, but Table C facilitates the method considerably. We illustrate this technique in Chapter 11.

The second procedure for obtaining $\hat{\alpha}$ from equation 9.60 is useful in some kinds of data for which the x_ν are equal for all values of ν within the range from zero to Ω. In other words, we can sometimes choose Ω so that $x_\nu = x$ for $\nu = 0, 1, 2, \cdots, \Omega$. Under this condition we can factor out $x_\nu = x$ and get from equations 9.60 and 9.63

(9.67)
$$\frac{D_1}{x} = \sum_{\nu=0}^{\Omega} \frac{\nu \hat{\alpha}^\nu q_0}{1 - \hat{\alpha}^\nu q_0}.$$

We denote the function on the right by $G(\hat{\alpha}, q_0, \Omega)$; it is an example of the function

(9.68)
$$G(\alpha, \beta, \Omega) = \sum_{\nu=0}^{\Omega} \frac{\nu \alpha^\nu \beta}{1 - \alpha^\nu \beta},$$

which we present in Table D. We may then write

(9.69)
$$G(\hat{\alpha}, q_0, \Omega) = D_1/x.$$

From a set of data, we can compute D_1/x and then use Table D to obtain the nearest value of $\hat{\alpha}$ for the known values of q_0 and Ω. In Chapter 10 we illustrate the procedure.

The third procedure for determining the estimate $\hat{\alpha}$ is an approximation for the case of $q_0 = 1$ which is especially useful when $\hat{\alpha}$ is near unity. We expand $\hat{\alpha}^\nu$ in a power series around $\hat{\alpha} = 1$:

(9.70)
$$\hat{\alpha}^\nu = [1 - (1 - \hat{\alpha})]^\nu = 1 - \nu(1 - \hat{\alpha}) + \cdots.$$

The annoying factor in equations 9.60 and 9.64 then becomes

(9.71)
$$\frac{\hat{\alpha}^\nu}{1 - \hat{\alpha}^\nu} = \frac{1 - \nu(1 - \hat{\alpha}) + \cdots}{\nu(1 - \hat{\alpha}) - \cdots}.$$

We drop terms beyond the linear one shown and have then, from equation 9.64

(9.72)
$$D_1 \cong \sum_{\nu=0}^{\Omega} x_\nu \frac{1 - \nu(1 - \hat{\alpha})}{1 - \hat{\alpha}}$$

$$= \frac{\sum\limits_{\nu=0}^{\Omega} x_\nu}{1 - \hat{\alpha}} - \sum_{\nu=0}^{\Omega} \nu x_\nu.$$

But D_1 is defined by equation 9.63, that is,

(9.73)
$$D_1 = \sum_{\nu=0}^{\Omega} \nu N_\nu - \sum_{\nu=0}^{\Omega} \nu x_\nu.$$

Combining the last two equations gives

(9.74)
$$\hat{\alpha} \cong 1 - \frac{\sum\limits_{\nu=0}^{\Omega} x_\nu}{\sum\limits_{\nu=0}^{\Omega} \nu N_\nu}.$$

This approximate formula may be used to estimate α directly from the data without the use of tables. Even when $\hat{\alpha}$ is not very near unity, the foregoing formula is useful in obtaining a preliminary value of $\hat{\alpha}$. Having such a preliminary value shortens the exact computations described above.

9.11 PROCEDURE FOR COMPUTING $\hat{\alpha}$ AND $\hat{q}_0$

When we wish to obtain the simultaneous maximum likelihood estimates of α and q_0 for the problem discussed in Section 9.9, we must solve a pair of equations (9.62). This ordinarily involves a great deal of computational labor. We provide a short-cut procedure for only one special case—when the x_ν are independent of ν over the range of values of ν, that is, when $x_\nu = x$ for $\nu = 0, 1, \cdots, \Omega$.

From the data we obtain two statistics D_1 and D_2 defined by

(9.75)
$$D_1 = \sum_{\nu=0}^{\Omega} \nu(N_\nu - x_\nu),$$

$$D_2 = \sum_{\nu=0}^{\Omega} (N_\nu - x_\nu).$$

In Table D we give the functions

(9.76)
$$G(\alpha, \beta, \Omega) = \sum_{\nu=0}^{\Omega} \frac{\nu \alpha^\nu \beta}{1 - \alpha^\nu \beta},$$

$$F(\alpha, \beta, \Omega) = \sum_{\nu=0}^{\Omega} \frac{\alpha^\nu \beta}{1 - \alpha^\nu \beta}.$$

From equations 9.62 we then have

(9.77)
$$G(\hat{\alpha}, \hat{q}_0, \Omega) = D_1/x,$$

$$F(\hat{\alpha}, \hat{q}_0, \Omega) = D_2/x.$$

Having computed x, D_1, and D_2 from the data, we may use Table D to find the nearest pair of values of the functions F and G and thereby obtain the estimates $\hat{\alpha}$ and $\hat{q}_0$. Because interpolation is usually necessary we illustrate the procedure in Chapter 10.

9.12 VARIANCE OF THE ESTIMATE $\hat{\alpha}$

We now return to the case described in Section 9.10 for which q_0 is known and we compute the estimate $\hat{\alpha}$. As indicated in Section 9.7, it is desirable to have an estimator which has a small variance, and so we inquire about the variance of our likelihood estimate $\hat{\alpha}$.

At the end of Section 9.8 we stated a well-known theorem about the asymptotic variance, $\sigma^2(\theta)$, of an estimate θ of some parameter θ. In our problem the parameter is α and the estimate is $\hat{\alpha}$. Thus we need to compute the second derivative of log P of equation 9.58 with respect to α. From the first derivative given by equation 9.59 we get

(9.78)
$$\frac{\partial^2}{\partial\alpha^2}\log P = \sum_{\nu=0}^{\Omega}\left\{-\frac{\nu}{\alpha^2}\left[(N_\nu - x_\nu) - x_\nu\frac{\alpha^\nu q_0}{1-\alpha^\nu q_0}\right]\right.$$
$$\left. -\frac{\nu}{\alpha}x_\nu\left[\frac{\nu\alpha^{\nu-1}q_0}{1-\alpha^\nu q_0} + \frac{(\alpha^\nu q_0)(\nu\alpha^{\nu-1}q_0)}{(1-\alpha^\nu q_0)^2}\right]\right\},$$

or

(9.79)
$$\frac{\partial^2}{\partial\alpha^2}\log P = -\sum_{\nu=0}^{\Omega}\frac{\nu}{\alpha^2}\left\{(N_\nu - x_\nu) - x_\nu\frac{\alpha^\nu q_0}{1-\alpha^\nu q_0}\right.$$
$$\left. + \nu x_\nu\frac{\alpha^\nu q_0}{(1-\alpha^\nu q_0)^2}\right\}.$$

We now need the expected value of this second derivative. We see from equation 9.50 that

(9.80) $$E(x_\nu) = N_\nu(1 - q_\nu) = N_\nu(1 - \alpha^\nu q_0).$$

Thus we get

(9.81)
$$-E\left(\frac{\partial^2}{\partial\alpha^2}\log P\right) = \sum_{\nu=0}^{\Omega}\frac{\nu}{\alpha^2}N_\nu\left\{1 - (1-\alpha^\nu q_0) - \alpha^\nu q_0 + \frac{\nu\alpha^\nu q_0}{1-\alpha^\nu q_0}\right\}$$
$$= \sum_{\nu=0}^{\Omega}\frac{\nu^2}{\alpha^2}N_\nu\frac{\alpha^\nu q_0}{1-\alpha^\nu q_0}.$$

From the theorem then we have, for the asymptotic variance of $\hat{\alpha}$,

(9.82) $$\sigma^2(\hat{\alpha}) = \frac{\alpha^2}{\displaystyle\sum_{\nu=0}^{\Omega}\nu^2 N_\nu\frac{\alpha^\nu q_0}{1-\alpha^\nu q_0}}.$$

We may estimate $\sigma^2(\hat{\alpha})$ by replacing α with its estimate $\hat{\alpha}$ in the right side of this equation. That N_ν and Ω random variables has been neglected.

The expression just obtained for the asymptotic variance of $\hat{\alpha}$ leads to a considerable amount of computational labor, but when $q_0 = 1$ we may use the approximation introduced in Section 9.10. Using equation 9.71 in the sum, we have

$$
\sum_{\nu=0}^{\Omega} \nu^2 N_\nu \frac{\alpha^\nu}{1 - \alpha^\nu} \cong \sum_{\nu=0}^{\Omega} \nu^2 N_\nu \frac{1 - \nu(1 - \alpha)}{\nu(1 - \alpha)}
$$

(9.83)

$$
= \frac{\sum_{\nu=0}^{\Omega} \nu N_\nu}{1 - \alpha} - \sum_{\nu=0}^{\Omega} \nu^2 N_\nu.
$$

The asymptotic variance is then given approximately by

$$
(9.84) \qquad \sigma^2(\hat{\alpha}) \cong \frac{\alpha^2(1 - \alpha)}{\sum\limits_{\nu=0}^{\Omega} \nu N_\nu - (1 - \alpha) \sum\limits_{\nu=0}^{\Omega} \nu^2 N_\nu} \qquad (q_0 = 1).
$$

This approximate formula is much easier to use than equation 9.82, as we see in Chapter 11.

9.13   GOODNESS-OF-FIT CONSIDERATIONS

In applying the mathematical system to a particular experimental problem, that is, in constructing a model for that problem, there are three major considerations: (1) identifications, (2) estimation of parameters, and (3) goodness-of-fit. We have already discussed the first two considerations, and in this section we comment on the question of goodness-of-fit. How well does the model account for the data?

Several statistical techniques are available for testing for goodness-of-fit, but most of these are not appropriate for the analyses in the following chapters. The major reason is that our model implies a distribution of probabilities on each trial. Consider first the most common criterion, the Pearson chi-square (χ^2) test. If we have a model that predicts that in N observations we have Np successes and $N(1 - p)$ failures, and if we observe x successes and $N - x$ failures, the Pearson test statistic is (without Yates' continuity correction)

$$
(9.85) \qquad \chi^2 = \frac{[x - Np]^2}{Np} + \frac{[(N - x) - N(1 - p)]^2}{N(1 - p)} = \frac{(x - Np)^2}{Np(1 - p)}.
$$

If this statistic is larger than some critical value, found in a χ^2 table, the fit is considered unsatisfactory, that is, the hypothesis that the true probability of a success is p is rejected. Now suppose that the model predicted that the sample of N observations was stratified in such a way that N_1 had a probability p_1 of a success and N_2 had a probability p_2 of a success, where $N_1 + N_2 = N$. Furthermore, suppose that we did not know which observations were associated with p_1 and which with p_2 so that we knew only the total number of observed successes, x. We might be tempted to replace p in the foregoing equation for χ^2 with the mean probability

$$(9.86) \qquad \bar{p} = \frac{N_1 p_1 + N_2 p_2}{N_1 + N_2} = \frac{N_1 p_1 + N_2 p_2}{N}.$$

But this would be wrong as we now show. Suppose $p_1 = 0$ and $p_2 = 1$; then the model predicts precisely N_1 successes, that is, $x = N_1$. If $x \neq N_1$, the model is wrong, and nothing further need be said, for the probability of observing anything but N_1 successes is zero, assuming that the model is correct. In other words, the distribution of the number of observed successes is discrete with unity density at $x = N_1$ and zero density elsewhere. The distribution of the quantity on the right side of equation 9.85 is not the χ^2 distribution, and so the Pearson criterion is not appropriate for this problem.

Now consider a less extreme case than the one just discussed. Suppose that p_1 were small and p_2 were large. The variance of the observed number of successes is small compared to $N\bar{p}(1 - \bar{p})$. The variance of the observed number of successes is, in fact,

$$(9.87) \quad \sigma^2(x) = N_1 p_1 (1 - p_1) + N_2 p_2 (1 - p_2) = N\bar{p} - (N_1 p_1{}^2 + N_2 p_2{}^2).$$

The variance of the p values is

$$(9.88) \qquad \sigma^2(p) = \frac{N_1 p_1{}^2 + N_2 p_2{}^2}{N} - \bar{p}^2.$$

Therefore

$$
\begin{aligned}
\sigma^2(x) &= N\bar{p} - [N\sigma^2(p) + N\bar{p}^2] \\
&= N\bar{p}(1 - \bar{p}) - N\sigma^2(p).
\end{aligned}
$$
$$(9.89)$$

We see that when $p_1 = 0$ and $p_2 = 1$, then $\sigma^2(x) = 0$, whereas when $p_1 = p_2 = \bar{p}$, $\sigma^2(p) = 0$, and so $\sigma^2(x) = N\bar{p}(1 - \bar{p})$. Now the χ^2 test for goodness-of-fit, given by equation 9.85, has the variance of the observed number of successes in the denominator, and as we have just seen this variance is not $N\bar{p}(1 - \bar{p})$ except when $\sigma^2(p) = 0$.

In Chapters 10 and 12 we find exceptions to the foregoing argument about the inappropriateness of a χ^2 test for goodness-of-fit. In Chapter 10, there will be a large collection of observations for which the same p value is appropriate, whereas in Chapter 12, all subjects have the same p value on a given trial. Under these conditions the Pearson criterion can be applied.

A possible test statistic is suggested by the discussion above. Our model predicts that on trial n there will be $NV_{1,n}$ responses of type A_1 and that the second raw moment of the p-value distribution is $V_{2,n}$.

Analogous to equation 9.87, the expression for the variance of the observed number x_n of A_1 responses is

(9.90)
$$\sigma^2(x_n) = N(V_{1,n} - V_{2,n}).$$

The proposed test statistic is then

(9.91)
$$u = \frac{(x_n - NV_{1,n})^2}{N(V_{1,n} - V_{2,n})}.$$

If the data for all trials were used, we suggest the test statistic

(9.92)
$$U = \sum_n \frac{(x_n - NV_{1,n})^2}{N(V_{1,n} - V_{2,n})}.$$

Inventing reasonable test statistics is easy, but the problem is to determine the distribution of such a statistic. It seems plausible that the statistic U has approximately the χ^2 distribution. The reasoning is that if x is a normally distributed variable with mean m and variance σ^2, then $(x - m)^2/\sigma^2$ has the χ^2 distribution. For large samples it is reasonable that x_n would be approximately normally distributed and that single terms of U would therefore be approximately distributed according to χ^2 with one degree of freedom. The sum of independent χ^2 values has a χ^2 distribution with degrees of freedom equal to the sum of the component degrees of freedom. We would have to pay back some degrees of freedom if the population values were estimated. But the crucial point concerns the independence of the separate terms. If on each trial we had a brand new set of organisms there would be no difficulty, but this is preposterous from a practical point of view. At best then the use of the U statistic is only suggestive, but it does provide an index for comparing the goodness-of-fit obtained by two or more methods.

A goodness-of-fit test that we will use is the run test [10, 11, 12]. It is an extremely simple test to use. We note on each trial whether the observed proportion of A_1's is above (+) or below (−) the theoretical mean $V_{1,n}$. We observe a sequence of +'s and −'s such as the following:

$$+ + - + + + - - + - - - - + + +$$

We then count the number of +'s, n_1, the number of −'s, n_2, and the number of *runs*, d. In the example above, $n_1 = 9$, $n_2 = 7$, and $d = 7$. We then ask if the number of runs is too large or too small, granted that the model is correct. Swed and Eisenhart [10] have computed tables of the probability of obtaining runs of various lengths for n_1 and n_2 from 1 to 20. For larger values of n_1 and n_2 we can determine the expected number of runs from the formula

(9.93)
$$E(d) = \frac{2n_1 n_2}{n_1 + n_2} + 1,$$

and the variance from

$$(9.94) \qquad \sigma_d^2 = \frac{2n_1 n_2 (2n_1 n_2 - n_1 - n_2)}{(n_1 + n_2)^2 (n_1 + n_2 - 1)}.$$

We then compute a normal deviate

$$(9.95) \qquad z = \frac{d - E(d)}{\sigma_d},$$

and consult a normal table to test for significance [12].

We have been discussing the problem of testing for goodness-of-fit in a rather technical way without making it very clear just what we expect of our general model. As we have said elsewhere in this book, our goal is to describe data *adequately*. This notion becomes more or less precise only when we specify what we mean by adequate. Another way of saying the same thing is that we want to account for most of the variability in learning data with our general model. But then what is "most" of the variability? Is 95 percent necessary? Answers to these questions are deeply rooted in a person's basic philosophy of scientific method. What degree of perfection is required, and for what purpose? Suppose that a model accounted for only 50 percent of the variability in a set of data. Would we therefore reject the model as useless? Our answer is negative because later we might account for most of the remaining 50 percent of variability by considerations outside the scope of the model. For example, 40 percent of the variability might be a result of individual subject differences, in which event the model would have been excellent for identical subjects. Furthermore, the model may provide a framework —a base-line—for analyzing individual differences, and this alone would be useful. In other words, we do not insist upon a narrow acceptance region in model building. Models of the sort we are studying make hundreds of predictions about a single set of data. When the predictions for many properties are accurate, while inaccurate for others, we do not have a simple acceptance-rejection problem that falls into the pattern of classical tests of significance, that is, we do not make one test of significance at the 5 percent level and let the model stand or fall by this result. Science does not move this way. Rather, the goodness-of-fit tests give us information about the satisfactory and unsatisfactory aspects of the model.

One final point on the broader question of goodness-of-fit: We do not expect any general model such as ours to describe adequately *all* experiments in learning. Many experiments cannot easily be made to fit into our basic paradigm of mutually exclusive alternatives. Others may fit the pattern in principle but may lead to serious mathematical complications. But even some of the experiments which are easily handled by our general model may not be well described by it.   If this occurred only infrequently we would not feel that the general model was doomed and henceforth useless.   We would immediately ask why a particular set of data led to poor agreement, and this might lead to significant further experimentation.   Without the model we could not even have asked the question!   A model which does adequately describe a fairly wide range of data may be used as a device for determining conditions under which an extended model or a different model is needed.

9.14 SUMMARY

The three major problems in applying the mathematical system of Part I to the analysis of particular experiments are discussed in this chapter. First we consider questions in *identifying* elements of the mathematical system with observable aspects of the behavior and environment of an organism.   We discuss possible correspondence between our general model and two main psychological theories, and how model parameters may depend upon some important experimental variables.   Next we indicate the nature of the general problem of *estimating* parameters of the model from experimental data, and we present several procedures to be used in the following chapters.   Finally we discuss some methods for measuring the *goodness-of-fit* of a model to particular data.

REFERENCES

1. Postman, L.   The history and present status of the law of effect.   *Psychol. Bull.*, 1947, **44**, 489–563.
2. Hull, C. L.   *The principles of behavior.*   New York: Appleton-Century-Crofts, 1943.
3. Spence, K. W.   Theoretical interpretations of learning.   *Handbook of experimental psychology*, S. S. Stevens, ed., New York: Wiley, 1951.
4. Guthrie, E. R.   *The psychology of learning.*   New York: Harper, 1935.
5. Guthrie, E. R.   Association and the law of effect.   *Psychol. Rev.*, 1940, **47**, 127–148.
6. Kendall, M. G.   *The advanced theory of statistics, Vol. I.*   London: J. B. Lippincott Co., 1943, pp. 178–180.
7. Wilks, S. S.   *Mathematical statistics.*   Princeton: Princeton University Press, 1943, pp. 136–142.
8. Mood, A. M.   *Introduction to the theory of statistics.*   New York: McGraw-Hill, 1950, pp. 152–154.
9. Bush, R. R., and Mosteller, F.   A stochastic model with applications to learning. *Annals of math. Stat.*, 1953, **24**, 559–585.
10. Swed, F. S., and Eisenhart, C.   Tables for testing randomness of grouping in a sequence of alternatives.   *Annals of math. Stat.*, 1943, **14**, 66–87.
11. Hoel, P. G.   *Introduction to mathematical statistics.*   New York: Wiley, 1947, pp. 177–182.
12. Mood, A. M., *loc. cit.*, pp. 390–394.
13. Savage, L. J.   *The foundations of statistics.*   New York: Wiley, 1954, pp. 220–245.

CHAPTER 10

Free-Recall Verbal Learning

10.1 THE EXPERIMENTS

An important area of experimental psychology during the last several decades has been that of verbal learning [1]. Many experiments have involved memorizing a list of words or nonsense syllables, and psychologists have intensively studied serial effects, for example, the comparative ease of memorizing words at the ends and middle of a list. Because of this interest in serial effects, the order of the words in the list was maintained, from trial to trial, in most of the experiments.

These experiments on serial rote learning have been of great importance and have led to a number of useful concepts [2]. From the point of view of quantitative models for learning, however, the problem of serial learning or the chaining of responses is a complex one. A much simpler empirical problem results when the order of the words being memorized is randomized on each presentation of the list. In this way, the serial effects are eliminated or at least minimized, but the basic learning or memorization process remains. Such experiments have been conducted by Bruner, Miller, and Zimmerman [3]. The model developed by Miller and McGill [4] was tailored to analyze data of this kind. Furthermore, they demonstrated the correspondence between their model and the model given in this book. The reader should find Miller and McGill's more intensive discussion rewarding.

In this chapter we are concerned with this one type of free-recall verbal learning experiment. The experiments are conducted as follows. A list of N monosyllabic words is read aloud to a subject. The subject is then instructed to write down all the words that he can recall. The experimenter gives him no indication of how well he has performed. Then the order of the words is randomized, and the procedure is repeated. The experiment is continued in this way for many trials until the proportion of words recalled nearly reaches an asymptote. The experiments conducted by Bruner, Miller, and Zimmerman used lists of 4, 8, 16, 32, and

64 words, and the experiments were continued in several cases for as many as 32 trials. In this chapter we analyze data obtained from one such experiment.

10.2 IDENTIFICATIONS AND ASSUMPTIONS

A basic assumption made by Miller and McGill in analyzing data from the experiments just described is that the N words on the list are independent of one another. This means that the subject's ability to recall a particular word does not depend upon his past or present performance with respect to any of the other words. We number the words by $i = 1$, $2, \cdots, N$. These are merely labels; they do not represent the position of the word in the list as these positions change from trial to trial.

Consider the ith word. On each trial this word either is recalled (written down) or not recalled. Denote correct recall of a word by response A_1, and non-recall by response A_2. The probability that the ith word is recalled on trial n is $p_{i,n}$. In the data we observe a sequence of A_1's and A_2's (recalls and non-recalls) for this ith word. The probability $p_{i,n}$ depends upon this sequence up to trial n, and by assumption does not depend upon the sequences of A_1's and A_2's for other words. We have N sequences of A_1's and A_2's, one for each word on the list. Since the words are assumed to be independent, we can think of these N sequences of responses as generated by N different subjects.

We do assume, however, that all words have the same initial probability of recall, p_0, and that all words have the same learning parameters λ_j and α_j. This is analogous to assuming that we have a group of N "identical" subjects. We are willing to make such a drastic assumption because the experimenters have gone to considerable trouble to arrange for the words to have approximate equivalence. All words were English and monosyllabic, the words were read aloud at a uniform rate, they had been equated for their "articulation value," and the order was scrambled between each presentation. This does not mean, of course, that the recall probabilities, $p_{i,n}$, are the same for all words on trial n for $n > 0$. On the other hand, it does imply that if the sequence of A_1's and A_2's for two words happen to be identical up to trial n, then the recall probabilities for these two words on trial n will be equal. Roughly speaking, this assumption means that the words are equally difficult and that position on the list is irrelevant. These assumptions are not obviously correct by any means, but they represent idealizations and simplifications which permit a relatively simple analysis of the data to be made. The test, of course, is how well the model reproduces or simulates the data.

The next basic assumption made by Miller and McGill is that a non-recall of the ith word does not change its probability of being recalled on the next trial. Thus the operator Q_2, which is applied when A_2 occurs, is the identity operator, that is, $\alpha_2 = 1$, and so

$$(10.1) \qquad p_{i,n+1} = Q_2 p_{i,n} = p_{i,n}.$$

In the next section we examine some data to see if assumption 10.1 is reasonable. When the word is recalled, that is, when A_1 occurs, we apply an operator Q_1 of the general form

$$(10.2) \qquad p_{i,n+1} = Q_1 p_{i,n} = \alpha_1 p_{i,n} + (1 - \alpha_1)\lambda_1.$$

This equation shows that when the recall probability is less than λ_1, recall will increase that probability. Hence we are simply assuming that there is a "practice effect" from writing down a word.

The two classes of events, recall and non-recall, change the recall probabilities as defined by equations 10.1 and 10.2. We assume that no other events alter these probabilities. Hence we can apply the analysis given in Section 8.5 for the special case of one identity operator. The alternative analysis given by Miller and McGill [4] could be used instead, but for consistency and continuity we draw from the results given in Chapter 8.

Until Section 10.7, we make one further assumption which limits the range of applicability of the model. We assume that the subject can learn the list of words perfectly, that is, that the proportion of words recalled approaches an asymptote of unity. Thus, we take $\lambda_1 = 1$. The operator Q_1, of equation 10.2, then becomes

$$(10.3) \qquad p_{i,n+1} = Q_1 p_{i,n} = \alpha_1 p_{i,n} + (1 - \alpha_1).$$

This restriction will simplify the estimation procedures given later. Only two quantities, the initial probability p_0 and the parameter α_1, remain to be estimated from the data.

10.3 THE DATA

The particular data we analyze in detail was obtained by Bruner, Miller, and Zimmerman [3] and was discussed by Miller and McGill [4]. A subject was read a list of 32 monosyllabic words and afterwards was asked to write down all the words he could recall. The order of the words was changed and the procedure was repeated. In this way the experiment gives a sequence of recalls and non-recalls for each of the 32 words. (Between some trials, the subject was given an "articulation test"—the words were read over a telephone system with noise introduced, and after each word was read the subject stated what he had heard. We were advised by Miller and McGill that their detailed analysis of the data showed that these tests did not influence the learning data in any detectable way.)

We find it convenient to represent the data by a set of random variables

$x_{i,n}.$ We let $x_{i,n} = 1$ if the ith word is recalled on trial n and let $x_{i,n} = 0$ if that word is not recalled on trial n. Thus for each word we obtain a sequence of 1's and 0's. In Fig. 10.1 we show the proportion of words correctly recalled on each trial. For purposes of analyzing the data we need to know the number k of previous recalls of a word on each trial, and so we provide these data in Table 10.1. (The value of $x_{i,n}$ for trial n

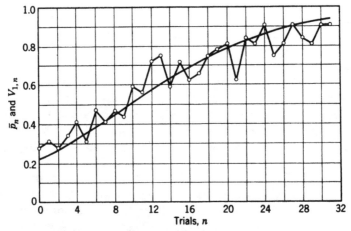

Fig. 10.1. Observed proportions, $\bar{p}_n$, of recalls on each trial from the Bruner-Miller-Zimmerman data, and curve of the means, $V_{1,n}$, computed from equation 8.51, with $\alpha_1 = 0.86$ and $p_0 = 0.22$.

can be obtained by subtracting an entry in column n from the entry in column $n + 1$. For example, for the first word, there was no recall on trial 9 because both trials 9 and 10 have 8 previous recalls, but there was a recall on trial 10 because trial 11 shows 9 previous recalls.)

Before we estimate the initial probability p_0 and the recall parameter α_1 we show that the assumption that $\alpha_2 = 1$ is consistent with the data. To do this, we look only at words with zero previous recalls and inquire about the proportion of them recalled for the first time on each of the first six trials. On trial 0 there are, of course, 32 words with zero previous recalls; from Table 10.1 we see that 9 of these are recalled on trial 0. Of the remaining 23 words, 8 are recalled for the first time on trial 1; of the remaining 15, 3 are recalled for the first time on trial 2, etc. In Table 10.2 we show these numbers for trials 0, 1, 2, 3, 4, 5. Now according to our null hypothesis, equation 10.1, the proportion of recalls will be the same on all trials. Thus the expected number of recalls in Table 10.2 is 27/98 times the number of words with zero previous recalls on each trial. We now can test for goodness-of-fit by the conventional Pearson χ^2 test.

TABLE 10.1. Number of previous recalls of each word for each trial. "Previous" means that the outcome for the trial numbered at the head of a column is not included in the entry in that column.

Trial n

Word	0	1	2	3	4	5	6	7	8	9	10	11	12	13	14	15	16	17	18	19	20	21	22	23	24	25	26	27	28	29	30	31	32
1 are	0	0	1	2	3	4	5	6	7	8	8	9	10	11	12	13	14	15	15	16	17	18	19	20	21	22	23	24	25	26	27	28	29
2 bad	0	0	0	0	0	0	1	1	1	1	1	1	1	2	2	2	2	2	2	4	4	6	6	7	8	9	10	11	12	12	13	14	14
3 bait	0	0	1	2	3	3	3	4	5	6	7	8	8	10	11	12	13	14	15	16	17	18	18	19	19	20	21	22	23	24	25	26	27
4 bask	0	0	0	0	0	0	0	1	1	1	1	1	2	3	4	5	6	7	8	9	10	11	12	13	14	15	15	16	17	17	18	20	20
5 bounce	0	0	0	0	2	2	2	2	2	2	2	3	4	4	5	5	5	6	7	7	7	7	7	7	7	8	9	9	10	11	12	13	14
6 box	0	0	0	1	1	2	3	3	4	4	5	5	6	7	8	9	10	11	12	13	14	15	16	17	18	19	20	21	22	23	24	25	26
7 crash	0	0	0	0	1	1	1	1	1	1	1	1	1	1	1	1	2	2	3	3	3	3	3	3	4	5	5	6	7	8	8	9	10
8 cleanse	0	0	0	0	1	1	1	1	1	1	1	2	2	3	4	5	5	6	6	7	7	7	7	8	8	9	9	9	9	9	10	10	10
9 clove	0	1	1	1	1	2	3	3	3	4	5	6	7	8	9	9	10	11	12	13	14	15	16	17	18	19	20	21	22	23	24	25	26
10 death	0	1	1	1	1	2	2	3	4	5	6	6	7	8	9	10	11	12	13	14	15	16	17	18	19	20	21	22	23	24	25	26	27
11 carl	0	0	0	0	0	1	2	3	4	5	6	6	7	8	9	9	10	10	11	12	13	14	15	16	17	18	19	20	21	22	23	24	25
12 fraud	0	0	0	0	0	1	1	1	1	1	1	2	2	2	2	2	2	2	2	3	4	5	6	7	8	9	10	11	12	13	13	14	15
13 grove	0	0	0	0	0	1	1	1	1	1	1	1	1	1	1	3	3	3	3	3	3	4	4	4	4	4	4	5	6	7	8	9	10
14 heap	0	0	0	0	1	1	1	1	1	1	1	1	2	2	2	3	3	3	3	4	5	5	5	6	7	7	7	7	8	8	8	9	9
15 hive	0	0	0	0	0	0	1	1	1	1	1	2	2	3	4	4	4	4	5	6	6	7	7	8	9	10	10	11	12	13	13	14	15
16 is	0	1	2	3	4	5	6	7	8	9	9	10	11	12	13	14	15	16	17	18	19	20	21	22	23	24	25	26	27	28	29	30	31
17 nook	0	0	1	1	2	2	3	3	3	3	3	3	5	6	7	8	9	9	10	11	12	13	13	13	13	14	15	15	16	17	18	19	20
18 not	0	0	1	1	1	1	2	2	2	3	3	4	5	6	6	7	8	9	10	11	12	13	13	14	15	16	17	18	19	20	21	22	23
19 pants	0	1	1	1	2	3	4	5	6	7	8	9	10	11	12	13	13	14	15	16	17	18	19	20	21	22	23	24	25	26	27	27	28
20 pests	0	1	1	2	2	2	2	2	2	3	3	5	5	6	6	7	8	9	9	9	10	13	14	14	14	15	16	17	18	19	19	19	20
21 plush	0	0	1	1	1	2	2	3	3	4	4	5	5	6	7	7	8	8	9	9	10	11	12	13	14	15	16	17	18	19	20	21	22
22 rag	0	0	0	0	0	0	0	0	0	0	0	0	0	0	0	0	0	1	1	1	2	4	4	5	5	5	5	5	6	7	8	9	10
23 rat	0	1	1	1	1	2	3	4	4	4	4	5	6	7	8	9	10	11	12	13	13	14	15	16	17	18	20	21	21	21	23	24	25
24 rise	0	0	0	0	0	0	0	0	0	1	1	2	2	4	4	5	6	7	7	7	7	7	7	8	9	10	11	12	13	13	15	16	17
25 rub	0	0	1	1	1	1	1	1	1	1	1	1	1	1	2	3	4	4	5	6	7	7	7	8	9	10	11	12	13	14	15	16	17
26 scythe	0	0	1	1	1	1	1	2	2	3	3	3	3	4	4	4	4	5	5	5	6	7	7	8	9	10	11	11	12	13	14	15	16
27 hoe	0	1	2	3	4	4	5	6	7	8	9	10	11	12	13	14	15	15	16	17	18	19	20	21	22	23	24	25	26	27	28	29	30
28 smile	0	1	1	1	2	3	3	4	4	5	6	7	8	8	9	10	11	12	13	14	15	16	17	18	19	20	21	22	23	24	25	26	27
29 start	0	1	1	1	1	1	1	1	2	3	4	5	6	7	8	9	10	11	11	12	13	14	15	16	17	18	19	20	20	21	22	23	24
30 toe	0	0	1	1	1	1	1	2	2	2	2	3	3	4	5	6	7	8	9	10	11	12	13	14	15	16	17	18	18	18	18	19	20
31 use	0	0	0	0	0	0	0	0	0	0	1	2	3	4	5	6	7	8	9	10	11	12	13	14	15	16	17	17	18	18	19	19	20
32 vast	0	0	0	0	0	1	1	2	2	2	2	2	2	2	2	3	4	4	5	6	7	8	8	9	10	11	11	12	13	13	14	15	16

We obtained $\chi^2 = 2.87$; the number of degrees of freedom is 5, and the probability obtained from a χ^2 table is about 0.72. Therefore, we consider the fit entirely adequate, and hence we have found evidence to support the assumption that $\alpha_2 = 1$.

In the following three sections we discuss some procedures for estimating the parameters p_0 and α_1 from the data. Perhaps the ideal way to present the estimation procedure would be to choose one method of estimating and carry it through for the particular data. If we did this, we might leave the reader with the mistaken impression that there is just one way to estimate. Instead of proceeding dogmatically, we present various approaches to the estimation problem for these data and discuss each briefly.

10.4 ESTIMATION OF p_0

First consider the problem of estimating the initial recall probability p_0. The obvious estimate of p_0 is the proportion of words recalled on trial 0. This proportion, which we call x_0, is given by

$$(10.4) \qquad x_0 = \frac{1}{N} \sum_{i=1}^{N} x_{i,0}.$$

We can think of the initial experimental trial as equivalent to N binomial trials with a probability p_0 of a success. An analogy is flipping N identical coins, each having a probability p_0 of falling heads on a single trial. The proportion of coins, x_0, which come up heads is an estimate of p_0. This estimate is unbiased, and is the maximum likelihood estimate of p_0 when only the data on the initial experimental trial are considered. From the data given in Table 10.1 we find that

$$(10.5) \qquad x_0 = 9/32 = 0.281.$$

The variance of this estimate is simply the binomial variance given by

$$(10.6) \qquad \sigma^2(x_0) = \frac{p_0(1 - p_0)}{N}.$$

We can estimate this theoretical variance of x_0 by replacing p_0 with its estimate x_0. If we do this for our data we obtain $\sigma^2(x_0) \doteq (0.281)(0.719)/32 = 0.0063$. The standard deviation, $\sigma(x_0)$, is then estimated by 0.079.

The variance of the estimate x_0 is appreciable, and so we would like to find a better estimate of p_0. Only the data from the initial trial were used in obtaining the estimate x_0. We can obtain a somewhat better estimate without introducing serious complications, however. From the basic assumptions being made in this chapter, we know that if an A_2 occurs on trial 0 (non-recall), the recall probability on trial 1 is still p_0. (From Table 10.2 we see that trial 1 gives an estimate $p_0 \triangleq 8/23 = 0.348$.) Similarly, if an A_2 occurs on trial 1, the recall probability on trial 2 remains p_0. (Again, Table 10.2 gives $p_0 \triangleq 3/15 = 0.200$.) We may readily use the data for each word on trials up through the trial on which recall first occurs to estimate p_0. This procedure will utilize considerably more data than the separate trial estimates. For each word we have

TABLE 10.2

Tabulations used in testing assumption that the non-recall parameter $\alpha_2 = 1$.

Trial (n)	Number of Words with Zero Previous Recalls	Number of Those Recalled on Trial n	Expected Number Recalled on Trial n
0	32	9	8.82
1	23	8	6.34
2	15	3	4.13
3	12	2	3.31
4	10	4	2.76
5	6	1	1.65
Totals	98	27	27.01

merely to record the number of trials preceded by zero recalls and sum these for all words; we denote this sum by N_0. On every such trial the probability of recall is p_0, and so the proportion of recalls on these N_0 word-trials is an estimate of p_0. Since each word can be recalled for the first time just once in the sequence, the number of recalls during the N_0 word-trials is simply N. Thus our new estimate, $\hat{p}_0$, of p_0 is

(10.7) $$\hat{p}_0 = N/N_0.$$

For the data given in the last section we have $N = 32$ and $N_0 = 120$. Thus

(10.8) $$\hat{p}_0 = 32/120 = 0.267.$$

We next investigate the variance and bias of this estimate. We can compute the expected number of word-trials, N_0, which is preceded by zero recalls. The probability of obtaining a value N_0 is given by the negative binomial distribution [5]:

(10.9) $$f(N_0) = \binom{N_0 - 1}{N - 1} (1 - p_0)^{N_0 - N} p_0^N.$$

The expected value of N_0 is

(10.10) $$E(N_0) = N/p_0.$$

We now demonstrate that $\hat{p}_0$ is the maximum likelihood estimate of p_0. We use the standard procedure described in Section 9.8. First we take the logarithm L of the expression in equation 10.9, omitting the constant coefficient, to obtain

$$(10.11) \qquad L = (N_0 - N) \log (1 - p_0) + N \log p_0.$$

If we maximize this logarithm, we shall have maximized the likelihood of obtaining the observed data. Hence we differentiate L with respect to p_0:

$$(10.12) \qquad \frac{\partial L}{\partial p_0} = - \frac{N_0 - N}{1 - p_0} + \frac{N}{p_0}.$$

When this derivative is set equal to zero we obtain for the maximum likelihood estimate of p_0

$$(10.13) \qquad \hat{p}_0 = N/N_0,$$

in agreement with equation 10.7. (To this point, the derivation is identical with that for the ordinary binomial distribution.) Next we compute the asymptotic variance of this estimate from the theorem given at the end of Section 9.8. Taking the second derivative of L gives

$$(10.14) \qquad \frac{\partial^2 L}{\partial p_0{}^2} = - \frac{N_0 - N}{(1 - p_0)^2} - \frac{N}{p_0{}^2}.$$

The negative of the expected value of this second derivative is, from equation 10.10,

$$(10.15) \qquad -E\left(\frac{\partial^2 L}{\partial p_0{}^2}\right) = \frac{(N/p_0) - N}{(1 - p_0)^2} + \frac{N}{p_0{}^2} = \frac{N}{p_0{}^2(1 - p_0)}.$$

Therefore, the asymptotic variance of $\hat{p}_0$ is

$$(10.16) \qquad \sigma^2(\hat{p}_0) = \frac{p_0{}^2(1 - p_0)}{N}.$$

The reader may note that this variance is p_0 times the variance of our older estimate x_0 (see equation 10.6). This seems reasonable because the expected number of non-recall trials is N/p_0; therefore we might expect to have the first trial sample size, N, replaced by N/p_0, as it is. For the data discussed above we get the estimate

$$(10.17) \qquad \sigma^2(\hat{p}_0) \cong \frac{(0.267)^2(0.733)}{32} = 0.0016,$$

and so

$$(10.18) \qquad \sigma(\hat{p}_0) \cong 0.04.$$

This standard deviation is just half of that obtained for the estimate x_0.

We are mildly concerned about the bias of the estimate $\hat{p}_0$. We could compute the expected value of p_0 and determine whether or not it was p_0. This computation is tedious, however, so we draw upon a result obtained by Girshick, Mosteller, and Savage [6]. They demonstrated that a unique unbiased estimate, $(p_0)_u$, of p_0 when N is fixed and N_0 is the only observed statistic, is given by

(10.19) $$(p_0)_u = \frac{N-1}{N_0-1}.$$

We do not repeat the proof here, but the reader can show that $(p_0)_u$ is indeed unbiased by using the distribution function $f(N_0)$ given by equation 10.9, and computing the expected value of $(p_0)_u$. From our data we get

(10.20) $$(p_0)_u = 31/119 = 0.261.$$

By comparing this unbiased estimate (equation 10.19) with the maximum likelihood estimate $\hat{p}_0$ (equation 10.13) we see that when N and N_0 are large the two estimates are nearly identical since the -1 terms are negligible. This establishes that $\hat{p}_0$ is unbiased when N gets arbitrarily large. However, $\hat{p}_0$ is necessarily biased for finite N. Moreover, for large values of N, since $\hat{p}_0 \cong (p_0)_u$ the distributions of these two estimates must be nearly identical. Therefore the variance of the unbiased estimate $(p_0)_u$ is given approximately by equation 10.16 for large N.

Since the variances of all the estimates of p_0 given above are not small, especially for small values of N, the reader may wonder why we cannot use still more of the data to estimate p_0. In principle this can be done, but we know from our basic operator, Q_1 (equation 10.3) that as soon as a word has been recalled once its probability of recall depends on the parameter α_1. We have already exhausted all the data which depend only upon p_0. The data on trials beyond the trial of the first recall depend upon both p_0 and α_1, and up to this point α_1 is completely unknown. In using these data on later trials, we have the choice of trying to estimate p_0 and α_1 simultaneously, or of estimating α_1 on the assumption that p_0 is known. The latter procedure is used in the next section.

10.5 ESTIMATION OF α_1

In this section we assume that the initial recall probability, p_0, is known; it may be estimated by the procedure given in the last section or its value may be assumed. We wish to estimate the parameter α_1 from the data, taking $p_0 = 0.26$.

In Section 9.6 we saw that the mean total number, T_2, of A_2 occurrences (non-recalls) may be used to estimate α_1. Equation 9.21 gives

(10.21)
$$T_2 \doteq \frac{-\log p_0}{1 - \alpha_1},$$

and so

(10.22)
$$\hat{\alpha}_1 = 1 - \frac{-\log p_0}{T_2}.$$

From Table 10.1 we obtain $T_2 = 11.59$, and this value along with $p_0 = 0.26$ gives

(10.23)
$$\hat{\alpha}_1 = 0.884.$$

Thus we have obtained an estimate of α_1 very cheaply. Nevertheless we use the more elegant maximum likelihood procedure described in Sections 9.9 and 9.10 to estimate α_1.

Equation 9.60 is appropriate for this problem, provided that we let ν stand for the number of previous recalls of a word, x_ν denote the number of words recalled at least $\nu + 1$ times, and N_ν the number of word-trials required for the x_ν words to be recalled $\nu + 1$ times after each has been recalled ν times. For example, with three words and six trials each, we might observe the sequences shown below:

$$
\begin{array}{ll}
\text{Word 1} & \text{0 1 1 0 0 1} \\
\text{Word 2} & \text{1 0 0 1 0 0} \\
\text{Word 3} & \text{0 0 1 1 1 0}
\end{array}
$$

(A 1 denotes a recall and a 0 a non-recall.) Then, for $\nu = 2$, words 1 and 3 are recalled $\nu + 1 = 3$ times; hence $x_\nu = 2$. Word 1 required 3 trials for the third recall after the second recall, whereas word 3 required one such trial. Thus, $N_\nu = 4$.

We consider α in equation 9.60 to be α_1 in the present model. We need the statistic D_1 defined by equation 9.63:

(10.24)
$$D_1 = \sum_{\nu=0}^{\Omega} \nu(N_\nu - x_\nu).$$

In Table 10.3 we tabulate the values of N_ν for $\nu = 0, 1, \cdots, 8$. These numbers were obtained from Table 10.1; N_ν is simply the number of times the integer ν appears in the entire table. For the above values of ν, the quantity x_ν has the value 32, the number of words on the list, for we see in Table 10.1 that every word that has 8 previous recalls is recalled at least once more. (Word number 14 has the smallest number of recalls.) By stopping at $\nu = 8$ we have chosen the upper limit Ω of the sum in the last equation to be 8. We make this choice of Ω so that we can use the second procedure given in Section 9.10, that is, we have chosen $\Omega = 8$ in order that $x_\nu = x = 32$.

TABLE 10.3
Tabulations used in estimating parameters.

ν	N_ν	x_ν	$\hat{p}_\nu$	$N_\nu - x_\nu$	$\nu(N_\nu - x_\nu)$
0	120	32	0.2605	88	0
1	127	32	0.2460	95	95
2	76	32	0.4133	44	88
3	62	32	0.5082	30	90
4	49	32	0.6458	17	68
5	49	32	0.6458	17	85
6	34	32	0.9394	2	12
7	57	32	0.5536	25	175
8	38	32	0.8378	6	48
9	43	31	0.7143	324	661
10	29	27	0.9286		
11	30	27	0.8966		
12	28	27	0.9630		
13	35	27	0.7647		
14	26	25	0.9600		
15	28	23	0.8148		
16	21	21	1.0000		
17	20	19	0.9474		
18	23	19	0.8182		
19	22	19	0.8571		
20	16	15	0.9333		
21	15	15	1.0000		
22	14	14	1.0000		
23	13	13	1.0000		
24	12	12	1.0000		
25	10	10	1.0000		

From Table 10.3 we see that $D_1 = 661$. From equation 9.69 we then have

$$(10.25) \qquad G(\hat{\alpha}_1, q_0, \Omega) = (D_1/x) = (661/32) = 20.66.$$

The function G is given in Table D. For our present problem, $q_0 = 0.74$ and $\Omega = 8$, and we wish to obtain $\hat{\alpha}_1$. From Table D,

$$(10.26) \qquad \begin{aligned} G(0.86, 0.74, 8) &= 19.49, \\ G(0.88, 0.74, 8) &= 23.08. \end{aligned}$$

Since we want the value of $\hat{\alpha}_1$ for which $G = 20.66$, we then interpolate to obtain

$$(10.27) \qquad\qquad\qquad \hat{\alpha}_1 = 0.867.$$

This estimate is quite close to the one given by equation 10.23.

We can estimate the variance $\sigma^2(\hat{\alpha}_1)$, of the estimate just obtained, from equation 9.82. The computation is straightforward since we know the values of N_ν from Table 10.3 and we replace α with the above estimate, 0.867, and take q_0 to be 0.74. The numerical computations yield $\sigma^2(\hat{\alpha}_1)$ $\doteq 0.00016$ ($\sigma(\hat{\alpha}_1) \doteq 0.013$). This is probably an underestimate of the variance of $\hat{\alpha}_1$ since q_0 is not actually known, as was assumed in the computation of that variance.

10.6 SIMULTANEOUS ESTIMATION OF q_0 AND α_1

In Section 10.4 we estimated the initial probability p_0 from that portion of the data which was independent of α_1, and then we used this value of p_0 to obtain estimates of α_1 in Section 10.5. We proceeded in this way

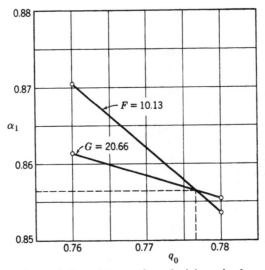

Fig. 10.2. Interpolation diagram for obtaining simultaneous maximum likelihood estimates of q_0 and α_1.

mainly for heuristic purposes. As already suggested in Section 9.11, we can obtain simultaneous maximum likelihood estimates of q_0 and α_1 by using Table D. We obtain these estimates now.

We need the statistic D_1 previously obtained and the statistic

$$(10.28) \qquad D_2 = \sum_{\nu=0}^{\Omega} (N_\nu - x_\nu).$$

From Table 10.3 we see that $D_1 = 661$ and $D_2 = 324$. The estimation equations (9.77) then become

$$(10.29) \qquad \begin{aligned} G(\hat{\alpha}_1, \hat{q}_0, 8) &= (D_1/x) = 20.66, \\ F(\hat{\alpha}_1, \hat{q}_0, 8) &= (D_2/x) = 10.13. \end{aligned}$$

The functions F and G are given in Table D, but we need to make a double interpolation to obtain the values of $\hat{\alpha}_1$ and $\hat{q}_0$. We assume that such interpolation schemes are not too well known, and so we give the details of one method here.

The procedure involves plotting α_1 versus q_0 for $F = 10.13$ and again for $G = 20.66$. The two curves will intersect at the point corresponding to the desired maximum likelihood estimates $\hat{\alpha}_1$ and $\hat{q}_0$. First consider the function F. We see that for $q_0 = 0.76$, the value $F = 10.13$ falls between $\alpha_1 = 0.86$ and $\alpha_1 = 0.88$. We interpolate and find that $F = 10.13$ for $\alpha_1 = 0.871$. For $q_0 = 0.78$, we interpolate between $\alpha_1 = 0.84$ and $\alpha_1 = 0.86$ and find that $F = 10.13$ for $\alpha_1 = 0.854$. Then we repeat the procedure for G and find that when $q_0 = 0.76$ we get $G = 20.66$ for $\alpha_1 = 0.861$, and that when $q_0 = 0.78$ we get $G = 20.66$ for $\alpha_1 = 0.855$. We then plot these values as shown in Fig. 10.2 and draw straight lines between the points. The line for $F = 10.13$ crosses the line for $G = 20.66$ at the point $(0.856, 0.778)$ and so we have finally for our maximum likelihood estimates

$$\hat{p}_0 = 0.222,$$
(10.30)
$$\hat{\alpha}_1 = 0.856.$$

These estimates can be compared to our previous estimates as summarized in Table 10.4.

TABLE 10.4

Comparison of the several estimates of the parameters p_0, α_1, λ_1. The numbers in parentheses in a particular row are values assumed in obtaining the estimates shown in that row.

Procedure	$\hat{p}_0$	$\hat{\alpha}_1$	$\hat{\lambda}_1$
Proportion recalls on trial 0	0.281	—	—
Maximum likelihood (early trials)	0.267	—	—
Unbiased (early trials)	0.261	—	—
Mean total number of non-recalls	(0.26)	0.884	(1.0)
Maximum likelihood	(0.26)	0.867	(1.0)
Simultaneous maximum likelihood	0.222	0.856	(1.0)
Minimum chi-square	0.228	0.850	0.982

Having estimated the parameters α_1 and p_0, we are now in a position to make some comparisons between the model and the data. We used the values $\alpha_1 = 0.86$ and $p_0 = 0.22$ and computed values of $V_{1,n}$ from equation 8.51 of Chapter 8. The computed curve, along with the experimental points, is shown in Fig. 10.1.

To make further comparisons, we ran 32 Monte Carlo computations for 32 trials each, as described in Section 6.2. The parameter values $\alpha_1 = 0.86$ and $p_0 = 0.22$ were used in making those computations. We computed several statistics from these runs, and the comparisons with those obtained from the data are shown in Table 10.5.

TABLE 10.5

Comparison of statistics of the Bruner-Miller-Zimmerman data and the Monte Carlo runs, computed with $p_0 = 0.22$, $\alpha_1 = 0.86$, $\lambda_1 = 1$.

	Data	Monte Carlo
Mean trial of first recall	2.75	3.66
Mean trial of second recall	6.72	7.19
Mean total number of non-recalls	11.59	11.22
Standard deviation of number of non-recalls	6.74	6.07
Estimate of $(p_0)_u$ from equation 10.19	0.261	0.209

10.7 IMPERFECT LEARNING

In previous sections we have regarded the list as ultimately being recalled perfectly. It is not absolutely necessary that we so regard this list-learning task. Furthermore, for other purposes it may be valuable to have a procedure for estimating the limit point λ_1 because not all tasks will ultimately lead to 100 percent performance, even after quite extended practice.

If the list of words is not learned perfectly, it is necessary to introduce a parameter λ_1 in the recall operator Q_1 to correspond to the upper limit of the recall proportion. The operators under such imperfect recall are

$$Q_1 p = \alpha_1 p + (1 - \alpha_1)\lambda_1,$$
(10.31)
$$Q_2 p = p,$$

where Q_1 is associated with recall, and Q_2 with non-recall. When a word has been recalled exactly ν times, its probability of recall has reached

$$(10.32) \qquad p_\nu = Q_1^\nu p_0 = \alpha_1^\nu p_0 + (1 - \alpha_1^\nu)\lambda_1.$$

As before, we let x_ν be the number of words recalled at least $\nu + 1$ times, and let N_ν be the number of word-trials required for those x_ν words to be recalled $\nu + 1$ times after each has been recalled ν times. (For small values of ν, x_ν will equal the total number of words, but some words are never recalled more than, say, eight times.) Analogous to equation 10.19, an unbiased estimate of p_ν is given by

(10.33)
$$\hat{p}_\nu = \frac{x_\nu - 1}{N_\nu - 1},$$

and

(10.34)
$$\sigma^2(\hat{p}_\nu) \cong \frac{p_\nu^2(1 - p_\nu)}{x_\nu}.$$

Knowledge of these quantities suggests the possibility of minimizing a χ^2-like quantity

(10.35)
$$\chi^2 = \sum_{\nu=0}^{\Omega} \frac{(p_\nu - \hat{p}_\nu)^2}{\sigma^2(\hat{p}_\nu)}$$
$$= \sum_{\nu=0}^{\Omega} W_\nu[\alpha_1{}^\nu p_0 + (1 - \alpha_1{}^\nu)\lambda_1 - \hat{p}_\nu]^2,$$

where

(10.36)
$$W_\nu = 1/\sigma^2(\hat{p}_\nu),$$

and Ω is the largest number of recalls retained for estimation purposes (for these data we use $\Omega = 25$). The W's involve the parameters being estimated, but it is worth noticing that the W's may be regarded as weights. Small changes in weights usually make relatively little difference in such problems, so we can obtain an approximation for the W's, and temporarily pretend they are constants. To obtain the minimizing equations we differentiate χ^2 partially with respect to p_0, λ_1, and α_1 and set the results equal to zero to obtain from p_0, λ_1, and α_1, respectively,

(10.37)
$$\sum_\nu W_\nu \alpha_1{}^\nu[\alpha_1{}^\nu p_0 + (1 - \alpha_1{}^\nu)\lambda_1 - \hat{p}_\nu] = 0,$$

(10.38)
$$\sum_\nu W_\nu(1 - \alpha_1{}^\nu)[\alpha_1{}^\nu p_0 + (1 - \alpha_1{}^\nu)\lambda_1 - \hat{p}_\nu] = 0,$$

(10.39)
$$\xi(\alpha_1) = \sum_\nu W_\nu \nu \alpha_1{}^{\nu-1}[\alpha_1{}^\nu p_0 + (1 - \alpha_1{}^\nu)\lambda_1 - \hat{p}_\nu] = 0,$$

where $\xi(\alpha_1)$ is introduced for convenience as a definition. Using the first of these equations in the second, and rewriting, we get

(10.40)
$$p_0 \sum_\nu W_\nu \alpha_1{}^\nu + \lambda_1 \sum_\nu W_\nu(1 - \alpha_1{}^\nu) = \sum_\nu W_\nu \hat{p}_\nu,$$

(10.41)
$$p_0 \sum_\nu W_\nu \alpha_1{}^{2\nu} + \lambda_1 \sum_\nu W_\nu \alpha_1{}^\nu(1 - \alpha_1{}^\nu) = \sum_\nu W_\nu \hat{p}_\nu \alpha_1{}^\nu,$$

(10.42)
$$\xi(\alpha_1) = p_0 \sum_\nu W_\nu \nu \alpha_1{}^{2\nu-1} + \lambda_1 \sum_\nu W_\nu \nu \alpha_1{}^{\nu-1}(1 - \alpha_1{}^\nu)$$
$$- \sum_\nu W_\nu \nu \alpha_1{}^{\nu-1} \hat{p}_\nu = 0.$$

We have already described how to obtain the $\hat{p}_\nu$. The W_ν can be estimated by choosing a fairly close fitting set of parameters and computing successive values of the variance for that set of parameters. A freehand fit to the curve of $\hat{p}_\nu$ against ν could be used to approximate the p_ν, but it would seem better to use a functional form that is of roughly the right shape. We chose $p_0 = 0.26$, $\alpha_1 = 0.85$, $\lambda_1 = 0.98$ for this purpose. As ν increased, the factor $(1 - p_\nu)$ decreased, but the number of words available also decreased, leaving the weights roughly equal for all ν (except $\nu = 0$, where the weight was twice the average). So, as a first approximation, it turns out that the W's can be dropped altogether.

For any given value of α_1, equations 10.40 and 10.41 are a pair of linear simultaneous equations that can be solved for p_0 and λ_1. When numerical values for all three parameters are then substituted into the third equation, we get a value of $\xi(\alpha_1)$, usually not zero. For example, for the word data we find:

α_1	$\xi(\alpha_1)$	p_0	λ_1
0.83	+0.0734	0.197198	0.966951
0.84	−0.0262	0.210704	0.976207
0.85	−0.1342	0.224892	0.986632

Interpolating on α_1 to make $\xi(\alpha_1)$ vanish, we find as estimates $p_0 \doteq 0.207$, $\alpha_1 \doteq 0.837$, $\lambda_1 \doteq 0.974$. This value of λ_1 is quite close to unity, as would be expected.

In obtaining these estimates we used the data of Table 10.3 for the first 25 recalls. By using so many recalls we are ultimately reduced to as few as 10 words for estimation purposes, and it seems wise to stop there. Of the 1024 observations, 997 were actually used (the cut-off at $\Omega = 25$, and the estimation procedure forced the discard of 27).

If the W's are taken to be equal, the summations not involving $\hat{p}$'s are readily obtained:

$$(10.43) \quad \Sigma_1 = \sum_{\nu=0}^{\Omega} \alpha_1^{\nu} = \frac{1 - \alpha_1^{\Omega+1}}{1 - \alpha_1},$$

$$(10.44) \quad \Sigma_2 = \sum_{\nu=0}^{\Omega} \alpha_1^{2\nu} = \frac{1 - \alpha_1^{2\Omega+2}}{1 - \alpha_1^2},$$

$$(10.45) \quad \Sigma_3 = \sum_{\nu=0}^{\Omega} \nu\alpha_1^{\nu-1} = \frac{1 - \alpha_1^{\Omega+1} - (1 - \alpha_1)(\Omega + 1)\alpha_1^{\Omega}}{(1 - \alpha_1)^2},$$

$$(10.46) \quad \Sigma_4 = \sum_{\nu=0}^{\Omega} \nu\alpha_1^{2\nu-1} = \frac{\alpha_1(1 - \alpha_1^{2\Omega+2}) - (1 - \alpha_1^2)(\Omega + 1)\alpha_1^{2\Omega+1}}{(1 - \alpha_1^2)^2},$$

$$(10.47) \quad \Sigma_5 = \sum_{\nu=0}^{\Omega} (1 - \alpha_1') = \Omega + 1 - \Sigma_1,$$

$$(10.48) \quad \Sigma_6 = \sum_{\nu=0}^{\Omega} \alpha_1'(1 - \alpha_1') = \Sigma_1 - \Sigma_2,$$

$$(10.49) \quad \Sigma_7 = \sum_{\nu=0}^{\Omega} \nu\alpha_1'^{\,\nu-1}(1 - \alpha_1') = \Sigma_3 - \Sigma_4.$$

Furthermore, when equal weights are used, the only tables that must be constructed are those for α_1' and $\nu\alpha_1'^{\,\nu-1}$, for each numerical value of α_1 used, and $\nu = 0, 1, \cdots, \Omega$. But when the W's are taken as unequal, it is necessary to construct tables of α_1', $\alpha_1'^{2\nu}$, $\nu\alpha_1'^{\,\nu-1}$, and $\nu\alpha_1'^{2\nu-1}$ for $\nu = 0, 1, \cdots, \Omega$, to facilitate computation of the summations involving the W's.

The first set of results obtained with equal W's was used to generate a new set of approximate p_ν's, $\nu = 0, 1, \cdots, 25$. From these p's new weights were obtained. The minimizing equations were solved using these unequal weights. The results were $\hat{p}_0 = 0.231$, $\hat{\alpha}_1 = 0.850$, $\hat{\lambda}_1 = 0.981$. Using these results a second set of unequal weights was computed, and this last iteration step gave finally

$$\hat{p}_0 = 0.228,$$

$$(10.50) \qquad\qquad \hat{\alpha}_1 = 0.850,$$

$$\hat{\lambda}_1 = 0.982.$$

These results are so close to the penultimate set, that the iterative process was abandoned at this point. Though it was worth keeping a third decimal for the purposes of iteration and computation, we know that the unreliability of the estimates is such that rounding to two decimals does them more than justice. It is worth noticing that the results for $\hat{p}_0$ and $\hat{\alpha}_1$ are extremely close to those obtained by the maximum likelihood procedure where λ_1 was assumed to be unity. This is no surprise now that we know λ_1 is so close to unity because minimum χ^2 is essentially maximum likelihood. However, we did use much more data in the minimum χ^2 approach, but, of course, the extra data from the later part of the experiment were made up for in the maximum likelihood procedure by the direct assumption of unit λ_1.

When the parameter values $\hat{p}_0 = 0.231$, $\hat{\alpha}_1 = 0.850$, $\hat{\lambda}_1 = 0.981$ were used to compute χ^2 in equation 10.35, the value turned out to be 56.3. This is a rather large value for 23 degrees of freedom (about four standard normal deviates out). In Fig. 10.3 values of $\hat{p}_\nu$ are plotted against ν. In this figure sources of the large value of χ^2 are apparent; for $\nu = 6$ and 7, the observed proportions differ from the computed ones by about 0.24 and 0.19, respectively. The weights associated with these points in the χ^2 computation are about 220. Thus these two points are contributing about 20.6 to the 56.3. There certainly is, therefore, more oscillation about the mean curve in the data than can be accounted for by the built-in variation in the model. If a subject learns words in clusters, rather than independently as assumed in this model, then one word in a cluster could

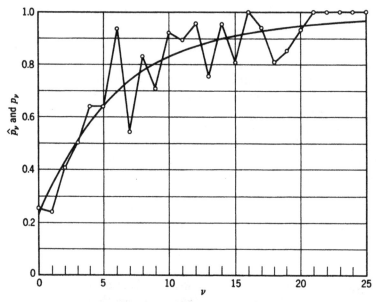

Fig. 10.3. Values of the estimates, $\hat{p}_\nu$, (circles joined by straight lines), of the recall probability after ν recalls, obtained from equation 10.33, and the values of p_ν, (smooth curve) computed from equation 10.32, with $p_0 = 0.231$, $\alpha_1 = 0.850$, and $\lambda_1 = 0.981$.

act as a cue for the whole cluster. Such behavior could introduce more variation than is provided in the models discussed in this chapter. Although we have not done so, it might be worth developing a model that would include such cluster-learning. The task looks difficult.

10.8 EVALUATION OF THE MODEL

Now that we have presented our first detailed application of the mathematical system described in Part I, we are in a position to evaluate its usefulness in more concrete terms than we have done heretofore. We delay a general evaluation until Chapter 15, but we now try to evaluate the model given in the present chapter. What has been accomplished with the model that could not have been done by routine analysis of the data? In attempting to answer this question we make three main points: (1) The model gives a more detailed description of the data than is ordinarily done, (2) the model yields a concise summary of the data which simplifies parametric studies, and (3) the model provides a baseline for studying more subtle effects, or differences in stimuli or subjects.

1. THE MODEL GIVES A DETAILED DESCRIPTION OF THE DATA. Most learning theories do not provide a framework for analyzing data at the level of single subjects and single trials. The model given in this chapter does give a probabilistic description of the data in terms of the probability of recall of a single word on each trial. By such a detailed analysis we can compare the relative effects of recall and non-recall of a word on the subsequent recall probabilities. We found that our assumption that non-recall had no effect was justified, that is, the data gave no evidence to reject this hypothesis. We also found that recall of a word reduced the probability of non-recall to about 85 percent of its previous value.

2. THE MODEL LEADS TO A CONCISE SUMMARY OF THE DATA. The data are completely specified by the model and the values of three parameters, p_0, α_1, and λ_1. From these parameters the model can generate a large number of statistics of the data. For example, the mean and variance of the number of recalls in a given number of trials, the mean and variance of the trial of the first recall, the mean and variance of the trial of the second recall, etc., can be computed from the model once we have estimated the three parameters. In this sense, the parameters p_0, α_1, and λ_1 are "basic" quantities. Furthermore, the psychological meaning of these parameters is not obscure. The parameter values may be considered summary statistics or properties of a particular set of data and may be used in making comparisons between various sets of data. Individual subjects can be compared easily, and the effects of changing the number of words or the speed of presentation of the words can be measured readily in terms of the parameters.

3. THE MODEL PROVIDES A BASELINE FOR STUDYING EFFECTS OUTSIDE THE MODEL. Several simplifying assumptions were made in constructing the model, and the validity of such assumptions depends on various experimental conditions. For example, we assumed that the words were equivalent for the subject—were equally difficult to learn. The comparisons between the data and the Monte Carlo runs, shown in Table 10.5, tend to support that assumption, but this is undoubtedly a result of the care taken by the experimenters to equate the words. But the model provides a way of measuring how well they succeeded! From Table 10.5 we note that the standard deviation of the number of non-recalls was 6.74 for the Bruner-Miller-Zimmerman data and 6.07 for the

Monte Carlo runs. Had the experimenters exercised less care and skill in equating the words, we would expect that the variance in the number of non-recalls would have been considerably larger. The experimenters scrambled the order of the words on each trial to eliminate serial effects, and our analysis indicates that they were successful. Experiments in which serial effects are present could be analyzed in the same way. In such cases, of course, the model would not be expected to reproduce the data accurately, but it could be used as a baseline for measuring the magnitude of the serial effects. Another possibility is that an experimenter might want to include some "loaded" or "emotionally toned" words in his list and find out if the subject had greater or less difficulty with such words. Again, the model provides a baseline for measuring the magnitude of such effects.

In the chapters which follow we have further opportunities for making evaluations such as those just given. The main points are the same, but the context is different in each case.

10.9 SUMMARY

An experiment on verbal learning with serial effects minimized is described, and the data are analyzed. It is assumed, following Miller and McGill [4], that non-recall of a word does not change the recall probability p. This assumption is supported by the data. Hence when non-recall occurs the operation applied is

$$Q_2 p = p.$$

When recall occurs, the operation is

$$Q_1 p = \alpha_1 p + (1 - \alpha_1)\lambda_1.$$

Thus, the analysis given in Section 8.5 is applied. Both the cases of the limit point $\lambda_1 = 1$ and $\lambda_1 < 1$ are considered. Several procedures for estimating parameters are described, and goodness-of-fit is measured.

REFERENCES

1. Hovland, C. I. Human learning and retention. *Handbook of experimental psychology*, S. S. Stevens, ed., New York: Wiley, 1951, pp. 618–624.
2. Hull, C. L., Hovland, C. I., Ross, R. T., Hall, M., Perkins, D. T., and Fitch, F. B. *Mathematico-deductive theory of rote learning.* New Haven: Yale University Press, 1940.
3. Bruner, J. S., Miller, G. A., and Zimmerman, C. Discriminative skill and discriminative matching in perceptual recognition. *J. exp. Psychol.*, 1955, **49**, in press.
4. Miller, G. A., and McGill, W. J. A statistical description of verbal learning. *Psychometrika*, 1952, **17**, 369–396.
5. Mood, A. M. *An introduction to the theory of statistics.* New York: McGraw-Hill, 1950, p. 61.
6. Girshick, M. A., Mosteller, F., and Savage, L. J. Unbiased estimates for certain binomial sampling problems with applications. *Annals of math. Stat.*, 1946, **17**, 13–23.

CHAPTER 11

Avoidance Training

11.1 INTRODUCTION

Since the original experiments of Bekhterev on the conditioned withdrawal response, reported in the early part of the century, a number of studies of avoidance conditioning have been made [1, 2]. In these experiments the response which is learned prevents the appearance of a noxious stimulus such as an electric shock. Responses such as withdrawal of hand or foot, running, and jumping have been conditioned in this way. In order to establish these responses, it was necessary of course to shock the animal on the first few trials and to present a warning (conditioned stimulus such as a buzzer or tone) on each trial.

The results of avoidance training experiments have been interpreted as a combination of "classical" and "instrumental" conditioning [3]. The electric shock is regarded as an unconditioned stimulus—the withdrawal response is a reflex—and the warning stimulus is considered to be a conditioned stimulus which acquires the ability to evoke the response through classical conditioning. The fact that the response causes a cessation of the noxious stimulus leads to the interpretation that instrumental conditioning is also involved. Escape from shock reinforces the withdrawal response during the early trials, whereas escape from the conditioned noxious stimulus reinforces that behavior on later trials.

Another interpretation of avoidance training has been given by Miller [4]. Fear is considered to be a learnable drive, and fear reduction is assumed to be a reward. The early trials of avoidance training presumably establish fear of the conditioned stimulus as a drive, whereas both escape and avoidance result in reinforcement of the behavior involved, escape because it leads to pain reduction and avoidance because it leads to fear or anxiety reduction.

We shall attempt to describe the results of an avoidance experiment in terms of a simple model, which is an application of our general model. On each trial the animal either avoids or does not avoid shock. Avoiding will be considered response A_1 and not-avoiding response A_2. Not-avoiding is necessarily followed by shock and ordinarily by escape from that shock. The occurrence of A_1 will have a specified effect on the probability p of response A_1 on a trial; our operator Q_1 describes this effect. Also, the occurrence of an A_2 will have another effect on p, and this effect is described by Q_2. Hence we consider this type of experiment as an example of two subject-controlled events. Before spelling out this application in detail, we describe an experiment on avoidance training. Later we estimate the parameters in the model from the data obtained from that experiment.

11.2　THE SOLOMON-WYNNE EXPERIMENT

In a study of "traumatic avoidance learning" Solomon and Wynne describe an experiment in which dogs learn to jump a barrier to avoid an intense electric shock [3]. The subjects were 30 "mongrel dogs of medium size" weighing 9 to 13 kilograms. The apparatus was a variation of the Miller-Mowrer shuttle box used for avoidance training of rats. The box consisted of two compartments separated by a barrier and a "guillotine-type gate," which could be raised or lowered. The barrier was adjusted to the height of each dog's back. The floor of the apparatus consisted of steel bars which were wired to the shock circuit.

The conditioned stimulus (CS) consisted of turning out the lights above the compartment the dog was in and simultaneously raising the gate. The other compartment was still illuminated. In ten pretest trials none of the 30 dogs jumped the barrier during a 2-minute exposure of the CS. During training the CS was presented for 10 seconds and was then followed by an intense electric shock applied through the floor to the dogs' feet. The voltage was the "highest possible without producing tetany of the dogs' leg muscles." The current was about 100 to 125 milliamperes for most dogs. The shock was left on until the dog escaped over the barrier into the illuminated compartment, where no shock was administered.* The gate was closed as soon as the dog jumped. If a dog jumped before the shock was turned on (during the 10-second period), the trial was recorded as an avoidance trial, whereas if the dog did not jump until it was shocked the trial was recorded as an escape or shock trial. The experiment was designed so that shock could be escaped or avoided only by jumping the barrier. Reaction latencies and various qualitative

---

* If a dog did not escape after 120 seconds of shock the trial was terminated and recorded as a shock trial with "infinite" latency. Out of 234 shock trials, only 8 were of this type. Rather than introduce a third operator and possibly a third response, we assume that these few trials had the same effect as trials on which escape actually occurred.

aspects of behavior were also recorded, but no use of these data will be made in the following analysis.

The experimental record of the 30 dogs is shown in Table 11.1. Trials on which shock occurred are indicated by an S; all other trials are avoidance trials. (Dog number 66 actually jumped within 10 seconds on

TABLE 11.1

Record of shocks for the first 25 trials on each of 30 dogs in the Solomon-Wynne experiment. Occurrence of shock is indicated by S, non-occurrence by a blank. (The number assignments to the dogs were made by Solomon and Wynne and do not imply that a selection has been made.)

Dog	Trial n																								
	0	1	2	3	4	5	6	7	8	9	10	11	12	13	14	15	16	17	18	19	20	21	22	23	24
13	S	S		S		S																			
16	S	S	S	S	S	S	S			S	S	S	S	S	S										
17	S	S	S	S	S			S			S		S		S										
18	S			S	S	S																			
21	S	S	S	S	S	S	S	S																	
27	S	S	S	S	S	S					S	S		S											
29	S	S	S	S	S	S	S		S	S	S	S	S					S							
30	S	S	S	S	S		S	S		S	S														
32	S	S	S	S	S		S			S				S	S	S						S			S
33	S	S	S	S		S	S																		
34	S	S	S	S	S	S	S	S	S	S						S									
36	S	S	S	S	S						S	S													
37	S	S	S				S		S	S															
41	S	S	S	S		S																			
42	S	S	S		S			S																	
43	S	S	S	S	S	S	S		S			S													
45	S		S		S	S	S					S													
47	S	S	S	S		S			S	S															
48	S	S	S	S	S	S		S	S	S		S													
46	S	S	S	S			S			S		S													
49	S	S	S			S											S								
50	S	S		S		S	S	S									S	S							
52	S	S	S	S	S	S	S	S				S	S	S											
54	S	S	S	S	S	S	S	S				S		S	S			S							
57	S	S	S	S	S	S		S					S	S											
59	S	S		S	S			S			S								S						
67	S	S	S	S		S													S						
66	S	S	S		S		S	S			S		S		S		S								
69	S	S	S	S				S	S			S		S		S									
71	S	S	S	S							S														
Total	30	27	26	25	18	18	14	12	10	7	9	11	4	8	4	3	3	2	1	0	0	1	0	0	1

the first experimental trial. However, this jump was not recorded as an avoidance but was considered an eleventh pretest trial for this dog. The next trial for this dog is labeled trial 0 in Table 11.1.) We are grateful to R. L. Solomon for making the raw data available to us and for numerous helpful suggestions.

11.3 THE MODEL

As already pointed out, the Solomon-Wynne experiment is taken as an example of two subject-controlled events. We identify avoidance of shock on an experimental trial as response A_1 (or event E_1) and non-avoidance (escape) as response A_2 (or event E_2). The probability of avoidance on trial n is denoted by p_n. When avoidance occurs on trial n we apply operator Q_1 to p_n to obtain p_{n+1}, and when non-avoidance occurs we apply Q_2 to obtain p_{n+1}. The probability that the dog is shocked on trial n is $q_n = 1 - p_n$. It is assumed that a dog jumps over the barrier on every trial, that is, the dog either avoids or escapes shock on every trial.

The operators Q_1 and Q_2 are, of course, of the form

(11.1) $$Q_i p = \alpha_i p + (1 - \alpha_i)\lambda_i,$$

but rather than maintain this general form we place some restrictions on these operators. First of all, we see from the data in Table 11.1 that all dogs ultimately learn to avoid. Data for trials beyond $n = 24$ obtained by Solomon and Wynne, but not shown in Table 11.1, show that every dog in the present experiment avoided without fail on trials beyond the twenty-fifth. (Some dogs had as many as 200 trials.) Therefore, the operator Q_1, which is applied to p when a dog avoids, must maintain p very near unity. For this reason we take $\lambda_1 = 1$.

We also take the limit point λ_2 equal to unity. Clearly, the administering of shock to a dog increases the probability of avoidance, at least during the early part of the training. Furthermore, if λ_2 were appreciably less than unity, and if a shock occurred late in training, the probability of avoidance would decrease, making shock more likely on the next trial. The data suggest that this does not occur, and so we take $\lambda_2 = 1$. These restrictions give us the operators defined by

(11.2)
$$Q_1 p = \alpha_1 p + (1 - \alpha_1) \quad \text{(avoidance)},$$
$$Q_2 p = \alpha_2 p + (1 - \alpha_2) \quad \text{(shock)}.$$

The probability of avoiding is p, of shock, $q = 1 - p$. These operators are of the type discussed in Section 8.2; they have equal limit points and therefore commute with one another. We make use of the analysis given in Sections 8.2 and 8.3.

It is convenient to use the complementary operators $\tilde{Q}_i$ which are applied to q, the probability of shock, and are defined by $\tilde{Q}_i q = 1 - Q_i p$. From equation 11.2 we get

(11.3)
$$\tilde{Q}_1 q = \alpha_1 q \quad \text{(avoidance)},$$
$$\tilde{Q}_2 q = \alpha_2 q \quad \text{(shock)}.$$

As we saw in Chapter 8, the probability of response A_2 (shock) on trial n, when k previous A_1's (avoidances) occurred, is

(11.4) $$q_{n,k} = \tilde{Q}_1{}^k \tilde{Q}_2{}^{n-k} q_0 = \alpha_1{}^k \alpha_2{}^{n-k} q_0,$$

where q_0 is the initial probability of shock.

We place one further restriction on the model: we assume that the initial probability of avoidance, p_0, is zero. We see from Table 11.1 that none of the 30 dogs avoided on trial 0. Moreover, as mentioned in Section 11.2, ten pretest trials were given each dog during which the CS was exposed for 2 minutes; none of the dogs jumped the barrier during these pretest trials. (We have already noted that dog 66 jumped on a trial considered to be an eleventh pretest trial. For analytic purposes, one jump in 301 trials cannot seriously change results based upon the assumption that $p_0 = 0$.) Since $p_0 = 0$ and $q_0 = 1$, equation 11.4 becomes

(11.5) $$q_{n,k} = \alpha_1{}^k \alpha_2{}^{n-k}.$$

Therefore, the model for this experiment contains but two parameters, α_1 and α_2. The next two sections discuss the procedures used in estimating these two parameters from the data. We note that α_1 is a measure of the effectiveness of an avoidance trial in increasing the probability of avoidance; if α_1 were unity, the trial would have no effect, and if α_1 were zero the probability of avoidance would go to unity at once. Similarly, α_2 is a measure of the effectiveness of a shock trial in teaching the dog to avoid. It is convenient to call α_1 the *avoidance parameter* and α_2 the *shock parameter*. We have more to say about these interpretations in Section 11.7.

11.4 ESTIMATION OF THE SHOCK PARAMETER, α_2

We first discuss the problem of estimating the shock parameter α_2 from the data given in Table 11.1. Again, we consider more than one method of estimating parameters. For α_2 we use these methods:

1. Information based on the results of trial 1; same for trial 2.
2. The mean number of trials before the first avoidance.
3. An approximate maximum likelihood estimate based on trials before the first avoidance to assist in obtaining estimate 4.
4. The preferred estimate, a maximum likelihood estimate based on the same trials as method 3.

Method 1 is an obvious method for introductory purposes. Method 2 was used in the Introduction to this book; the estimate is easy to obtain from tables we provide. Estimates from 4 are more trouble to get, but we think they have more precision than the others.

We use the symbol $\hat{\alpha}_2$ indiscriminately for all these estimates. The interpretation of the "hat" on α_2 then is to mean "*an* estimate of" rather than "*the* estimate of." If we were to try to distinguish each of the estimates by a different symbol, we feel that the reader would be little further ahead, and perhaps rather confused. The source of each estimate should be clear from the context.

METHOD 1. By assumption, the probability of shock, q_0, on trial 0 is unity, and so on trial 1 all dogs have the same probability of shock. From equation 11.5 we learn that this latter probability is

$$(11.6) \qquad\qquad q_{1,0} = \alpha_2.$$

We may estimate $q_{1,0}$ at once by the proportion of the 30 dogs that receive shock on trial 1. In Table 11.1 we see that 27 dogs are shocked on trial 1, and so as our first estimate of α_2 we obtain

$$(11.7) \qquad\qquad \hat{\alpha}_2 = (27/30) = 0.900.$$

Furthermore, we see that these 27 dogs have a probability of shock on trial 2 given by

$$(11.8) \qquad\qquad q_{2,0} = \alpha_2{}^2.$$

We can estimate $q_{2,0}$ by the proportion of the 27 dogs that receive shock on trial 2. From the data we get $q_{2,0} = 24/27$, and so as a second estimate of α_2 we get

$$(11.9) \qquad\qquad \hat{\alpha}_2 = \sqrt{24/27} = 0.943.$$

This estimate is not too close to the estimate given by equation 11.7 and so we would like to obtain further estimates and average them. We could, with no difficulty, obtain further estimates of α_2 from $q_{3,0}$, $q_{4,0}$, etc., and then combine them to obtain a grand estimate of α_2, but, instead, we consider two other procedures.

METHOD 2. In Section 9.6 we described the statistic F_1, defined as the mean number of trials before the first A_1 occurrence (avoidance). Equation 9.3 tells us that when $\lambda_2 = 1$ and $q_0 = 1$, as is the case in the present model,

$$(11.10) \qquad\qquad F_1 \triangleq \Phi(\alpha_2, 1),$$

where $\Phi(\alpha, \beta)$ is the function given in Table A. The number of trials before the first avoidance of each dog is obtained directly from the data in Table 11.1. The mean is 4.50, and the standard deviation of the 30 observations is 2.25. From Table A we see that $\Phi(\alpha_2, 1)$ is 4.39 for

$\alpha_2 = 0.92$ and 5.08 for $\alpha_2 = 0.94$. Therefore, by interpolation we get as an estimate of α_2,

$$(11.11) \qquad \hat{\alpha}_2 = 0.923.$$

We see that this estimate is between the two earlier estimates, 0.900 and 0.943.

In Section 8.3 we computed the variance of the number of trials before the first A_1 occurrence. Equation 8.32 gives for this variance when $q_0 = 1$,

$$(11.12) \qquad \sigma^2 = 2\Psi(\alpha_2, 1) + \Phi(\alpha_2, 1) - [\Phi(\alpha_2, 1)]^2,$$

where $\Psi(\alpha, \beta)$ is the function given in Table A. We may use the estimate $\hat{\alpha}_2$ just obtained, compute σ^2 from the above formula, and compare the result with the variance of the trials to first avoidance obtained from Table 11.1. For $\alpha_2 = 0.92$ we get from Table A, $\Phi = 4.39$ and $\Psi = 9.97$. Equation 11.12 then gives $\sigma^2 = 5.06$ and $\sigma = 2.25$. The observed standard deviation is 2.25. Such close agreement is pleasant to find.

METHOD 3. We have already obtained three estimates of α_2, but we now obtain another based upon the maximum likelihood procedure of Sections 9.9 and 9.10. We are concerned only with that portion of the data up through the first avoidance; hence $k = 0$ in equation 11.5, and we have

$$(11.13) \qquad q_{n,0} = \alpha_2{}^n \qquad (k = 0).$$

This equation is of the form of equation 9.40, and so the procedure developed in Section 9.9 is appropriate. The index ν of that section becomes trial number n and so the quantities N_ν will become N_n, the number of dogs on trial n with zero previous avoidances. Furthermore, $x_\nu = x_n$ is the number of the N_n dogs that avoid on trial n. The numbers N_n and x_n are readily tabulated from the data given in Table 11.1. We give the results in Table 11.2.

In Section 9.10 we presented three methods of obtaining the maximum likelihood estimate. The third procedure we use first. From equation 9.74 we have

$$(11.14) \qquad \hat{\alpha}_2 \cong 1 - \frac{\sum\limits_{n=0}^{\infty} x_n}{\sum\limits_{n=0}^{\infty} n N_n}.$$

In Table 11.2, we then see that

$$(11.15) \qquad \hat{\alpha}_2 = 1 - \frac{30}{447} = 0.933.$$

TABLE 11.2
Tabulations obtained from the data in Table 11.1.

n	N_n	x_n	nN_n	$n(N_n - x_n)$	n^2N_n
0	30	0	0	0	0
1	30	3	30	27	30
2	27	3	54	48	108
3	24	4	72	60	216
4	20	7	80	52	320
5	13	4	65	45	325
6	9	2	54	42	324
7	7	4	49	21	343
8	3	2	24	8	192
9	1	0	9	9	81
10	1	1	10	0	100
11	0	0	0	0	0
12	0	0	0	0	0
Totals		30	447	312	2039

METHOD 4. This approximate value of the maximum likelihood estimate can now be used to get a more exact value. We use the first procedure given in Section 9.10. The sum D_1, defined by

$$(11.16) \qquad D_1 = \sum_{n=0}^{\infty} n(N_n - x_n),$$

has the value 312 as can be seen from Table 11.2. We must solve numerically equation 9.66, which is in our present notation

$$(11.17) \qquad D_1 = \sum_{n=0}^{\infty} x_n g_n(\hat{\alpha}_2).$$

The function $g_n(\hat{\alpha}_2)$ can be obtained from Table C for given values of $\hat{\alpha}_2$. We proceed by trial-and-error as shown in Table 11.3. We first try $\hat{\alpha}_2 = 0.93$, since we just obtained $\hat{\alpha}_2 = 0.933$. Using the values of x_n given in Table 11.2 and the values of $g_n(\hat{\alpha}_2)$ from Table C, we perform the multiplications and summation to get a sum of 350.5. This value is too large since $D_1 = 312$ as we have seen, and so we try $\hat{\alpha}_2 = 0.92$. This value of $\hat{\alpha}_2$ leads to a sum of 297.5, which is quite near $D_1 = 312$. By interpolation we get finally

$$(11.18) \qquad \hat{\alpha}_2 = 0.923.$$

This happens to agree to three places with the estimate given by equation 11.11, which was obtained from the statistic F_1.

TABLE 11.3

Computations made for numerical solution of equation 11.17.

n	x_n	$g_n(0.93)$	$x_n g_n(0.93)$	$g_n(0.92)$	$x_n g_n(0.92)$
1	3	13.29	39.87	11.50	34.50
2	3	12.80	38.40	11.02	33.06
3	4	12.33	49.32	10.56	42.24
4	7	11.88	83.16	10.10	70.70
5	4	11.43	45.72	9.67	38.68
6	2	11.00	22.00	9.24	18.48
7	4	10.57	42.28	8.83	35.32
8	2	10.16	20.32	8.43	16.86
9	0	9.77	0	8.05	0
10	1	9.38	9.38	7.68	7.68
Totals			350.55		297.52

Now that we have obtained the maximum likelihood estimate of α_2 we estimate its variance by the method presented in Section 9.12. Since $q_0 = 1$, we can use the approximate formula 9.84, which becomes

$$(11.19) \qquad \sigma^2(\hat{\alpha}_2) \triangleq \frac{\hat{\alpha}_2^2(1 - \hat{\alpha}_2)}{\sum_{n=0}^{\infty} n N_n - (1 - \hat{\alpha}_2) \sum_{n=0}^{\infty} n^2 N_n}.$$

Using the value $\hat{\alpha}_2 = 0.92$ and the value of the sums in Table 11.2, we then get

$$(11.20) \qquad \sigma^2(\hat{\alpha}_2) \triangleq \frac{(0.92)^2(0.08)}{447 - (0.08)2039} = 0.00024,$$

and

$$(11.21) \qquad \sigma(\hat{\alpha}_2) \triangleq 0.015.$$

Only the data up through the first avoidance of each dog has been used in obtaining estimates of α_2. Data beyond the first avoidance depend on α_2 but also depend on the avoidance parameter α_1. Rather than try to use all the data to give simultaneous estimates of α_1 and α_2, we use the estimates of α_2 already obtained and proceed to estimate α_1 from the remainder of the data. In the next section we behave as if the true value of α_2 were 0.92.

11.5 ESTIMATION OF THE AVOIDANCE PARAMETER, α_1

Having estimated α_2, we now consider some estimates of the avoidance parameter α_1. First we consider the statistic T_2, defined in Section 9.6 to be the mean total number of A_2 occurrences (shocks). Equation 9.20

becomes for $q_0 = 1$,

$$(11.22) \qquad T_2 \triangleq \frac{-\log \dfrac{1-\alpha_2}{1-\alpha_1}}{\alpha_2 - \alpha_1} = T(\alpha_2, \alpha_1),$$

where $T(\alpha_2, \alpha_1)$ is the function given in Table B. From the data we get $T_2 = 7.80$. Using the previously obtained value $\alpha_2 = 0.92$, we obtain by interpolating in Table B,

$$(11.23) \qquad \hat{\alpha}_1 = 0.807.$$

We now develop another scheme for estimating α_1. First we consider only those data for which there is precisely one previous avoidance ($k = 1$). Equation 11.5 gives

$$(11.24) \qquad q_{n,1} = \alpha_1 \alpha_2^{n-1}.$$

Since we have assumed that $\alpha_2 = 0.92$ we need only estimate $q_{n,1}$ for a given value of n and then compute the estimate of α_1 from the last equation. An unbiased estimate of $q_{n,1}$ is readily obtained from the data. We merely count up the number of dogs on trial n that have just one previous avoidance; we call this number $N_{n,1}$. Then we note how many of those $N_{n,1}$ dogs avoided on trial n, and we call this number $x_{n,1}$. In Table 11.4 we present these tabulations. An unbiased estimate of $q_{n,1}$ is then $1 - (x_{n,1}/N_{n,1})$, and we have from equation 11.24,

$$(11.25) \qquad \hat{\alpha}_{1,n} = \frac{1 - \dfrac{x_{n,1}}{N_{n,1}}}{\alpha_2^{n-1}}.$$

For each trial n on which $N_{n,1}$ is not zero we obtain an estimate of α_1 from this equation, and so we denote the estimate for trial n by $\hat{\alpha}_{1,n}$.

Wishing to combine the several estimates $\hat{\alpha}_{1,n}$ in a sensible way, we take a weighted mean:

$$(11.26) \qquad \hat{\alpha}_1 = \frac{\sum\limits_n W_n \hat{\alpha}_{1,n}}{\sum\limits_n W_n}.$$

The problem is to determine the weights W_n. It is well known that the variance of the grand estimate $\hat{\alpha}_1$ is minimized when the weights W_n are inversely proportional to the variances of the $\hat{\alpha}_{1,n}$, provided that the $\hat{\alpha}_{1,n}$ are uncorrelated; and they are uncorrelated. The variance of the estimate, $1 - (x_{n,1}/N_{n,1})$, of $q_{n,1}$ is (from the binomial)

$$(11.27) \qquad \frac{q_{n,1}(1 - q_{n,1})}{N_{n,1}} = \frac{\alpha_1 \alpha_2^{n-1}(1 - \alpha_1 \alpha_2^{n-1})}{N_{n,1}}.$$

TABLE 11.4

Tabulations obtained from Table 11.1 and used to obtain the estimates of α_1 given in Table 11.5.

n	$N_{n,1}$	$x_{n,1}$	$N_{n,2}$	$x_{n,2}$	$N_{n,3}$	$x_{n,3}$	$N_{n,4}$	$x_{n,4}$	$N_{n,5}$	$x_{n,5}$	$N_{n,6}$	$x_{n,6}$
0	0	–	0	–	0	–	0	–	0	–	0	–
1	0	–	0	–	0	–	0	–	0	–	0	–
2	3	1	0	–	0	–	0	–	0	–	0	–
3	5	1	1	0	0	–	0	–	0	–	0	–
4	8	5	2	0	0	–	0	–	0	–	0	–
5	10	5	7	3	0	–	0	–	0	–	0	–
6	9	6	9	5	3	3	0	–	0	–	0	–
7	5	3	10	6	5	3	3	2	0	–	0	–
8	6	4	7	5	8	4	4	4	2	1	0	–
9	4	2	6	4	9	8	4	4	5	5	1	0
10	2	0	4	3	5	5	8	6	4	2	6	4
11	3	1	1	1	3	2	7	3	8	5	4	4
12	2	1	1	1	2	2	6	6	6	5	5	4
13	1	0	1	1	1	1	2	1	7	5	6	4
14	1	1	0	0	1	1	2	1	3	2	7	5
15	0	–	1	1	0	–	2	1	2	1	4	4
16	0	–	0	–	1	1	1	1	2	2	1	0
17	0	–	0	–	0	–	1	1	1	1	3	3
18	0	–	0	–	0	–	0	–	1	1	1	0
19	0	–	0	–	0	–	0	–	0	–	2	2
Sums	59	30	50	30	38	30	40	30	41	30	40	30

In equation 11.25 we see that $\hat{a}_{1,n}$ is obtained by dividing the estimate of $q_{n,1}$ by $\alpha_2{}^{n-1}$. Therefore, the variance of $\hat{a}_{1,n}$ is obtained by dividing the variance of the estimate of $q_{n,1}$ by the *square* of $\alpha_2{}^{n-1}$, that is,

(11.28)
$$\sigma^2(\hat{a}_{1,n}) = \frac{\alpha_1(1 - \alpha_1\alpha_2{}^{n-1})}{\alpha_2{}^{n-1}N_{n,1}}.$$

The common factor, α_1, does not matter, and so we can take the weights to be

(11.29)
$$W_n = \frac{\alpha_2{}^{n-1}N_{n,1}}{1 - \alpha_1\alpha_2{}^{n-1}}.$$

We then substitute these weights in equation 11.26. The numerator is

(11.30)
$$\sum_n W_n\hat{a}_{1,n} = \sum_n \frac{N_{n,1} - x_{n,1}}{1 - \alpha_1\alpha_2{}^{n-1}},$$

and the denominator is

$$(11.31) \qquad \sum_n W_n = \sum_n \frac{\alpha_2{}^{n-1}N_{n,1}}{1 - \alpha_1\alpha_2{}^{n-1}} .$$

Both of the last two sums contain α_1, the parameter being estimated, and so we replace α_1 by $\hat{\alpha}_1$ in the last two equations. We then get from equation 11.26

$$(11.32) \qquad \sum_n \frac{\hat{\alpha}_1\alpha_2{}^{n-1}N_{n,1}}{1 - \hat{\alpha}_1\alpha_2{}^{n-1}} = \sum_n \frac{N_{n,1} - x_{n,1}}{1 - \hat{\alpha}_1\alpha_2{}^{n-1}} .$$

This simplifies to

$$(11.33) \qquad \sum_n \frac{x_{n,1}}{1 - \hat{\alpha}_1\alpha_2{}^{n-1}} = \sum_n N_{n,1}.$$

This equation must be solved numerically to obtain $\hat{\alpha}_1$. The process is somewhat laborious, but not excessively so, since we have already obtained some estimates of α_1 and know the range it is in. The right side of equation 11.33 has the value 59 from the tabulations in Table 11.4. We have computed the left side for $\hat{\alpha}_1 = 0.73$ and $\hat{\alpha}_1 = 0.74$ and obtained 58.83 and 59.67, respectively. Interpolating, we then get

$$(11.34) \qquad \hat{\alpha}_1 = 0.732,$$

as the solution of equation 11.33 for the data in Table 11.4.

The variance of the last estimate of α_1 may be obtained from the relation

$$(11.35) \qquad \sigma^2(\hat{\alpha}_1) = \frac{\sum\limits_n W_n{}^2\sigma^2(\hat{\alpha}_{1,n})}{[\sum\limits_n W_n]^2} ,$$

though this is probably an underestimate because we have assumed α_2 to be known. Furthermore, the estimates of α_1 and α_2 are surely correlated. The weights are given by equation 11.29 and the variances $\sigma^2(\hat{\alpha}_{1,n})$ by equation 11.28. Some algebra leads to the result

$$(11.36) \qquad \sigma^2(\hat{\alpha}_1) = \alpha_1/\sum\limits_n W_n.$$

When we replace α_1 by its estimate 0.732, compute the weights from equation 11.29, and substitute these in the preceding equation, we have the estimate

$$(11.37) \qquad \sigma(\hat{\alpha}_1) \cong 0.095.$$

We use this result shortly.

The procedure just described for obtaining an unbiased estimate of α_1 may be extended to obtain unbiased estimates of α_1^k for $k = 2, 3, \cdots$. An unbiased estimate of $q_{n,k}$ of equation 11.5 is $1 - (x_{n,k}/N_{n,k})$, where $N_{n,k}$ is the number of dogs on trial n with precisely k previous avoidances, and $x_{n,k}$ is the number of those dogs that avoid on trial n. The development follows that given above for $k = 1$, and so we omit the details. Analogous to equation 11.33 we get

$$(11.38) \qquad \sum_n \frac{x_{n,k}}{1 - \widehat{\alpha_1}^k \alpha_2^{n-k}} = \sum_n N_{n,k},$$

and these must be solved for the estimates $\widehat{\alpha_1^k}$. Although the estimate of α_1^k is unbiased, its kth root, the corresponding estimate of α_1, will be somewhat biased for $k > 1$. We have computed these estimates of α_1 for $k = 2, 3, 4, 5$, and 6, and they are given in Table 11.5. When we take a simple average of the six estimates of α_1 we get finally

$$(11.39) \qquad \hat{\alpha}_1 = 0.797.$$

TABLE 11.5

Estimates of the avoidance parameter, α_1, made from equation 11.38.

k	$\widehat{\alpha_1^k}$	$\hat{\alpha}_1$
1	0.732	0.732
2	0.642	0.801
3	0.350	0.705
4	0.424	0.807
5	0.468	0.859
6	0.449	0.875
Mean		0.797

The variance of this grand estimate of α_1 may be computed—or at least estimated—in several ways. However, we use only a rather rough estimate. Equation 11.37 gives the standard deviation of the estimate of α_1 obtained for $k = 1$. We assume that the corresponding standard deviation for $k = 2, 3, 4, 5, 6$ is about this same value, and so we merely divide 0.095 by $\sqrt{6}$ since we have averaged six estimates. We then get

$$(11.40) \qquad \sigma(\hat{\alpha}_1) \sim 0.04,$$

as the standard deviation of the estimate given by equation 11.39. The estimate of 0.797 of equation 11.39 compares very well with our earlier estimate of 0.807 given by equation 11.23. For further discussion we use the estimate 0.80.

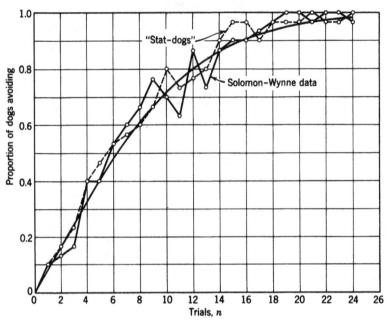

Fig. 11.1. Proportion of dogs that avoided on each trial in the Solomon-Wynne experiment (circles joined by solid lines), proportion of the stat-dogs that avoided on each trial (circles joined by dashed lines), and the theoretical means, $V_{1,n}$, computed from equation 11.41 (smooth curve).

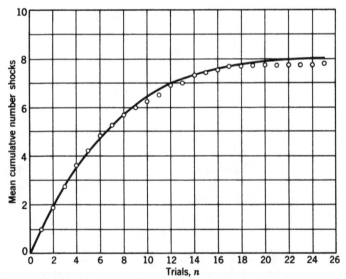

Fig. 11.2. Mean cumulative number of shocks received by the dogs in the Solomon-Wynne experiment (circles) and the cumulative curve computed with the aid of equation 11.41.

11.6   GOODNESS-OF-FIT

In the preceding two sections we computed several estimates of the parameters α_2 and α_1. The values we now use are $\alpha_2 = 0.92$ and $\alpha_1 = 0.80$. We use these values to compute some theoretical properties of the data and compare them with the observed properties. We are interested in the proportion of dogs that avoid on each trial; this proportion is compared to the theoretical means $V_{1,n}$ discussed in Section 8.2 and elsewhere.

The approximate explicit formula for the means is given by equation 8.8. For $\lambda = 1$, $p_0 = 0$, and the above values of α_1 and α_2, this equation becomes

$$(11.41) \qquad V_{1,n} \cong \frac{1}{2} \left\{ 0.33 + 1.67 \, \frac{0.67 e^{0.20n} - 1}{0.67 e^{0.20n} + 1} \right\}.$$

For $n = 0, 1, \cdots, 25$ we have computed the right side of this equation and show the results in Fig. 11.1, along with the proportions from the data given in Table 11.1.

Another way to present the data is to plot the mean cumulative number of shocks versus trial number. In Fig. 11.2 we show such a plot for the data in Table 11.1. The theoretical curve was obtained by integrating $(1 - V_{1,n})$ over n, using equation 11.41 to give $V_{1,n}$.

For further comparisons we have made thirty Monte Carlo computations (called "stat-dogs"), using the operators

$$(11.42) \qquad \begin{aligned} Q_1 p &= 0.80p + 0.20, \\ Q_2 p &= 0.92p + 0.08, \end{aligned}$$

and $p_0 = 0$. In Table 11.6 we show the resulting "data," and in Table 11.7 we compare several statistics of these data with the corresponding statistics of the Solomon-Wynne data. The close agreement between these two sets of statistics is evidence that the model gives a reasonably accurate description of the data. The reader may find it instructive to compare the dogs and stat-dogs not only by eye but also numerically on the basis of any statistic that occurs to him.

Now that we have obtained some numerical computations from the model we measure the goodness-of-fit in more formal ways. First, we make a run test, as described in Section 9.13. We note whether the computed $V_{1,n}$ of Fig. 11.1 is above (+) or below (−) the observed proportion of avoidances on each trial. We see in Fig. 11.1 that there are eight pluses and seventeen minuses; the number of runs is $d = 13$. The critical value for too few runs is 7, and the critical value for too many runs is 16, both at the 5 percent level. Hence, we conclude that the agreement between the model and the data is satisfactory as far as the run test is concerned.

TABLE 11.6

Record of "shocks" (occurrences of A_2) of 30 stat-dogs for 25 trials each, computed with the operators of equations 11.42.

Dog	— Trial n																								
	0	1	2	3	4	5	6	7	8	9	10	11	12	13	14	15	16	17	18	19	20	21	22	23	24
1	S	S	S	S	S	S	S	S																	
2	S	S	S	S	S		S	S	S		S	S		S	S	S	S			S					
3	S	S		S			S	S		S	S	S						S							
4	S	S		S				S		S	S														
5	S	S	S	S	S	S	S	S	S			S						S							
6	S	S				S			S																
7	S	S	S	S				S																	
8	S	S	S		S				S	S		S													
9	S	S	S	S		S	S		S	S		S		S											
10	S	S	S		S				S	S		S		S											
11	S	S	S	S		S		S				S	S	S				S							
12	S	S	S	S					S	S		S	S	S											
13	S	S	S	S	S	S			S	S															
14	S	S	S	S	S																				
15	S	S	S	S			S							S											
16	S	S	S		S					S		S		S											
17	S	S	S	S	S		S	S	S	S		S	S	S											
18	S	S	S	S					S	S											S		S		
19	S			S		S		S										S							
20	S	S	S	S		S	S																		
21	S	S		S		S					S			S									S		
22	S	S		S		S																			
23	S	S	S	S	S	S			S	S	S														
24	S	S		S	S	S			S	S		S		S											
25	S	S	S	S	S			S	S	S				S											
26	S	S	S	S	S	S	S	S			S	S		S				S						S	
27	S	S		S	S		S	S		S	S	S		S											
28	S			S		S	S					S	S	S				S							
29	S	S	S	S	S	S		S													S				
30	S	S	S		S		S	S			S		S	S											
Total	30	27	25	23	18	16	14	13	12	10	6	8	7	6	3	1	1	3	1	1	1	0	1	1	0

TABLE 11.7

Comparisons of the stat-dog "data" in Table 11.6 and the Solomon-Wynne data in Table 11.1.

	Stat-Dogs		Real Dogs	
	Mean	S.D.	Mean	S.D.
Trials before first avoidance	4.13	2.08	4.50	2.25
Trials before second avoidance	6.20	2.06	6.47	2.62
Total shocks	7.60	2.27	7.80	2.52
Trial of last shock	12.53	4.78	11.33	4.36
Alternations	5.87	2.11	5.47	2.72
Longest run of shocks	4.33	1.89	4.73	2.03
Trials before first run of four avoidances	9.47	3.48	9.70	4.14

Next we compute the statistic U defined in Section 9.13. We need an expression for the second raw moment $V_{2,n}$. From equation 8.7 with $\lambda = 1$ we get

$$(11.43) \quad V_{2,n} = \frac{(V_{1,n+1} - V_{1,n}) - (1 - \alpha_2)(1 - V_{1,n})}{\alpha_1 - \alpha_2} + V_{1,n}.$$

Thus the statistic U of equation 9.92 is

$$(11.44) \quad U = \frac{\alpha_2 - \alpha_1}{N} \sum_n \frac{(x_n - N V_{1,n})^2}{(V_{1,n+1} - V_{1,n}) - (1 - \alpha_2)(1 - V_{1,n})}.$$

Using the values of $V_{1,n}$ computed from equation 11.41 and the values $\alpha_2 = 0.92$, $\alpha_1 = 0.80$, and $N = 30$, we obtained $U = 18.6$. Assuming twenty-two degrees of freedom (twenty-five trials less three parameters estimated) we obtain $P = 0.67$ from a χ^2 table. We again conclude that the fit is satisfactory.

The assumed commutativity of the operators implies that the avoidance probability after say two avoidances and three shocks is independent of the order in which those avoidances and shocks occurred. This in turn implies, among other things, that the number of additional trials required for say the fifth avoidance is independent of that order. It was suggested to us that this prediction can be tested, in principle, by examining the data, but it turns out that the number of observations available for such a test is small, even with data on thirty dogs.

That we succeed so well in fitting the model to the data is undoubtedly a result, in part, of the great care taken by the experimenters in controlling conditions. Considerable effort was made to control the strength of shock and the height of the barrier relative to each dog, to eliminate disturbing influences during the course of the experiment, etc. This analysis of the data shows that there is little evidence for dog-to-dog differences, and this is testimony to the carefully controlled conditions in the experiment.

11.7 A THEORETICAL INTERPRETATION

In their theoretical paper, Solomon and Wynne discuss avoidance training in terms of a dual theory of learning. They conceive that two effects take place: (1) By classical conditioning, the conditioned stimulus (CS) acquires the essential property of the unconditioned stimulus (US), namely, ability to evoke the withdrawal response; and (2) by instrumental conditioning, the probability of the response with the CS becomes greater. Cessation of anxiety is considered to be the reinforcing agent, but anxiety is produced by the CS only after the classical conditioning has occurred.

In the remarks which follow we attempt to reconcile this theoretical position of Solomon and Wynne with the model discussed in the preceding sections.

In Chapter 2 we presented a set-theoretic description of instrumental conditioning and demonstrated that our linear operators could be deduced from some simple assumptions about stimulus sampling and conditioning (as proposed by W. K. Estes). Hence, we have already provided an interpretation of our operator Q_1 in terms of instrumental conditioning;

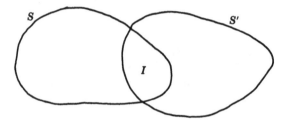

Fig. 11.3. Two sets of stimuli, S and S', which have an overlap or intersection I. The set S denotes the experimental situation with the conditioned stimulus (CS, light out and gate up), and the set S' represents the situation with the unconditioned stimulus (US, shock). The intersection I represents those stimuli common to the two situations.

in a given situation (experimental box with the CS), a response (jumping) has a probability p of occurrence and, when it occurs, the probability changes to $Q_1 p$.

Next, we interpret our operator Q_2 in terms of the classical conditioning effect described by Solomon and Wynne. We denote the experimental situation with the CS by a set S and the situation with the US by a set S'. These sets have an intersection I as shown in Fig. 11.3. We define an index η by

$$(11.45) \qquad \eta = \frac{\mathscr{M}(I)}{\mathscr{M}(S)},$$

where $\mathscr{M}(\ )$ denotes the measure of the set or subset named in the parentheses. The elements contained in S' are assumed to be completely conditioned to the jumping response, that is, the response has probability one of occurring on a trial on which the US appears. We assume that the similarity of S to S' increases each time the CS is followed by the US. In particular, we assume that the new index is

$$(11.46) \qquad Q_2\eta = \alpha_2\eta + (1 - \alpha_2).$$

Now if none of the elements contained in the complement of I in S are conditioned to the jumping response, the probability p of jumping in S

(avoidance) equals the index η. Therefore, under these conditions, equation 11.46 leads at once to our basic equation for $Q_2 p$ (equation 11.2).

There remains to be discussed how the two effects can be combined to lead to the same two operators Q_1 and Q_2. As a result of assumptions already made, Q_1 will have the form given in equations 11.2 even though a subset I of S is completely conditioned to jumping. Thus, for any similarity η, the first of equations 11.2 describes the effect of an avoidance on the probability of avoidance. However, after one or more avoidances

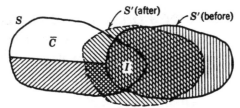

Fig. 11.4. The situations S and S' before and after a trial on which shock occurs. The shaded area indicates stimuli conditioned to jumping; it is assumed that all elements in S' (situation with the US) are conditioned to jumping and that a shock trial increases the measure of the intersection of S and S'.

have occurred we have elements conditioned to jumping in the manner shown in Fig. 11.4: a part of the complement of I in S is conditioned. Now when the US follows the CS, the new S' is as shown by the dotted line in Fig. 11.4. The index increases by an amount

$$(11.47) \qquad \Delta\eta = (1 - \alpha_2)(1 - \eta).$$

The probability q of not avoiding is

$$(11.48) \qquad q = \frac{\mathscr{M}(\bar{C})}{\mathscr{M}(S)},$$

where $\bar{C}$ is the non-conditioned part of S. This probability decreases by an amount

$$(11.49) \qquad \Delta q = \frac{\mathscr{M}(\bar{C})}{\mathscr{M}(S) - \mathscr{M}(I)}\, \Delta\eta.$$

Using equations 11.45, 11.47, and 11.48 in equation 11.49 gives

$$(11.50) \qquad \Delta q = (1 - \alpha_2)q.$$

Hence the new probability of not avoiding is

$$(11.51) \qquad \tilde{Q}_2 q = q - \Delta q = \alpha_2 q.$$

This is the assumed operator shown in equations 11.3.

We have just shown how our two basic operators of equations 11.2 and

11.3 can be deduced from a set-theoretic description of the two effects described by Solomon and Wynne. The instrumental part follows from the simple conditioning mechanism treated earlier, and the classical part involves an increasing index of similarity of the situation with the CS to the situation with the US. We do not consider this interpretation of the model to be essential or unique. Rather, we have indicated once again that the model *can* be interpreted in terms of current learning theories or principles.

11.8 EXPERIMENTS ON THE CS-US INTERVAL

Although we have estimated parameters from the data and succeeded in reproducing those data with the model, we might ask how the model helps in interpreting the experimental results. The model does provide a measure of the relative effects of shock and avoidance trials. We found that $\hat{\alpha}_2 = 0.92$ and $\hat{\alpha}_1 = 0.80$; since $(0.92)^{2.7} \simeq 0.80$, we can say that an avoidance trial is worth about three shock trials, that is, three successive shocks have the same effect on the probability of avoidance as a single avoidance trial. Furthermore, our analysis of the data suggests but little evidence for dog-to-dog differences. These conclusions seem to us to be useful ones and conclusions that would not have been reached without a model. However, a major value of the model is that it provides two summary statistics of the data which may be used in comparing dogs run under different experimental conditions. We now describe a series of experiments conducted by Solomon and his colleagues and present the results of our analysis of the data.

In the experiment described in Section 11.2, the time interval between the conditioned stimulus (CS, light out and gate raised) and the unconditioned stimulus (US, shock turned on) was 10 seconds. In more recent studies, Brush, Brush, and Solomon [5] have run five additional groups of dogs with CS-US intervals of 2.5, 5, 20, 40, and 80 seconds. There were eleven dogs in each group, and so our estimates of the shock and avoidance parameters α_1 and α_2 for those groups are not as reliable as the estimates for the 10-second dogs. In Table 11.8 we show the mean number of trials, F_1, before the first avoidance and the mean number, T_2, of total shocks received per dog for each group. From these two statistics, we estimated α_1 and α_2 by using equations 11.10 and 11.22. The results are also given in Table 11.8.

The estimates of the shock parameter α_2 indicate (1) that a shock has nearly the same effect with CS-US intervals of 2.5, 5, 10, and 20 seconds; (2) that it has a smaller effect with an interval of 40 seconds; and (3) that it has an appreciably larger effect when the interval is 80 seconds. The estimates of the avoidance parameter α_1 show that an avoidance trial has

TABLE 11.8

Data obtained from the experiment by Brush, Brush, and Solomon on the effect of the CS-US interval in avoidance training of dogs. The mean number of trials before the first avoidance is denoted by $\bar{F}_1$ and the mean number of total shocks per dog is denoted by $\bar{T}_2$. The estimates of the avoidance parameter α_1 and the shock parameter α_2 are given in the last two columns.

CS-US Interval	Number of Dogs	$\bar{F}_1$	$\bar{T}_2$	$\hat{\alpha}_1$	$\hat{\alpha}_2$
2.5	11	4.45	7.64	0.80	0.92
5	11	4.64	7.18	0.76	0.93
10	30	4.50	7.80	0.81	0.92
20	11	5.00	9.09	0.82	0.94
40	11	7.82	11.45	0.79	0.97
80	11	3.27	10.18	0.94	0.86

about the same effect when the CS-US interval is increased from 2.5 to 40 seconds, but that the effect decreases at 80 seconds. Neither estimated parameter, $\hat{\alpha}_1$ nor $\hat{\alpha}_2$, is a monotonic function of the CS-US interval, and this leads us to suspect that more data are needed to give a complete picture of the effect of changing the interval. The variances of the estimates $\hat{\alpha}_1$ and $\hat{\alpha}_2$ are appreciable, even for the thirty dogs in the 10-second group as we saw in Sections 11.4 and 11.5. This may be seen most easily from the variances of the estimates F_1 and T_2 from the data. We saw in Section 11.4 that the standard deviation of the observed number of trials before the first avoidance was 2.25. For thirty dogs, then, F_1 has a standard deviation of about 0.41. The standard deviation of the estimated mean number of total shocks, T_2, is about 0.46. These estimated standard deviations are for the thirty dogs in the 10-second group, and the corresponding estimates for the other groups would, be appreciably larger because the groups are smaller.

11.9 SUMMARY

A recent experiment of Solomon and Wynne [3] on the avoidance training of dogs is analyzed in terms of a model with $Q_1 p = \alpha_1 p + (1 - \alpha_1)$ and $Q_2 p = \alpha_2 p + (1 - \alpha_2)$. Alternative A_1 is identified with avoidance and alternative A_2 is identified with non-avoidance (shock). The initial avoidance probability is taken to be zero. We estimate the avoidance parameter α_1 and the shock parameter α_2 from the data and obtain $\hat{\alpha}_1 = 0.80$ and $\hat{\alpha}_2 = 0.92$. The computations made from the model in terms of those two parameters are in close agreement with the data. A theoretical interpretation of the experiment is described in Section 11.7, and data on the effects of the CS-US interval are discussed in Section 11.8.

REFERENCES

1. Hilgard, E. R., and Marquis, D. G. *Conditioning and learning.* New York: D. Appleton-Century, 1940, pp. 58–62.
2. Keller, F. S., and Schoenfeld, W. N. *Principles of psychology.* New York: Appleton-Century-Crofts, 1950, pp. 311–315.
3. Solomon, R. L., and Wynne, L. C. Traumatic avoidance learning: acquisition in normal dogs. *Psychol. Monogr.,* 1953, **67**, No. 4 (Whole No. 354).
4. Miller, N. E. Learnable drives and rewards. *Handbook of experimental psychology,* S. S. Stevens, ed., New York: Wiley, 1951, pp. 435–472.
5. Brush, F. R., Brush, E. S., and Solomon, R. L. Traumatic avoidance learning: the effects of CS-US interval with a delayed-conditioning procedure. *J. comp. physiol. Psychol.,* 1955, **48**, in press.

CHAPTER 12

An Experiment on Imitation

12.1 THE EXPERIMENT

Imitation has been considered a basic process in social behavior, and it has been shown experimentally that imitation can be learned, that is, that it can be controlled by rewards and punishments. Miller and Dollard have given a detailed analysis of imitative behavior in terms of reinforcement theory, and they describe a number of experiments on rats and on children [1]. These experiments demonstrate that one subject will learn to imitate another subject when it is rewarded for doing so, and that this imitative behavior generalizes to other stimulus situations.

Schein investigated imitation in small groups of army inductees in problem-solving situations [2], but his data show little evidence that imitation can be controlled by rewards and punishments. Later Shwartz conducted a simple imitation experiment (suggested by Schein's work) on grade school children as well as high school students [3]. Some of the data from her experiment are analyzed in this chapter.

In Shwartz's experiment, two children were brought into a room and told that they were to participate in a guessing game. Each child was to guess whether the experimenter was going to say "a" or "b" on each of fifty trials; first, child 1 said "a" or "b," then child 2 said "a" or "b," and then the experimenter said "a" or "b." During each block of 10 trials, the experimenter said what child 1 said 8 times. A schedule was drawn up in advance by randomizing within blocks of 10 trials. The same schedule was used on all pairs of subjects in the experimental groups. According to the design, child 1 was correct 8 out of 10 times, but child 2 had the option of "imitating" child 1 or not; if he did imitate he would be right 8 out of 10 times. Of course the subjects were not told that child 1 was right on 80 percent of the trials; they were told that they were used together to save time and "somebody had to be first."

Four main groups of subjects were used in Shwartz's experiment, with twenty pairs of subjects in each group. There were two experimental groups and two control groups; two age levels were used—fourth grade students and high school students. The control groups received the same treatment as the experimental groups except that child 1 was "rewarded" on only 5 out of 10 trials. The main results found were that the experimental grade school children imitated significantly more than their controls whereas there was no significant difference between the experimental and control groups at the high school level. Furthermore, both control groups imitated less than half the time.

In the next section we describe a model for analyzing Shwartz's data. In later sections we estimate the model parameters from the data, and finally we discuss the goodness-of-fit of the model.

12.2 THE MODEL

The behavior of child 2 is of major interest; he says "a" or "b" and is right or wrong, depending on whether or not he imitated child 1. In constructing the model for this experiment we could identify the responses "a" and "b" of child 2 with alternatives A_1 and A_2 in the mathematical system, but, since the purpose is to study the imitation behavior of child 2, we say that alternative A_1 occurred if child 2 made the same response as child 1 and that alternative A_2 occurred if child 2 made the opposite response. In other words, the response patterns aa and bb are identified with A_1 and the patterns ab and ba with A_2. As always, these identifications are not unique.

If it should happen that the first child always said "a" after a few trials, it would be difficult to argue that the second child was learning to imitate; we could just as well argue that the second child was learning to say "a" independent of the first child. However, in Shwartz's experimental grade school group, the number of "a" responses made by the first child averaged 26.25 for the 50 trials, and ranged from 23 to 33. Thus we infer that child 2 is not just learning to say "a" or "b."

The events E_1 and E_2 in the mathematical system are identified with the announcements of the experimenter following the guesses of the subjects. When the experimenter confirms the guess of child 1, event E_1 occurs. When the experimenter denies the guess of child 1, event E_2 occurs. If we choose, we may consider that E_1 rewards child 2 for imitating or punishes him for not imitating, depending on what child 2 actually did. We consider this a single event, independent of child 2's behavior. Similarly we may regard E_2 as reward of non-imitation or punishment of imitation. With these identifications, the events are *experimenter controlled*. Event E_1 occurs on a specified 8 trials in each block of 10. Hence, if p_n is the probability that child 2 imitates on trial n, the order of application of operators Q_1 and Q_2 to p_n is completely specified by the experimenter's schedule and is independent of the behavior of the subjects. Furthermore, if we assume that the subjects are "identical," that is, have the same initial probability of imitation and the same parameter values, then all subjects have the same probability p_n on trial n. We do not obtain a distribution of probabilities as we did in the applications presented in earlier chapters. This fact makes the analysis especially simple.

The operator Q_1, applied to p_n when event E_1 occurs, is assumed to be given by

$$(12.1) \qquad\qquad Q_1 p_n = \alpha_1 p_n + (1 - \alpha_1),$$

that is, $\lambda_1 = 1$. When the experimenter confirms what child 1 said on trial n we apply Q_1 to p_n, the probability that child 2 made the same response as child 1 on trial n. We take the operator Q_2 to be

$$(12.2) \qquad\qquad Q_2 p_n = \alpha_2 p_n,$$

that is, $\lambda_2 = 0$. This operator is applied to p_n when the experimenter denies child 1's response. Since we know the precise sequence of events used in the experiment we can compute the probabilities p_n on every trial in terms of the initial probability p_0 and the parameters α_1 and α_2. In later sections we estimate p_0, α_1, and α_2 from Shwartz's data.

An alternative model, but a more complex one, can be obtained as follows. We could still take the position that child 2 imitates child 1 (response A_1) or does not imitate child 1 (response A_2) on every trial. We could then say that there are two possible *outcomes* (not events): outcome O_1 would be confirmation, by the experimenter, of child 1's response, and outcome O_2 would be denial of that response. As a result, we would have four events listed below.

Child 2's Response	Outcome	Event
A_1 (imitation of child 1)	O_1 (confirmation of child 1's response)	E_{11}
A_1 (imitation of child 1)	O_2 (denial of child 1's response)	E_{12}
A_2 (non-imitation of child 1)	O_1 (confirmation of child 1's response)	E_{21}
A_2 (non-imitation of child 1)	O_2 (denial of child 1's response)	E_{22}

For the four events E_{jk} we would then have operators Q_{jk} defined by

$$Q_{jk} p = \alpha_{jk} p + (1 - \alpha_{jk})\lambda_{jk}.$$

Reasonable restrictions on the parameters can be made: (1) that $\lambda_{11} = 1$ and $\lambda_{12} = 0$, that is, that the probability p of imitations tends to unity or zero, depending on whether or not it is rewarded when it occurs, (2) that

$\lambda_{21} = 1$ and $\lambda_{22} = 0$ by making the analogous assumption about non-imitation. These restrictions give the operators,

$$Q_{11}\,p = \alpha_{11}\,p + (1 - \alpha_{11}),$$

$$Q_{12}\,p = \alpha_{12}\,p,$$

$$Q_{21}\,p = \alpha_{21}\,p + (1 - \alpha_{21}),$$

$$Q_{22}\,p = \alpha_{22}\,p.$$

The experimental design calls for confirmation of child 1's response a fixed proportion π of the trials, independent of the responses made by the two children. Thus we get the following event probabilities.

Event	Probability of Occurrence
E_{11}	$p\pi$
E_{12}	$p(1 - \pi)$
E_{21}	$(1 - p)\pi$
E_{22}	$(1 - p)(1 - \pi)$

We have set up a model which is an application of the case of experimenter-subject-controlled events described in Sections 3.13 and 4.6. Further assumptions about the four parameters α_{jk} can be made to particularize the model. One such set of assumptions leads to the model previously described, namely, we let $\alpha_{11} = \alpha_{21} = \alpha_1$, and $\alpha_{12} = \alpha_{22} = \alpha_2$. In effect these assumptions mean that confirmation after imitation or non-imitation has the same effect on child 2, and a similar statement holds for denial. In other words, outcome O_1 has the same effect on p whether it follows A_1 or A_2, and, similarly, outcome O_2 has the same effect after either response.

We wish to emphasize that other restrictions on the α_{jk} are possible; we might prefer to make no further restrictions and estimate all four parameters from the data. Moreover, still other identifications between the model alternatives and events could be made, leading to still other models.

12.3 THE DATA

We analyze in detail only the data obtained by Shwartz from the experimental grade school group. We represent the data by random variables $x_{i,n}$ defined by

$$(12.3) \quad x_{i,n} = \begin{cases} 1 & \text{if child 2 of the } i\text{th pair } (i = 1, 2, \cdots, N) \\ & \text{imitates child 1 of that pair on trial } n, \\ 0 & \text{otherwise.} \end{cases}$$

The number of imitations occurring on trial n is

$$(12.4) \qquad x_n = \sum_{i=1}^{N} x_{i,n}.$$

Table 12.1 gives the values of x_n obtained by Shwartz for the experimental grade school group on each of fifty trials. For that group, $N = 20$.

TABLE 12.1

The number of imitations x_n made on each of 50 trials for the grade school group studied by Shwartz. There were 20 pairs of subjects. The asterisk (*) beside the trial number indicates that event E_2 occurred on that trial.

n	x_n	n	x_n	n	x_n
0	7	17	6	34	9
*1	10	*18	15	35	10
2	7	19	9	36	13
3	8	20	12	37	14
4	5	21	11	38	14
5	9	22	15	*39	15
*6	10	*23	13	*40	9
7	8	24	9	41	8
8	7	*25	14	42	11
9	9	26	8	43	12
10	13	27	10	44	12
11	12	28	12	45	15
12	15	29	13	46	11
13	11	30	13	*47	17
14	17	*31	15	48	8
15	13	32	9	49	13
*16	15	33	12		

It is convenient to introduce another set of random variables y_n defined by

$$(12.5) \qquad y_n = \begin{cases} 1 & \text{if experimenter said what} \\ & \text{child 1 said on trial } n, \\ 0 & \text{otherwise.} \end{cases}$$

In other words, $y_n = 1$ if event E_1 occurred, and $y_n = 0$ if event E_2 occurred. In Table 12.1 asterisks (*) indicate trials on which event E_2 occurred.

In a later section we inquire about the homogeneity of the subjects, and so in Table 12.2 we give the number of imitations made by each of the 20 subjects for the first 25 trials and the last 25 trials.

TABLE 12.2

The number of imitations made by each subject during the first and last 25 trials of Shwartz's imitation experiment.

Subject	Trials 0–24	Trials 25–49	Total
1	14	21	35
2	21	24	45
3	11	15	26
4	17	16	33
5	19	13	32
6	15	6	21
7	13	15	28
8	12	20	32
9	12	20	32
10	10	17	27
11	15	18	33
12	9	12	21
13	9	11	20
14	8	8	16
15	19	23	42
16	8	9	17
17	11	8	19
18	10	8	18
19	16	11	27
20	17	22	39
Total	266	297	563
Mean	13.30	14.85	28.15

12.4 ESTIMATION OF α_1 AND α_2

The simplicity of the experimental design and the model chosen leads to a direct procedure for estimating the parameters α_1 and α_2. First consider α_2. When event E_2 occurs on trial n,

(12.6) $$p_{n+1} = \alpha_2 p_n.$$

For any trial n we can estimate p_n by the proportion of subjects that imitated on trial n, and we can estimate p_{n+1} by the corresponding proportion on trial $n + 1$. Hence we obtain estimates of α_2 without further ado. The estimate of p_n is x_n/N, and the estimate of p_{n+1} is x_{n+1}/N. Thus we can write

(12.7) $$x_{n+1} \triangleq \alpha_2 x_n.$$

We want to combine these estimates for all trials for which event E_2 occurs, that is, for which y_n of equation 12.5 is zero. Therefore we have

$$(12.8) \qquad \sum_n (1 - y_n)x_{n+1} \triangleq \alpha_2 \sum_n (1 - y_n)x_n.$$

This leads to the estimate

$$(12.9) \qquad \hat{\alpha}_2 = \frac{\sum\limits_n (1 - y_n)x_{n+1}}{\sum\limits_n (1 - y_n)x_n}.$$

In words, the estimate is obtained by counting the total number of imitations on trials immediately after E_2 occurs and dividing that number by the total number of imitations on trials on which E_2 occurs. From the data given in Table 12.1 we get

$$(12.10) \qquad \hat{\alpha}_2 = 81/133 = 0.609.$$

The procedure for estimating α_1 is much the same as that for α_2. We obtain from equation 12.1 and the relation $q_n = 1 - p_n$,

$$(12.11) \qquad q_{n+1} = \alpha_1 q_n.$$

This equation is appropriate when event E_1 occurs on trial n, that is, when $y_n = 1$. For such trials, q_n is estimated by $1 - (x_n/N)$ and q_{n+1} is estimated by $1 - (x_{n+1}/N)$. Therefore, analogous to equation 12.9 is the equation

$$(12.12) \qquad \hat{\alpha}_1 = \frac{\sum\limits_n y_n(N - x_{n+1})}{\sum\limits_n y_n(N - x_n)}.$$

The data from Table 12.1 give

$$(12.13) \qquad \hat{\alpha}_1 = 305/363 = 0.840.$$

Note that $\hat{\alpha}_1$ is considerably larger than $\hat{\alpha}_2$. This implies that events which increase imitation have a smaller effect than events which inhibit imitation. We might argue that this kind of copying is not approved in our society and hence the observed effect.

The same estimation procedures were also used on Shwartz's experimental high school group, and the values $\hat{\alpha}_2 = 0.925$ and $\hat{\alpha}_1 = 0.982$ were obtained. These parameter estimates are much nearer unity than the corresponding ones for the grade school group. This result agrees with Shwartz's finding that the grade school students showed much more reaction to the rewards and punishments for imitation than did the high school students.

12.5 ESTIMATION OF p_0

In the last section we estimated the two parameters α_1 and α_2, but we made no reference to the initial probability of imitation, p_0. We now consider the problem of estimating p_0 from the data.

The obvious estimate of p_0 is the proportion, x_0/N, of imitations observed on trial zero. Table 12.1 gives

$$(12.14) \qquad \hat{p}_0 = (x_0/N) = (7/20) = 0.350.$$

The variance is

$$(12.15) \qquad \sigma^2(\hat{p}_0) = \frac{p_0(1-p_0)}{N},$$

and may be estimated by replacing p_0 with $\hat{p}_0$; the standard deviation is then

$$(12.16) \qquad \sigma(\hat{p}_0) \triangleq 0.107.$$

The variation in $\hat{p}_0$ is large, and so we want to obtain a better estimate of p_0.

Table 12.1 shows that event E_1 occurred on trial 0; equation 12.1 gives

$$(12.17) \qquad p_1 = \alpha_1 p_0 + (1 - \alpha_1).$$

We have already estimated α_1, and we can estimate p_1 by the proportion x_1/N, and from this we obtain another estimate of p_0. The data give

$$(12.18) \qquad 0.500 \triangleq 0.840 p_0 + 0.160.$$

Solving for p_0, we get the estimate

$$(12.19) \qquad \hat{p}_0 = 0.405.$$

We could continue in this way and get an estimate of p_0 for every trial, but the problem remaining is how to combine the estimates. The obvious procedure is to take a weighted mean of the individual trial estimates. We denote the weights by W_n and the individual estimates by $\hat{p}_{0,n}$. The grand estimate is then

$$(12.20) \qquad \hat{p}_0 = \frac{\sum\limits_{n} W_n \hat{p}_{0,n}}{\sum\limits_{n} W_n}.$$

The only question is how to determine the weights W_n. To do this we will appeal to a χ^2 criterion.

We would like to choose p_0 so that we get the best possible fit to the data. The goodness-of-fit may be measured by the Pearson χ^2. The expected number of imitations on trial n is Np_n and the expected number of non-imitations is Nq_n; the observed numbers are x_n and $N - x_n$, respectively. Therefore, the Pearson χ^2 is

(12.21) $$\chi^2 = \sum_n \left\{ \frac{(x_n - Np_n)^2}{Np_n} + \frac{(N - x_n - Nq_n)^2}{Nq_n} \right\}.$$

Using the relation $q_n = 1 - p_n$, we have after simplifications

(12.22) $$\chi^2 = \sum_n \frac{(x_n - Np_n)^2}{Np_n(1 - p_n)}.$$

The smaller the value of this χ^2 the better the fit to the data. The problem is to choose p_0 so that χ^2 is minimized. Since p_n is a function of p_0 we need to differentiate χ^2 with respect to p_0 and set the derivative equal to zero. But unfortunately p_n appears in both the denominator and the numerator of each term in the sum; this complicates greatly the resulting equations.

To avoid the complications just mentioned we shall use Neyman's modification of the Pearson χ^2 criterion. This modification, which might be called the "observed χ^2," is an application of the "best asymptotically normal" estimate developed by Neyman [4]. It is obtained simply by replacing the expected quantity in the denominator of the Pearson χ^2 by the observed quantity:

(12.23) $$\chi^2{}_{\text{ob}} = \sum_n \frac{N(x_n - Np_n)^2}{x_n(N - x_n)}.$$

Neyman has shown that minimizing such a statistic leads to estimates that have essentially all the nice properties of maximum likelihood estimates. The important pleasant feature is that p_n appears only in the numerator, whereas in an ordinary χ^2 both numerator and denominator contain p_n. Differentiating the observed χ^2 with respect to p_0, we get

(12.24) $$\frac{\partial \chi^2{}_{\text{ob}}}{\partial p_0} = \sum_n \frac{-2N^2(x_n - Np_n)}{x_n(N - x_n)} \frac{\partial p_n}{\partial p_0}.$$

We need p_n as a function of p_0. For every trial n we can write

(12.25) $$p_n = A_n + B_n p_0,$$

where A_n and B_n are functions of α_1 and α_2 alone and depend upon the precise sequence of events. For example,

(12.26) $$p_1 = Q_1 p_0 = (1 - \alpha_1) + \alpha_1 p_0,$$

and so $A_1 = 1 - \alpha_1$ and $B_1 = \alpha_1$. Table 12.1 shows that E_2 occurs on trial 1, giving

(12.27) $$p_2 = Q_2 Q_1 p_0 = \alpha_2(1 - \alpha_1) + \alpha_2 \alpha_1 p_0.$$

Therefore, $A_2 = \alpha_2(1 - \alpha_1)$ and $B_2 = \alpha_2 \alpha_1$. Event E_1 occurs on trial 2, giving

(12.28) $$p_3 = Q_1 Q_2 Q_1 p_0 = \alpha_1 \alpha_2(1 - \alpha_1) + (1 - \alpha_1) + \alpha_2 \alpha_1^2 p_0.$$

In this way we can obtain A_n and B_n in terms of α_1 and α_2 for all values of n. Using expression 12.25 in equation 12.24 we get

(12.29) $$\frac{\partial \chi^2_{ob}}{\partial p_0} = \sum_n \frac{-2N^2(x_n - NA_n - NB_n p_0)}{x_n(N - x_n)} B_n.$$

Setting this derivative equal to zero gives the equation

(12.30) $$\sum_n \frac{x_n - NA_n}{x_n(N - x_n)} B_n = \hat{p}_0 \sum_n \frac{NB_n^2}{x_n(N - x_n)}.$$

This equation can be used to solve for the minimum observed χ^2 estimate $\hat{p}_0$. However, we note that the individual trial estimates $\hat{p}_{0,n}$ which

TABLE 12.3

Computations used in obtaining the estimate $\hat{p}_0$ from equations 12.20, 12.31, and 12.33.

n	A_n	B_n	x_n	$\hat{p}_{0,n}$	W_n	$W_n \hat{p}_{0,n}$
0	0	1.000	7	0.350	0.2198	0.0769
1	0.1600	0.8400	10	0.405	0.1411	0.0571
2	0.0976	0.5124	7	0.493	0.0577	0.0284
3	0.2420	0.4304	8	0.367	0.0386	0.0142
4	0.3633	0.3615	5	−0.313	0.0349	−0.0109
5	0.4651	0.3037	9	−0.050	0.0186	−0.0009
6	0.5507	0.2551	10	−0.199	0.0130	−0.0026
7	0.3359	0.1556	8	0.412	0.0050	0.0021
8	0.4422	0.1307	7	−0.705	0.0038	−0.0027
9	0.5314	0.1098	9	−0.741	0.0024	−0.0018
10	0.6064	0.0922	13	0.474	0.0019	0.0009
11	0.6692	0.0774	12	−0.894	0.0013	−0.0012
12	0.7221	0.0650	15	0.429	0.0011	0.0005
13	0.7665	0.0546	11	−3.965	0.0006	−0.0024
14	0.8040	0.0459	17	1.002	0.0008	0.0008
15	0.8353	0.0386	13	−4.801	0.0003	−0.0014
16	0.8617	0.0324	15	−3.448	0.0003	−0.0010
					0.5412	0.1560

entered equation 12.20 may be written as

$$(12.31) \qquad \hat{p}_{0,n} = \frac{(x_n/N) - A_n}{B_n},$$

and so equation 12.30 may be written in the form

$$(12.32) \qquad \sum_n \frac{NB_n{}^2}{x_n(N - x_n)} \hat{p}_{0,n} = \hat{p}_0 \sum_n \frac{NB_n{}^2}{x_n(N - x_n)}.$$

Hence the weights W_n in equation 12.20 are

$$(12.33) \qquad W_n = \frac{NB_n{}^2}{x_n(N - x_n)}.$$

With these weights and the individual trial estimates of equation 12.31 we can compute the grand estimate $\hat{p}_0$ from equation 12.20. Table 12.3 shows the computations for the estimation of p_0. The coefficients A_n and B_n of equation 12.25 were computed for trials 0 to 16, using the estimates $\hat{\alpha}_1 = 0.840$ and $\hat{\alpha}_2 = 0.609$ obtained in the last section. From Table 12.3 we finally get

$$(12.34) \qquad \hat{p}_0 = 0.288.$$

In the next section these estimates are used in measuring goodness-of-fit.

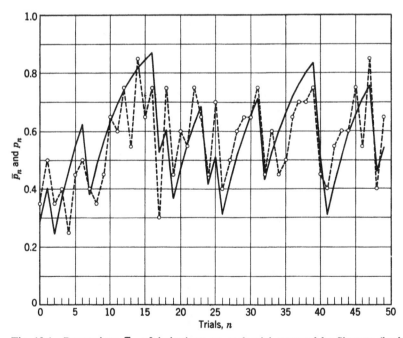

Fig. 12.1. Proportions, $\bar{p}_n$, of imitations on each trial observed by Shwartz (broken lines) and computed probability, p_n, of imitation (solid lines).

12.6   GOODNESS-OF-FIT

Having estimated the three parameters, we can compute the probabilities p_n of imitation on each trial. We use

$$Q_1 p_n = 0.840 p_n + 0.160,$$

(12.35) $$\qquad Q_2 p_n = 0.609 p_n,$$

$$p_0 = 0.288.$$

The results are shown in Table 12.4 along with the observed proportions of imitations on each trial. These are also shown in Fig. 12.1.

TABLE 12.4

Computed probability, p_n, of imitation and observed proportion, $\bar{p}_n$, of imitations on each trial for Shwartz's data.

n	p_n	$\bar{p}_n$	Diff.	n	p_n	$\bar{p}_n$	Diff.
0	0.288	0.350	—	25	0.511	0.700	—
1	0.402	0.500	—	26	0.311	0.400	—
2	0.246	0.350	—	27	0.421	0.500	—
3	0.366	0.400	—	28	0.514	0.600	—
4	0.467	0.250	+	29	0.592	0.650	—
5	0.553	0.450	+	30	0.657	0.650	+
6	0.624	0.500	+	31	0.712	0.750	—
7	0.380	0.400	—	32	0.434	0.450	—
8	0.480	0.350	+	33	0.524	0.600	—
9	0.563	0.450	+	34	0.600	0.450	+
10	0.633	0.650	—	35	0.664	0.500	+
11	0.692	0.600	+	36	0.718	0.650	+
12	0.741	0.750	—	37	0.763	0.700	+
13	0.782	0.550	+	38	0.801	0.700	+
14	0.817	0.850	—	39	0.833	0.750	+
15	0.846	0.650	+	40	0.507	0.450	+
16	0.871	0.750	+	41	0.309	0.400	—
17	0.530	0.300	+	42	0.419	0.550	—
18	0.606	0.750	—	43	0.512	0.600	—
19	0.369	0.450	—	44	0.590	0.600	—
20	0.470	0.600	—	45	0.656	0.750	—
21	0.555	0.550	+	46	0.711	0.550	+
22	0.626	0.750	—	47	0.757	0.850	—
23	0.686	0.650	+	48	0.461	0.400	+
24	0.418	0.450	—	49	0.547	0.650	—

The goodness-of-fit between the computed probabilities p_n and the observed proportions $\bar{p}_n$ may be determined by computing the Pearson χ^2 given by equation 12.22. This was done for the first 25 trials and the last 25 trials separately. The values obtained are given in Table 12.5. The

TABLE 12.5

Results of chi-square computations for testing goodness-of-fit of model to Shwartz's data.

	Trials 0–24	Trials 25–49	Trials 0–49
χ^2	36.00	25.34	61.34
d.f.	22*	22*	47
P	0.03	0.28	0.08

* The number of degrees of freedom was taken to be 22 in order to be conservative (see text).

number of degrees of freedom was taken to be 47 for all 50 trials since three parameters were estimated from the data. To be conservative, we assumed that there were 22 degrees of freedom for the two halves of the data. The probabilities P obtained from χ^2 tables show that there is a significant difference between the computed and observed figures during the first 25 trials but not for the last 25 trials. One might conjecture that child 2 is not alerted to imitation as much in the first 25 trials as in the last 25, and hence that the model is more appropriate for the second half of the data. When all 50 trials are considered, the disagreement is not great—there is an 8 percent chance of a fit at least as bad as the obtained one, assuming the model is correct.

The goodness-of-fit between the computed and observed proportions of imitations can be tested in another way. We can apply the "run test," described in Section 9.13, to see if the differences follow a systematic pattern. In Table 12.4 we have noted whether the difference $p_n - \bar{p}_n$ was positive or negative; 22 are positive and 28 are negative. We then counted up the number of runs of pluses and minuses. (A run is a consecutive series of pluses or a consecutive series of minuses.) A total of 23 runs is observed in Table 12.4. Referring to Section 9.13, we find the expected number of runs is 25.64 and that the variance of the number of runs is 11.89. This leads to a standard deviate of 0.77. Therefore we conclude that there is no significant systematic difference between the computed and the observed proportions.

The expected total number of imitations during the 50 trials can be compared to the observed mean, though it is not much of a check on the model. In terms of the random variables $x_{i,n}$ defined by equation 12.3, the total number of imitations for the ith subject is

$$(12.36) \qquad\qquad T_i = \sum_{n=0}^{49} x_{i,n}.$$

These variables $x_{i,n}$ are independent binomial observations, and so the the expected value of T_i is

$$(12.37) \qquad\qquad E(T_i) = \sum_n E(x_{i,n}) = \sum_n p_n.$$

In other words, to obtain the expected number of imitations made by a single child, sum the probabilities p_n over all trials. Table 12.4 gives

$$(12.38) \qquad\qquad E(T_i) \triangleq 28.535.$$

In Table 12.2 we have tabulated the number of imitations made by each child. The mean of these is

$$(12.39) \qquad\qquad \bar{T} = 28.15,$$

agreeing closely with the expected value just estimated.

We may further compute the expected variance in the numbers T_i from equation 12.36. The $x_{i,n}$ are each independently drawn from a different binomial distribution. Hence,

$$(12.40) \qquad\qquad \sigma^2(T_i) = \sum_n \sigma^2(x_{i,n}),$$

but the variance of $x_{i,n}$ is simply the binomial variance $p_n(1 - p_n)$ and so

$$(12.41) \qquad\qquad \sigma^2(T_i) = \sum_n p_n(1 - p_n).$$

From the values of p_n given in Table 12.4 we get

$$(12.42) \qquad\qquad \sigma^2(T_i) \triangleq 11.02.$$

From the values of T_i given in Table 12.2 for the 20 subjects we computed the variance and obtained

$$(12.43) \qquad\qquad \hat{\sigma}^2(T_i) = 68.33.$$

This observed variance is appreciably larger than the expected variance. Our interpretation is that the subjects are not "identical"—that they did not have the same values of p_0, α_1, and α_2. It is quite possible, however, that a more complicated model would account for the data better than the model used here.

12.7 SUMMARY

A simple experiment on imitation is described and the data analyzed in terms of experimenter-controlled events. The schedule of events is the same for all subjects, giving a single probability (rather than a distribution) for each trial. Three parameters are estimated from the data and the goodness-of-fit of the model considered. It is found that events which inhibit imitation have a greater effect than events which increase it.

REFERENCES

1. Miller, N. E., and Dollard, J. *Social learning and imitation.* New Haven: Yale University Press, 1941.
2. Schein, E. *The effect of reward on imitation.* Ph.D. Thesis, Harvard University, 1952.
3. Shwartz, N. *An experimental study of imitation: the effects of reward and age.* Senior honors thesis, Radcliffe College, 1953.
4. Neyman, J. Contribution to the theory of the χ^2 test. *Proceedings of the Berkeley symposium on mathematical statistics.* Berkeley: University of California Press, 1949, pp. 239–273.

CHAPTER 13

Symmetric Choice Problems

13.1 INTRODUCTION

This chapter deals with a class of experimental problems involving a choice between two or more alternatives on each trial, alternatives with a certain type of symmetry. Examples are the T-maze in which a rat chooses to turn right or left, and a "two-armed bandit" in which a person pushes one of two buttons. The symmetry can be recognized by the fact that interchanging or renaming the alternatives seems to change nothing of psychological importance. For example, whether "left" or "right" is the favorable side is of little interest. Indeed, many experimenters will randomize the favorable side from rat to rat to protect against initial position preferences. The outcomes of the possible choices, however, ordinarily will not be the same. One may result in reward and the other not, or one may pay off a certain proportion of the time and the other a different proportion of the time.

13.2 T-MAZE EXPERIMENTS

An experimental arrangement that has received considerable attention is the T-maze [1, 2]. It is one of the simplest mazes used with rats, because only a single choice point is involved. Although modifications and generalizations of the experimental design have been used with rats and with other subjects, we first focus attention on the simple type of open maze used with rats.

A diagram of the apparatus is shown in Fig. 13.1. A rat is placed at the starting position, s, and it runs to the choice point, c. The rat then goes to one of two goal boxes, A_1 and A_2, where it may or may not receive food. This sequence of behavior constitutes one experimental trial. The procedure is ordinarily repeated many times with the same rat. The total behavior of a rat on an experimental trial is a rather complex course of action. Moreover, from the point of view of stimulus conditioning, discussed in Chapter 2, the problem is not simple. The rat is in a varying stimulus situation as it traverses the maze, and after passing the choice point is in one of two possible stimulus situations. This total behavior on a trial can be broken up, of course, and appropriate measures or indices used for each part. For example, we might ask about the latency of the starting position, s, or the running speed between s and the choice point c. It seems to us, however, that the portion of the rat's behavior peculiar to this experiment is the behavior at the choice point, c. Which of the two alternatives does the rat choose?

In the analysis which follows we ignore all aspects of a rat's behavior on a trial except the way a rat turns at the choice point. On each complete experimental trial, the rat arrives at the choice point where the population of stimulus elements is held constant from trial to trial. Two classes of responses are defined corresponding to the goal box reached, A_1 or A_2. On each experimental trial one and only one of these response classes occurs. An experimental trial, then, corresponds to a trial as defined in Chapter 1—an opportunity for choosing among mutually exclusive and exhaustive alternatives. According to the model, the state of the organism on a particular trial is completely specified by a probability p

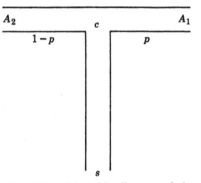

Fig. 13.1. Schematic diagram of the T-maze used by Brunswik [1] and by Stanley [2]. The starting box is denoted by s, the choice point by c, and the goal boxes by A_1 and A_2. The probabilities of going to A_1 and A_2 are p and $1 - p$ respectively.

that the rat will go to goal box A_1 and a probability $q = 1 - p$ that it will go to goal box A_2. When these probabilities are known for every trial we have complete information about the model of the learning process. These probabilities can be estimated from the proportion of turns to goal box A_1 made by a single rat on a number of trials or by the proportion of a population of rats which go to goal box A_1 on a particular trial.

In the experiments of Brunswik and Stanley various schedules of rewards and non-rewards were used on the two sides of the maze. For example, food was placed on one side of the maze on 75 percent of the trials and on the other side on 25 percent. In this example the reward probabilities, 0.75 and 0.25, add to unity, but they need not and in some experiments did not. Stanley ran one group, for example, with reward on one side for 50 percent of the trials and reward on the other side for 0 percent of the trials. (In the following sections we denote such experimental conditions by 75 : 25, 50 : 0, etc., where the first number gives the reward probability on the more favorable side.) Brunswik and Stanley did not actually use complete randomness in setting up the reward schedules, but did use restricted randomness within blocks of trials.

13.3   EXPERIMENTS WITH HUMAN SUBJECTS

The experimental literature contains reports on numerous studies of human behavior in simple two-choice situations. Several of these were modifications and extensions of an experiment described in 1939 by Humphreys [3]. In those experiments, a subject was seated before a panel containing two light bulbs. On each trial, the left light was turned on; a few seconds later, on some trials the right bulb would come on, whereas on other trials the right bulb would not come on. The subject was to guess whether or not the right light would come on during each trial. In Humphreys' original experiment, three schedules were used: continuous reinforcement (right light turned on during each trial), partial reinforcement (right light on during half of trials), and extinction (right light on during none of the trials). If we identify event E_1 with turning on of the right light and event E_2 with not turning it on, we may describe the schedule by the two probabilities π_1 and π_2 of events E_1 and E_2, respectively. Humphreys' schedules were then 100 : 0, 50 : 50, and 0 : 100. (The first number gives the percent of E_1 occurrences.)

Experiments similar to Humphreys' were reported by Grant, Hake, and Hornseth, and, in one of these, schedules of 0 : 100, 25 : 75, 50 : 50, 75 : 25, and 100 : 0 were used [4]. (In other studies, the intertrial interval was varied, but no significant differences from this source were found [5].) Note that in all groups the two event probabilities necessarily summed to unity—the light either came on or it did not.

Jarvik reported an experiment in which subjects predicted on each trial whether the experimenter was going to say the word "check" or the word "plus" [6]. Schedules of 60 : 40, 67 : 33, and 75 : 25 were used in that study. More recently, Hake and Hyman described an experiment in which subjects predicted whether a horizontal or a vertical set of lights would come on [7]. Their schedules were 50 : 50 and 75 : 25, but, in addition, a Markov dependence was introduced into the sequences used with two groups of subjects.

In all the human experiments described so far, one of two events occurred on each trial. If we care to think of confirmation of a guess or prediction as reinforcement of that guess, then one and only one of the two possible guesses was reinforced on each trial. This restriction was not present in the Brunswik and Stanley rat experiments described in the preceding section, for in those experiments the reward probabilities for the two responses could be varied independently. During the spring of

1951 we designed and Henry Gerbrands constructed a "two-armed bandit" to replicate the Brunswik-Stanley experiments, and Jacqueline Jarrett Goodnow carried out the replications. The apparatus consisted of a panel of lights and two push-button electrical switches as shown in Fig. 13.2. A large center light indicated the start of a trial. The subject inserted a poker chip to activate the machine and then pushed one of the two buttons; a light came on above that button. By means of an electromagnet one or more poker chips could be delivered to the subject

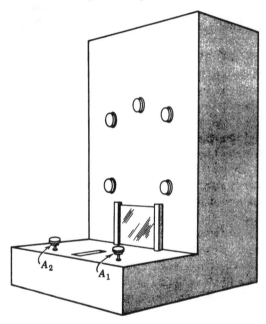

Fig. 13.2. Sketch of the two-armed bandit used by Goodnow, Robillard, and the authors. When the machine is in operation, the three upper lights are on and a soft buzz may be heard. The subject pushes button A_1 or button A_2, and this turns on the light directly above the button pushed. Poker chips are dropped into the region behind the glass.

in a box at the center of the apparatus. If a chip was delivered he "won," if not he "lost." The pay-offs were not automatic but were controlled by the experimenter who sat in an adjacent room; the apparatus appeared to be automatic to the subjects. By appropriate pay-off schedules, various partial reinforcement conditions were introduced. Schedules used by Goodnow and Laval Robillard include 100 : 0, 50 : 0, 75 : 25, 100 : 50, 30 : 0, 80 : 0, 80 : 40, 60 : 30.* (This apparatus was also used by David

---

* The optimum decision rule (the "best strategy") for playing the two-armed bandit for n trials seems not to be known.

Carver to collect data for a senior honors thesis at Harvard. He studied the effects of differential amounts of reward on the two sides under partial reinforcement.)

Detambel [8] and Neimark [9] had subjects push one of two telegraph keys to turn on a light. In Detambel's experiment, one light bulb was used. On each trial, each key was either correct or incorrect; when a correct key was pressed the light came on. In one of Neimark's experiments, the same procedure was used except that there was a light bulb above each key. In both these studies, as well as in the two-armed bandit experiments, the outcome probabilities for the two responses could be manipulated independently.

More recently, M. M. Flood has developed a simple punch-board technique of collecting data. Several columns of holes appeared on a punch-board, and the subject was told to punch one hole in each row, working from the top down. Behind the hole punched out was a special mark to indicate a win, or a different mark to indicate a loss. In one of Flood's original punch-board experiments, nine columns were used (nine-choice situation), and later Flood and Mosteller used these punch-boards for two-choice experiments. (With these boards a subject can punch at his own rate, and so trials may be strongly massed.)

Another variation of the experimental design was developed by Bush, R. L. Davis, and G. L. Thompson. A subject was presented with two ordinary playing cards, face down, and was asked to turn over one of them. If the card turned over was red, the subject was given a nickel; if the card was black, the subject received nothing. The procedure was repeated for many trials. By "stacking" the cards in advance, various partial reinforcement schedules were obtained. This procedure was used with six Santa Monica (California) high-school students during the summer of 1952, and with ten Harvard undergraduates and ten Cambridge (Massachusetts) high-school students during the fall of 1952 as part of a course in experimental social psychology at Harvard.

The procedural differences among the many experiments mentioned above is rather important, we believe, in deciding how to analyze the data. In the studies by Humphreys, by Grant, Hake, and Hornseth, by Jarvik, by Hake and Hyman, and others, one of two environmental changes occurred on every trial independent of the subject's response. As a result, this procedure has been called *non-contingent* [9]: a light came on or it did not, a horizontal or vertical set of lights came on, or the experimenter said "check" or "plus." These procedures suggest that we identify these two possible environmental changes with two events E_1 and E_2 in the model and use the case of experimenter-controlled events first introduced in Section 3.9. A different procedure was used in the studies by

Brunswik and by Stanley, in the two-armed bandit experiments, in the study by Detambel, and in the punch-board and playing-card experiments. In those experiments, the environmental change depended upon the subject's response. The probability of presentation of food or of a poker chip depends upon which choice is made by the subject. Therefore, this procedure has been called *contingent*. It suggests at once the case of experimenter-subject-controlled events first discussed in Section 3.13. The two possible *outcomes* would then be reward and non-reward, or "light on" and "no light on." As a result, there would be four events corresponding to the possible response-outcome pairs.

In the experiments performed by Neimark [9] both procedures were used. In the non-contingent groups, the schedules of appearances and non-appearances of the lights were independent of the subject's responses, whereas in the contingent groups, a light came on only when a subject pressed the appropriate key. This study permits a direct comparison of the two procedures. In addition, Neimark ran several groups of subjects with *three* keys and three lights. We discuss this three-choice problem in a later section.

We have pointed out that the non-contingent procedure suggests the case of experimenter-controlled events, whereas the contingent procedure seems to require the case of experimenter-subject-controlled events. However, both procedures could be analyzed using the case of experimenter-subject-controlled events. We might argue that pressing key 1 and having light 2 come on was not the same event as pressing key 2 and having light 2 come on. Therefore, we would prefer to have an operator for each possibility and to determine from the data whether the two occurrences had the same effect on behavior. Experimenter-subject-controlled events are always more general than experimenter-controlled events. Nevertheless, for practical analysis of data, we find it convenient to reduce the number of parameters in the model, and so we assume that experimenter-controlled events are appropriate for experiments that use the non-contingent procedure.

13.4   A MODEL WITH EXPERIMENTER-CONTROLLED EVENTS

We employ experimenter-controlled events and equal alphas for describing data from the non-contingent experiments. For most of these experiments, one of two environmental changes occurs on each trial; we identify these changes with events E_1 and E_2, which correspond to operators Q_1 and Q_2, respectively. The two possible responses of the subject are identified with responses A_1 and A_2. Furthermore, we take the limit points of the operators such that if E_i occurs on all trials, response A_i

will occur with a probability which tends to unity. The operators are then given by

(13.1)
$$Q_1 p = \alpha p + (1 - \alpha)$$
$$Q_2 p = \alpha p.$$

Event E_1 occurs with probability π_1 and event E_2 with probability $\pi_2 = 1 - \pi_1$.

In Section 5.3 we discussed such a model. In that section, the asymptotic mean was found to be

(13.2) $$V_{1,\infty} = \pi_1.$$

This result implies that, after a large number of trials, the proportion of A_1 responses is equal to the proportion of E_1 occurrences. In the next section we examine data from a number of experiments to see whether human subjects behave according to this prediction of the model.

Although the main interest is in the value of the asymptote, we now describe a procedure for estimating the parameter α. In Section 5.3, the mean on trial n was found to be

(13.3) $$V_{1,n} = \pi_1 - (\pi_1 - V_{1,0})\alpha^n.$$

This equation suggests a simple method of estimating α. Sum this expression from trial 0 to trial $N - 1$ to get

(13.4) $$\sum_{n=0}^{N-1} V_{1,n} = N\pi_1 - (\pi_1 - V_{1,0})\frac{1 - \alpha^N}{1 - \alpha}.$$

If we have a way of estimating the left side of this equation and can also estimate $V_{1,0}$, we can then compute an estimate of α. We next demonstrate how the sum on the left side can be estimated readily from data.

Consider a random variable $x_{i,n}$ which is unity if the ith subject makes response A_1 on trial n and is zero otherwise. For each subject we then obtain a sequence of 1's and 0's. The expected value of $x_{i,n}$, if nothing is known about the ith subject's responses previous to trial n, is

(13.5) $$E(x_{i,n}) = Pr\{x_{i,n} = 1\} = V_{1,n}.$$

Next, define a statistic T_i as the total number of A_1 responses by the ith subject during the first N trials, that is,

(13.6) $$T_i = \sum_{n=0}^{N-1} x_{i,n}.$$

The expected value is

(13.7) $$E(T_i) = \sum_{n=0}^{N-1} E(x_{i,n}) = \sum_{n=0}^{N-1} V_{1,n}.$$

Hence we have, from combining equations 13.4 and 13.7,

$$(13.8) \qquad E(T_i) = N\pi_1 - (\pi_1 - V_{1,0}) \frac{1 - \alpha^N}{1 - \alpha}.$$

The expected value of T_i can be estimated by the mean number T of A_1 responses of K subjects during trials 0 to $N - 1$, that is,

$$(13.9) \qquad T = \frac{1}{K} \sum_{i=1}^{K} T_i.$$

This leads to the estimation equation

$$(13.10) \qquad T \triangleq N\pi_1 - (\pi_1 - V_{1,0}) \frac{1 - \alpha^N}{1 - \alpha}.$$

We are often able to choose N large enough to make α^N negligible compared to unity. When this is so, we can solve for α in the foregoing equation to get

$$(13.11) \qquad \alpha \triangleq 1 - \frac{\pi_1 - V_{1,0}}{N\pi_1 - T}.$$

All that remains is to estimate $V_{1,0}$. This initial mean may be estimated from data obtained during the first few trials, or it can be assumed that $V_{1,0} = 0.5$, for this simply means that the *group* of subjects has no initial preference or bias.

Both α and $V_{1,0}$ can be estimated from equation 13.10 by choosing two values of N. Denote these trial numbers by N_1 and N_2, and let T_1 and T_2 be the corresponding values of T. From the two equations of the form 13.10, we can eliminate $V_{1,0}$ to get

$$(13.12) \qquad \frac{N_1 \pi_1 - T_1}{N_2 \pi_1 - T_2} \triangleq \frac{1 - \alpha^{N_1}}{1 - \alpha^{N_2}}.$$

This equation can be solved numerically for the estimate of α, and then $V_{1,0}$ can be estimated from

$$(13.13) \qquad V_{1,0} \triangleq \pi_1 - (N_1 \pi_1 - T_1) \frac{1 - \alpha}{1 - \alpha^{N_1}}.$$

We use these two equations in the next section.

The model just described is adequate for describing most experiments that use the non-contingent procedure. However, Neimark introduced a modification of the procedure in some of her groups [9], and she modified the model accordingly. Her subjects were required to guess which of two lights would come on during each trial, but on some trials neither light came on. These "blank trials," as she called them, constitute a third class of events which we label E_3.   If we assume an identity operator for such events, we have the set of operators given by

$$Q_1 p = \alpha p + (1 - \alpha) \qquad [\pi_1]$$

(13.14) $\qquad Q_2 p = \alpha p \qquad\qquad\quad [\pi_2]$

$$Q_3 p = p \qquad\qquad\qquad [1 - \pi_1 - \pi_2].$$

The probabilities of application of these three operators are shown in square brackets beside each equation.   We now have a case of three experimenter-controlled events.   From Section 4.4, equations 4.30, we see that

$$\bar{a} = \pi_1(1 - \alpha),$$

(13.15) $\qquad \bar{\alpha} = 1 - (\pi_1 + \pi_2)(1 - \alpha),$

$$V_{1,\infty} = \frac{\pi_1}{\pi_1 + \pi_2}.$$

From equation 4.28 we see that

(13.16) $\qquad\qquad V_{1,n} = V_{1,\infty} - (V_{1,\infty} - V_{1,0})\bar{\alpha}^n.$

Therefore the estimation procedures described above may be used to estimate $\bar{\alpha}$, and then the corresponding estimate of α can be obtained from the second of equations 13.15.   Equations 13.12 and 13.13 can be used by replacing α with $\bar{\alpha}$ and π_1 with $V_{1,\infty}$:

$$\frac{N_1 V_{1,\infty} - T_1}{N_2 V_{1,\infty} - T_2} \triangleq \frac{1 - \bar{\alpha}^{N_1}}{1 - \bar{\alpha}^{N_2}},$$

(13.17)

$$V_{1,0} \triangleq V_{1,\infty} - (N_1 V_{1,\infty} - T_1) \frac{1 - \bar{\alpha}}{1 - \bar{\alpha}^{N_1}}.$$

13.5 DATA FROM EXPERIMENTS USING THE NON-CONTINGENT PROCEDURE

A number of experiments involving the non-contingent procedure are briefly described in Section 13.3.   Data from some of these experiments are now examined, and the results are interpreted in terms of the model described in the preceding section.

In Humphreys' experiment, the schedules used were 100 : 0, 50 : 50, and 0 : 100.   The data presented in his paper [3] indicate that the proportions of guesses of "lights on" approached asymptotes of 1.0, 0.5, and 0, respectively.   These empirical asymptotes agree with those predicted by equation 13.2.   Similar agreement is found in the data of Grant, Hake, and Hornseth [4], who used schedules of 0 : 100, 25 : 75, 50 : 50, 75 : 25, and 100 : 0. Each group of subjects approaches a performance asymptote nearly equal to the probability that the light will come on.

Two of the groups in the Hake and Hyman study [7] were presented with independent sequences of symbols with reward probabilities of 50 : 50 and 75 : 25. These groups tended towards asymptotes of 0.5 and 0.75, respectively, in agreement with equation 13.2. The other two groups were presented with Markov sequences of symbols; these groups also tended towards asymptotes equal to the proportions of symbols of one kind in the sequence. This result agrees with the model predictions in Section 5.11 which described the effects of a Markov sequence of operators with the equal-alpha restriction imposed.

TABLE 13.1

Data obtained by Jarvik [6] in an experiment in which subjects predicted whether the experimenter would say "check" or "plus" on each trial. Three groups of subjects were run with the values of π_1, the proportions of "check" announcements by the experimenter, shown. The proportion of "check" predictions are shown for each trial block. Also the estimates of α obtained from equation 13.11 are given. (These data are reproduced from the *Journal of Experimental Psychology* with the permission of the American Psychological Association.)

Trial Block	Proportions		
	$\pi_1 = 0.60$	$\pi_1 = 0.67$	$\pi_1 = 0.75$
0–10	0.456	0.459	0.456
11–21	0.458	0.518	0.564
22–32	0.461	0.641	0.667
33–43	0.536	0.675	0.804
44–54	0.615	0.611	0.798
55–65	0.622	0.626	0.703
66–76	0.555	0.579	0.771
77–86	0.573	0.674	0.759
Number of subjects	29	21	28
$\hat{\alpha}$	0.982	0.974	0.952

In the Jarvik experiment [6], subjects predicted whether the experimenter would say "check" or "plus" on each trial. The data obtained are presented in Table 13.1 and Fig. 13.3. As can be seen, the proportions of "check" guesses are tending towards the asymptotes predicted by equation 13.2. Also, in Table 13.1 we show for each group the estimates of α obtained from equation 13.11. These estimates indicate that the speed of learning increases with increasing proportions of "checks" in the experimenter's sequence.

Four of the sixteen groups of subjects studied by Neimark [9] received

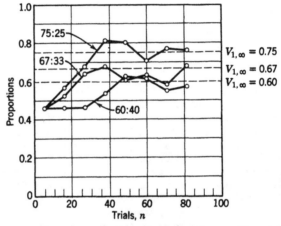

Fig. 13.3. The prediction data obtained by Jarvik [6]. The theoretical asymptotes are shown by the dotted lines.

a non-contingent procedure with two keys. The reward probabilities were 100 : 0, 66 : 0, 66 : 34, and 66 : 17. In Table 13.2 we show the

TABLE 13.2

Results obtained by Neimark [9] from three non-contingent groups in her two-key experiment. The event probabilities, π_1 and π_2, the theoretical asymptotic means, $V_{1,\infty}$, the observed mean number, $\overline{T}_1$, of A_1 responses during the first 30 trials, the observed mean number, $\overline{T}_2$, of A_1 responses during all 100 trials, and the computed estimates of α and $\hat{V}_{1,0}$ are shown. There were 20 subjects in each group.

Group	π_1	π_2	$V_{1,\infty}$	$\overline{T}_1$	$\overline{T}_2$	$\hat{\alpha}$	$\hat{V}_{1,0}$
66 : 0	0.66	0	1.00	21.00	88.75	0.921	0.415
66 : 17	0.66	0.17	0.795	16.60	69.05	0.953	0.378
66 : 34	0.66	0.34	0.66	15.05	56.50	0.982	0.457

results obtained by Neimark from three of these groups along with the estimates of α and $V_{1,0}$ for each, computed from equations 13.17. Note that the estimates of α increase as the difficulty of the discrimination increases, as we would expect. In Fig. 13.4 we show the data for blocks of five trials from the 66 : 0, 66 : 34, and 66 : 17 groups. The smooth

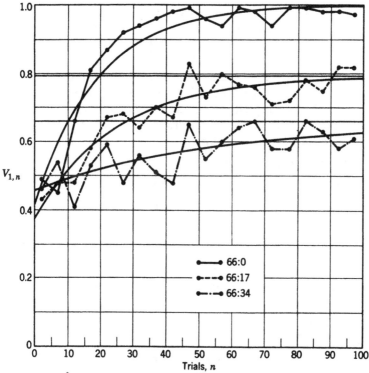

Fig. 13.4. Data obtained by Neimark [9] with the non-contingent procedure, and theoretical means, $V_{1,n}$, computed from equations 13.15 and 13.16 with the parameter values shown in Table 13.2. The heavy horizontal lines are the theoretical asymptotes.

curves were plotted from equations 13.15 and 13.16, using the estimates of α and $V_{1,0}$ given in Table 13.2.

All experiments mentioned in this section used the non-contingent procedure, and in all cases the empirical estimates of the asymptotic means are in close agreement with the prediction of the model described in the preceding section. Furthermore, we are aware of no similar experiments that yield different results. However, we presented no measures of goodness-of-fit of the model to the trial-by-trial data. We merely estimated the parameter α and then compared the computed mean curves with the data.

13.6 A MODEL WITH EXPERIMENTER-SUBJECT-CONTROLLED EVENTS

In Section 13.3 it was pointed out that the T-maze experiments with rats, as well as several human experiments, used a contingent procedure; the outcomes of trials were contingent upon the subjects' responses. A model for these experiments uses experimenter-subject-controlled events discussed in Sections 3.13, 4.6, and 5.8. The two alternatives A_1 and A_2 are identified with the two available choices, for example, turning right or left or pushing the right or left button. The two outcomes O_1 and O_2 correspond to reward and non-reward or correct and incorrect, respectively. (For simplicity, we speak of right and left turns and rewards and non-rewards; appropriate modifications for other experiments are required.) An event then is an alternative chosen *and* an outcome; we have four possible events as tabulated.

Event	Alternative	Outcome	Identification
E_{11}	A_1	O_1	right turn + reward
E_{12}	A_1	O_2	right turn + no reward
E_{21}	A_2	O_1	left turn + reward
E_{22}	A_2	O_2	left turn + no reward

Corresponding to each event E_{jk} there is an operator Q_{jk} defined by

$$(13.18) \qquad Q_{jk}p = \alpha_{jk}p + (1 - \alpha_{jk})\lambda_{jk}.$$

We now place some restrictions on these four operators.

The first restrictions result from the symmetry of the situation. We assume that reward of a right turn has the same effect on p as reward of a left turn has on $q = 1 - p$, so we require that

$$(13.19) \qquad \begin{aligned} \alpha_{21} &= \alpha_{11} = \alpha_1, \\ 1 - \lambda_{21} &= \lambda_{11} = \lambda_1. \end{aligned}$$

For the same reasons we wish non-reward to have a symmetric effect and so we take

$$(13.20) \qquad \begin{aligned} \alpha_{22} &= \alpha_{12} = \alpha_2, \\ 1 - \lambda_{22} &= \lambda_{12} = \lambda_2. \end{aligned}$$

These four restrictions reduce our four operators to

$$(13.21) \qquad \begin{aligned} Q_{11}p &= \alpha_1 p + (1 - \alpha_1)\lambda_1, \\ Q_{12}p &= \alpha_2 p + (1 - \alpha_2)\lambda_2, \\ Q_{21}p &= \alpha_1 p + (1 - \alpha_1)(1 - \lambda_1), \\ Q_{22}p &= \alpha_2 p + (1 - \alpha_2)(1 - \lambda_2). \end{aligned}$$

In an experiment in which reward always follows A_1 and never follows A_2, the subject chooses A_1 with almost certainty after a number of trials. This means that operator Q_{11} is applied repeatedly after learning has occurred, and we want this operator to maintain p at or very near unity. Hence, we choose $\lambda_1 = 1$. Similarly, if the non-reward operator, Q_{12}, were applied repeatedly, the probability p should tend towards zero. This implies that $\lambda_2 = 0$. These restrictions reduce our four operators to

$$
\begin{aligned}
Q_{11}p &= \alpha_1 p + (1 - \alpha_1) & &[p\pi_1] \\
Q_{12}p &= \alpha_2 p & &[p(1 - \pi_1)] \\
Q_{21}p &= \alpha_1 p & &[(1 - p)\pi_2] \\
Q_{22}p &= \alpha_2 p + (1 - \alpha_2) & &[(1 - p)(1 - \pi_2)].
\end{aligned}
$$

(13.22)

Only two parameters, α_1 and α_2, remain. We call α_1 the reward parameter and α_2 the non-reward parameter.

These four operators characterize a reinforcement-extinction model because reward improves and non-reward reduces the probability of going to the chosen side. But if non-reward occurs in the presence of stimuli associated with reward, secondary reinforcement can occur. If the right-hand sides of the second and fourth listed operators are interchanged, a secondary-reinforcement model results. This model has two absorbing barriers; asymptotically nearly all sequences terminate at zero or unity. Though we do not make essential use of either model here, we regard both as important.

As in Section 3.13, the conditional probabilities of the outcomes, given a particular response, are constant. The probability of outcome O_k, given alternative A_j, is denoted by π_{jk}. We drop the double subscript for simplicity and let

$$
\begin{aligned}
\pi_1 &= \pi_{11} = 1 - \pi_{12}, \\
\pi_2 &= \pi_{21} = 1 - \pi_{22}.
\end{aligned}
$$

(13.23)

The event probabilities are as shown in brackets to the right of equation 13.22.

We next consider two alternative additional restrictions on the model First we consider the equal alpha condition ($\alpha_1 = \alpha_2$) and then we discuss the condition $\alpha_2 = 1$ already considered in Section 8.6.

The instructions given a human subject in a two-choice experiment may lead him to believe that non-reward of one response necessarily implies that the other response would have been rewarded if it had been made. Such an assumption by the subject may result from direct instruction or implication, or may be made without the experimenter suggesting it to him. The schedule used in the experiment may be such that the assumption is correct, or it may not be, but the subject often has no opportunity for testing the assumption in an unambiguous way. When this assumption is made by a subject, it seems plausible to assume that a rewarded A_1 response has nearly the same effect upon future behavior as a non-rewarded A_2 response and vice versa. This in turn implies the equal alpha condition,

$$(13.24) \qquad \alpha_1 = \alpha_2 = \alpha.$$

The operators of equations 13.22 then become

$$(13.25) \qquad \begin{aligned} Q_{11}p &= \alpha p + (1 - \alpha), \\ Q_{12}p &= \alpha p, \\ Q_{21}p &= \alpha p, \\ Q_{22}p &= \alpha p + (1 - \alpha). \end{aligned}$$

In Section 5.9 it was shown that these operators lead to the following explicit formula for the means:

$$(13.26) \quad V_{1,n} = V_{1,\infty} - (V_{1,\infty} - V_{1,0})[1 - (2 - \pi_1 - \pi_2)(1 - \alpha)]^n,$$

where

$$(13.27) \qquad V_{1,\infty} = \frac{1 - \pi_2}{2 - \pi_1 - \pi_2}.$$

Note that $\pi_1 + \pi_2$ need not be unity. This equation for the asymptotic mean provides us with a direct way of telling whether the condition $\alpha_1 = \alpha_2$ is appropriate for any given set of data. In the following two sections we present several sets of data which clearly do not satisfy this condition.

When we regard the outcomes as reward and non-reward rather than as information about the pay-off schedules, there is no intuitive reason for assuming that the reward and non-reward parameters α_1 and α_2 are equal. Certainly, when we think of rat experiments, we have little basis for discussing assumptions made by the subject. We do not care to think of the rat as collecting data and making decisions accordingly or as reasoning that the other choice would have paid off when he failed. Instead, we prefer to think of the rat as undergoing a simple conditioning process with reward leading to an increase in response strength.

Estimating both α_1 and α_2 from a set of data leads to difficult computational problems, and we have not found a wholly satisfactory procedure for obtaining such estimates. However, the data we present in the following two sections give some support to the assumption that α_2 is very near unity. A closed expression for the asymptotic mean in terms of α_1 and α_2 is not available, but the expected operator approximation is suggestive (cf. Section 6.4). From equation 6.24, we can show that the expected-operator asymptotic mean is given by the formula

(13.28) $V_{1,\infty}$

$$\cong \frac{(\pi_1 - \pi_2) - 2(1 - \pi_2)x + \sqrt{(\pi_1 - \pi_2)^2 + 4(1 - \pi_2)(1 - \pi_1)x^2}}{2(\pi_1 - \pi_2)(1 - x)},$$

where

(13.29) $$x = \frac{1 - \alpha_2}{1 - \alpha_1} \neq 1.$$

For the 50 : 0 case ($\pi_1 = 0.5$, $\pi_2 = 0$) this reduces to

(13.30) $$V_{1,\infty} \cong \frac{1 - 4x + \sqrt{1 + 8x^2}}{2(1 - x)}.$$

When $\alpha_2 = 1$, we have $x = 0$ and $V_{1,\infty} \cong 1$. But when $x = 0.5$, the last formula gives $V_{1,\infty} \cong 0.732$. Moreover, we already know that for $\alpha_1 = \alpha_2$, that is, when $x = 1$, that $V_{1,\infty} = 0.667$ for the 50 : 0 case. (Even if $_{\infty 2} = 1$, $V_{1,\infty}$ is not generally exactly unity unless $\pi_2 = 0$, because the model guarantees that some p-value sequences are "absorbed" at $p = 0$. This implies that a small proportion of animals will stabilize on the unfavorable side of the maze unless $\pi_2 = 0$.)

Some of the 50 : 0 data described in the next two sections show that $V_{1,\infty}$ can be at least as high as 0.9. To obtain such a large value of $V_{1,\infty}$ would require a value of x in the foregoing formula as small as 0.1. As a result we assume that $x = 0$, that is, that $\alpha_2 = 1$.

If $\alpha_2 = 1$ is used, equations 13.22 simplify to

(13.31)
$$Q_{11}p = \alpha_1 p + (1 - \alpha_1),$$
$$Q_{12}p = p,$$
$$Q_{21}p = \alpha_1 p,$$
$$Q_{22}p = p.$$

The problem then is to estimate the remaining parameter α_1 from data. And having estimated α_1, we wish to compute the means $V_{1,n}$ to compare with the trial-by-trial means from the data.

In Section 8.6 we derived the approximate formula

(13.32) $$V_{1,n} \cong \frac{V_{1,0}}{V_{1,0} + (1 - V_{1,0})e^{-(\pi_1 - \pi_2)(1 - \alpha_1)n}}.$$

From this equation we can compute the mean $V_{1,n}$ for any trial n in terms of the initial mean, the known values of π_1 and π_2, and the parameter α_1. We use this equation in the following two sections.

The one remaining task before proceeding to an analysis of data is to develop a method for estimating the reward parameter α_1. We use a technique employed previously in other problems. The total number of errors (choices of alternative A_2) on trials $0, 1, \cdots, K-1$ can be estimated by the integral

$$(13.33) \quad \int_0^K (1 - V_{1,n})\, dn \cong \int_0^K \left\{ 1 - \frac{V_{1,0}}{V_{1,0} + (1 - V_{1,0})e^{-(\pi_1 - \pi_2)(1 - \alpha_1)n}} \right\} dn.$$

The result of the integration is

$$(13.34) \quad \int_0^K (1 - V_{1,n})\, dn \cong - \frac{\log\left[V_{1,0} + (1 - V_{1,0})e^{-(\pi_1 - \pi_2)(1 - \alpha_1)K}\right]}{(\pi_1 - \pi_2)(1 - \alpha_1)}.$$

For large numbers of trials K we have, approximately,

$$(13.35) \quad \int_0^K (1 - V_{1,n})\, dn \cong - \frac{\log V_{1,0}}{(\pi_1 - \pi_2)(1 - \alpha_1)}.$$

If we denote the total number of errors made by a subject (or the mean for a group of subjects) by T_2, we have the estimation equation

$$(13.36) \quad T_2 \triangleq - \frac{\log V_{1,0}}{(\pi_1 - \pi_2)(1 - \alpha_1)},$$

or

$$(13.37) \quad \hat{\alpha}_1 = 1 - \frac{-\log V_{1,0}}{(\pi_1 - \pi_2)T_2}.$$

When K is not large enough to permit us to use the large-trial approximation, we substitute T_2 for the integral on the left side of equation 13.34 and solve numerically for the estimate of α_1.

For most of the experiments to be analyzed, it is satisfactory to take $V_{1,0} = 1/2$. The symmetry of the experimental responses makes this reasonable when groups of subjects are considered, at least. Nevertheless, we may have occasion to examine the data to see if 0.5 is a good estimate of $V_{1,0}$. This is most easily done by plotting the data on semi-log paper; the way to make the plot may be seen by using equation 13.32 above. Let $V_{1,n}$ be estimated by $\bar{p}_n$, the proportion of subjects that make response A_1 on trial n, and obtain from equation 13.32

$$(13.38) \quad \frac{1 - \bar{p}_n}{\bar{p}_n} \triangleq \frac{1 - V_{1,0}}{V_{1,0}} e^{-(\pi_1 - \pi_2)(1 - \alpha_1)n}.$$

Taking logarithms, we obtain

$$(13.39) \quad \log\left[\frac{1 - \bar{p}_n}{\bar{p}_n}\right] \triangleq \log\left[\frac{1 - V_{1,0}}{V_{1,0}}\right] - (\pi_1 - \pi_2)(1 - \alpha_1)n.$$

Thus, if we plot $(1 - \bar{p}_n)/\bar{p}_n$, obtained directly from the data on the log scale, versus n on the linear scale of semi-log paper, a straight line with intercept $(1 - V_{1,0})/V_{1,0}$ at $n = 0$ should be obtained. We suggest that only the data from the early trials be plotted and that a straight line be drawn by eye through the plotted points to obtain the intercept. The initial mean $V_{1,0}$ can be obtained directly from the value of the intercept.

13.7 THE STANLEY T-MAZE DATA

The T-maze experiments with rats, conducted by Brunswik [1] and Stanley [2], were described in Section 13.2. In this section we analyze some of the data obtained by Stanley. Seven rats were used in each group, and we are concerned with the 100 : 0, 50 : 0, and 75 : 25 groups. In Table 13.3 we give the observed proportion of turns to the favorable side for each of these groups during each day of the experiment. Eight trials were run on each day.

Stanley controlled the number of rewards for each rat instead of the total number of trials, and so the number of rats remaining in the experiment decreases as shown in Table 13.3. For the 100 : 0 group, Stanley ran each rat until the animal went two days, not necessarily consecutive, without errors. A rat in each of the other groups was matched by the litter-mate technique to a rat in the 100 : 0 group. Stanley ran a 50 : 0 rat, for example, until it received the same number of rewards as its litter-mate in the 100 : 0 group. Hence the unequal numbers of trials for the various animals. We describe this procedure in some detail because we wish to use the data beyond the point where the first rat drops out of the experiment in each group. We would expect that proportions of favorable turns given in Table 13.3 would have been larger on the later days of the experiment if Stanley had kept all rats in the experiment to the end. Certainly in the 100 : 0 group the animals that learned fastest dropped out of the experiment first, and so the proportions beyond day 5 shown in Table 13.3 for that group are clearly biased in the downward direction. Whether or not a similar bias exists in the 50 : 0 and 75 : 25 groups depends on whether differences between litters are large or not. If the matching technique used were in fact effective and if some litters were "smarter" than others, surely a bias would result. On the other hand, if the matching technique had little effect or if the observed differences were the result of statistical "accidents" such as those observed in our "stat-rats" (cf. Section 6.2), then there would be no reason to expect an appreciable bias. We do not try to settle this question but merely point out the arguments on both sides. The important point is that the direction of bias, if one exists, is such that the proportions given in Table 13.3 are too small after some rats have dropped out ($N < 7$).

(A lower bound of 0.955 for the learning parameter in the 100 : 0 group was obtained by recomputing as if rats reaching the criterion continued with perfect performance instead of dropping out.)

TABLE 13.3

Summary of data obtained by Stanley [2] on three groups of seven rats each. There were eight trials each day. The proportions of turns to the favorable side by each group on each day are shown. (As discussed in the text, some rats were run through more trials than others.) The estimates of α_1 were obtained from equations 13.34 and 13.37.

Day	100 : 0 Group		50 : 0 Group		75 : 25 Group	
	N	Proportion	N	Proportion	N	Proportion
1	7	0.571	7	0.518	7	0.500
2	7	0.696	7	0.553	7	0.554
3	7	0.714	7	0.661	7	0.589
4	7	0.821	7	0.696	7	0.589
5	7	0.804	7	0.643	7	0.643
6	6	0.833	7	0.661	7	0.714
7	6	0.937	7	0.714	7	0.750
8	4	0.844	7	0.821	6	0.792
9	4	0.875	7	0.786	6	0.792
10	4	0.875	7	0.893	4	0.844
11	3	0.875	6	0.833	4	0.875
12	3	0.917	6	0.875	4	0.813
13			6	0.854	4	0.782
14			6	0.854	3	0.834
15			5	0.850	3	0.917
16			4	0.875		
17			4	0.906		
18			4	0.875		
19			4	0.937		
20			4	0.937		
21			3	0.917		
22			3	0.917		
23			3	1.000		
24			3	0.958		
$\hat{\alpha}_1$	0.961		0.962		0.964	

The equal-alpha condition described in Section 13.6 is clearly not met by Stanley's 50 : 0 and 75 : 25 groups. The equal-alpha asymptotes are

0.667 and 0.750, respectively, for those two groups. On days beyond the sixth, the 50 : 0 proportions exceed 0.667 on all days. Similarly, on all days beyond the seventh, the 75 : 25 proportions exceed 0.750. Furthermore, the data in Table 13.3 strongly suggest that all groups are approaching an asymptote of nearly 1.00. Therefore, we apply the model with $\alpha_2 = 1$.

The initial mean $V_{1,0}$ is taken to be 0.5 for Stanley's 100 : 0 and 50 : 0 groups. The data for the 75 : 25 group suggest that $V_{1,0}$ is slightly less than 0.5 and so we plotted $(1 - \bar{p}_n)/\bar{p}_n$ versus n on semi-log paper as

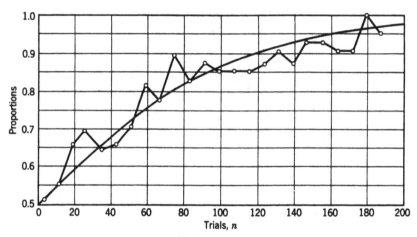

Fig. 13.5. The 50 : 0 data obtained by Stanley [2] from seven rats. The smooth curve was computed from equation 13.32 with $\pi_1 = 0.5$, $\pi_2 = 0$, $V_{1,0} = 0.5$, and $\alpha_1 = 0.96$.

described at the end of the preceding section. This led to the estimate $\hat{V}_{1,0} = 0.48$ for the 75 : 25 group.

The reward parameter α_1 was estimated for each group by using equations 13.37 and 13.34 of the last section. An approximate value of $\hat{\alpha}_1$ was first obtained from equation 13.37, and then equation 13.34 was solved numerically for each of the three groups. The quantity T_2 was defined to be the mean total number of errors per subject, but caution must be used in computing this quantity when the number of subjects is not the same on all trials. The simplest procedure is to estimate the integral on the left side of equations 13.34 and 13.35 by summing the observed proportions of errors on each day and multiplying by 8, since there were eight trials per day. In this way we obtained the estimates of α_1 from Stanley's data shown in Table 13.3. These three estimates are remarkably close to one another, and within the framework of the model this implies that the effect of a reward is independent of the reinforcement schedule used. In the next section we see that this result does not hold for data obtained from human subjects.

Using the estimated values of α_1, we computed the approximate values of the trial means for each group from equation 13.32. In Figs. 13.5 and 13.6 we show the results for the 50 : 0 and 75 : 25 groups along with the experimental points.

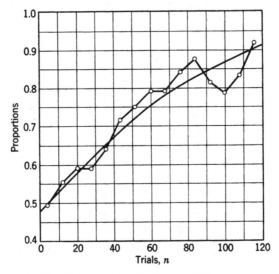

Fig. 13.6. The 75 : 25 data obtained by Stanley [2] from seven rats. The smooth curve was computed from equation 13.32 with $\pi_1 = 0.75$, $\pi_2 = 0.25$, $V_{1,0} = 0.48$, and $\alpha_1 = 0.96$.

13.8 HUMAN EXPERIMENTS USING THE CONTINGENT PROCEDURE

In this section we discuss several experiments using the contingent procedure on human subjects in two-choice situations. A set of experiments conducted by Jacqueline Jarrett Goodnow in the spring of 1951 was designed by her and the authors to replicate the rat experiments by Stanley described in the preceding section. The subjects were six groups of five Harvard undergraduates. The reward probabilities used were 100 : 0, 50 : 0, 100 : 50, and 75 : 25. For the 100 : 0 and 50 : 0 conditions, two different procedures were used: "pay to play" and "play free." In the pay-to-play procedure, the subject's net pay was computed by subtracting his losses from his wins. In the play-free procedure, a subject's net pay was simply the number of chips won. The chips were worth one cent. Each subject in the two 100 : 0 groups was run for 75 trials whereas all other subjects were run for 150 trials of acquisition.

In addition, each subject received 50 trials of extinction, but we delay discussion of those data until a later section.

Table 13.4 gives the proportions of choices of the more favorable side made by all subjects in each group during blocks of 10 trials. As in the last section, we see that the equal alpha condition is not satisfied in these data; for the 50 : 0 groups the equal alpha asymptote is 0.67, and for the

TABLE 13.4

"Two-armed bandit" data obtained by Goodnow from six groups of five subjects each. The reward probabilities and the two procedures, "pay to play" and "play free," identify each group. For each block of ten trials the group proportions of choices of the favorable side are given. The estimates of α_1 were obtained from equations 13.34 and 13.37.

Trial Block	100 : 0 Pay to Play	100 : 0 Play Free	50 : 0 Pay to Play	50 : 0 Play Free	100 : 50 Pay to Play	75 : 25 Pay to Play
0–9	0.54	0.38	0.44	0.44	0.54	0.42
10–19	0.84	0.66	0.70	0.50	0.54	0.68
20–29	1.00	0.90	0.60	0.70	0.64	0.54
30–39	1.00	0.88	0.70	0.68	0.66	0.78
40–49	1.00	0.90	0.70	0.58	0.84	0.64
50–59	1.00	0.88	0.82	0.74	0.96	0.68
60–69†	1.00	0.99	0.86	0.60	0.90	0.66
70–79			0.90	0.68	1.00	0.80
80–89			0.88	0.76	1.00	0.80
90–99			0.80	0.70	1.00	0.88
100–109			0.94	0.76	1.00	0.92
110–119			0.78	0.68	1.00	0.90
120–129			0.78	0.84	1.00	0.92
130–139			0.82	0.76	1.00	0.82
140–149			0.90	0.74	1.00	0.86
$\hat{\alpha}_1$	0.888	0.951	0.962	0.979	0.928	0.967

† For the 100 : 0 groups this includes trials 60–74.

75 : 25 group it is 0.75. The data in Table 13.4 for these groups clearly exceed those asymptotes. (The 100 : 0 and 100 : 50 groups should have an asymptote of 1.00 for either the equal alpha condition or the condition that $\alpha_2 = 1$ discussed in Section 13.6.)

For all six groups we assumed that $V_{1,0} = 0.5$ and $\alpha_2 = 1$, and we estimated α_1 from equations 13.34 and 13.37. We obtained the estimates shown in the last row of Table 13.4. In Fig. 13.7 we show the data for the 75 : 25 group along with a curve of $V_{1,n}$ computed from equation 13.32, with $V_{1,0} = 0.5$ and $\alpha_1 = 0.967$.

Two inferences may be drawn from the estimates $\hat{\alpha}_1$ for the six groups. First, reward had a larger effect on the probability of choosing the more favorable side ($\hat{\alpha}_1$ was smaller) for subjects who paid to play than for subjects who played free. This is true for both the 100 : 0 and 50 : 0 conditions. Second, the effect of a reward was much greater in the continuous reinforcement cases (100 : 0) than in the corresponding

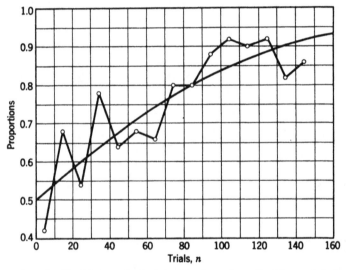

Fig. 13.7. The 75 : 25 data obtained by Goodnow from five Harvard students, using the two-armed bandit with the pay-to-play condition. The smooth curve was computed from equation 13.32 with $\pi_1 = 0.75$, $\pi_2 = 0.25$, $V_{1,0} = 0.5$, and $\alpha_1 = 0.967$.

partial reinforcement cases. For the pay-to-play procedure the order of decreasing effect of reward was 100 : 0, 100 : 50, 50 : 0, 75 : 25. This is the order of our intuitive guesses about the order of increasing "ambiguity" of the situations, that is, the 100 : 0 seems to be the easiest discrimination to learn and the 75 : 25 the most difficult. In a sense, the 50 : 0 and 100 : 50 conditions are symmetric in pay-off and non-pay-off, but the subject is presumably searching for the more favorable side and so experiences the consistent success side more often than the consistent failure side.

Another set of experiments using the two-armed bandit was conducted by Laval Robillard as part of a senior honors thesis at Harvard in 1952–

1953. He used seven groups of 10 Harvard freshmen as subjects. The play-free procedure was used for all groups but no chip was used to activate the machine. The first part of Robillard's set of experiments was a study of the effect of the amount of reward in the 50 : 0 situation. Subjects in one group were told to win as many chips as possible but were not told that the chips won would be exchanged for money. Another group was told that each chip was worth one cent, and a third group was told that each chip was worth five cents. The chips were exchanged for money at the end of the experiment. Each subject was run for 100 trials of acquisition. Four additional groups were run without monetary value placed on the chips and with reward probabilities of 30 : 0, 80 : 0, 80 : 40, and 60 : 30. Extinction trials were run on all 70 subjects, but we delay discussion of the extinction data until a later section.

TABLE 13.5

"Two-armed bandit" data obtained by Robillard from seven groups of ten subjects each. The reward probabilities and monetary value of each chip are shown for each group. For each block of ten trials the group proportion of choices of the favorable side is given. The estimates of α_1 were obtained from equations 13.34 and 13.37.

Trial Block	50 : 0 (0¢)	50 : 0 (1¢)	50 : 0 (5¢)	30 : 0 (0¢)	80 : 0 (0¢)	80 : 40 (0¢)	60 : 30 (0¢)
0–9	0.52	0.51	0.49	0.56	0.59	0.50	0.49
10–19	0.63	0.54	0.58	0.57	0.77	0.59	0.58
20–29	0.69	0.67	0.67	0.55	0.88	0.71	0.62
30–39	0.63	0.59	0.59	0.63	0.88	0.64	0.51
40–49	0.64	0.66	0.63	0.60	0.86	0.63	0.51
50–59	0.75	0.66	0.64	0.66	0.91	0.63	0.61
60–69	0.76	0.77	0.71	0.65	0.92	0.53	0.57
70–79	0.85	0.70	0.73	0.65	0.89	0.71	0.57
80–89	0.87	0.83	0.72	0.65	0.88	0.73	0.65
90–99	0.90	0.83	0.81	0.66	0.89	0.70	0.55
$\hat{\alpha}_1$	0.958	0.969	0.973	0.967	0.943	0.975	0.975

Table 13.5 summarizes the acquisition data obtained by Robillard. Proportions of choices of the more favorable side for all subjects in each group are given for blocks of 10 trials. We assumed $V_{1,0} = 0.5$ and estimated α_1 for each group from equation 13.34. The estimates obtained

are shown in the last row of Table 13.5. In Fig. 13.8 we give an example of the data from Robillard's experiments along with the curve of $V_{1,n}$ versus n computed from equation 13.32.

Again, two inferences may be drawn about how the effectiveness of a success trial depends on certain experimental conditions. First, we see that success has the greatest effect when the stakes are lowest. A comparison of Robillard's three 50 : 0 groups shows that success is most effective when the chips have no monetary value and is least effective when each chip is worth five cents. The differences are small and probably not significant, however. The second inference has to do with the effectiveness

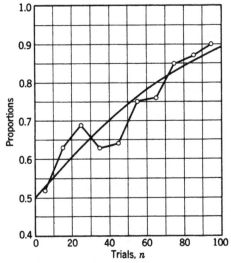

Fig. 13.8. The 50 : 0 data obtained by Robillard from ten Harvard freshmen, using the two-armed bandit. This group received no money in exchange for the chips. The smooth curve was computed from equation 13.32 with $\pi_1 = 0.5$, $\pi_2 = 0$, $V_{1,0} = 0.5$, and $\alpha_1 = 0.958$.

of success in different partial reinforcement schedules. Robillard confirmed the conclusions from the Goodnow data that a single reward is less effective when the situation is a difficult discrimination. For the procedure involving no money value, the order of decreasing effectiveness of success is 80 : 0, 50 : 0, 30 : 0, 80 : 40, 60 : 30. (The last two groups gave identical estimates of α_1.)

A similar experiment used the playing card technique described in Section 13.3. Briefly, a subject was presented with two playing cards face down on each trial and asked to choose the red card; if he succeeded he was paid five cents. Twenty subjects were run by students in a course in experimental social psychology at Harvard in October, 1952. The

50 : 0 condition was used, and each subject was run for 100 trials. One group of 10 subjects were Cambridge high-school students, and another group of ten subjects were Harvard undergraduates majoring in mathematics, physics, or astronomy. Table 13.6 summarizes the data obtained,

TABLE 13.6

Data obtained from the playing-card experiment with ten subjects in each group. The proportion of choices of the favorable side by each of the two groups is shown for each block of ten trials. The estimates of α_1 were obtained from equations 13.34 and 13.37.

Trial Block	High-School Students	Harvard Students
0– 9	0.47	0.59
10–19	0.63	0.72
20–29	0.64	0.77
30–39	0.71	0.75
40–49	0.74	0.76
50–59	0.71	0.82
60–69	0.77	0.83
70–79	0.76	0.87
80–89	0.89	0.94
90–99	0.91	0.96
$\hat{\alpha}_1$	0.958	0.934

and Fig. 13.9 shows the data for the high-school group along with a computed curve of $V_{1,n}$ versus n. As in the previous two sets of experiments, the data exceed the equal-alpha asymptote of 0.67 during all except the early blocks of trials.

The mean initial probability was taken to be 0.5 for both groups. The number of correct choices made by each group during the first five trials was 21 out of a possible 50, but since the standard deviation of the expected number of successes in 50 binomial observations with $p = 0.5$ is about 3.5, we saw no reason for not taking $V_{1,0} = 0.5$. Furthermore, there is no evidence that the two groups of subjects had different mean initial probabilities of choosing the favorable card. The success parameters α_1 were estimated by using equation 13.34, and the results are given in Table 13.6. The data show that the college undergraduates "learned faster" than the high-school students.

In Neimark's study [9], four groups of subjects were run with the contingent procedure in a two-choice situation. The experimental method was described in Section 13.3. With reward probabilities of 66 : 0 and 66 : 17, the group performance exceeded the equal-alpha asymptotes, as in other investigations described in this section. However, Neimark's 66 : 34 group seemed to be approaching the equal-alpha asymptote of 0.66. This agrees with the results from Detambel's 50 : 0 group [8], which also seemed to approach the equal-alpha asymptote of 0.67.

Except for the data from the Neimark 66 : 34 group and the Detambel

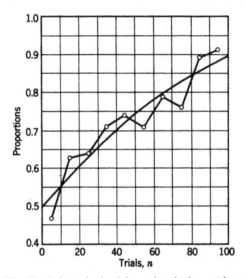

Fig. 13.9. The 50 : 0 data obtained from the playing-card experiment with ten Cambridge high-school students. The smooth curve was computed from equation 13.32 with $\pi_1 = 0.5$, $\pi_2 = 0$, $V_{1,0} = 0.5$, and $\alpha_1 = 0.958$.

50 : 0 group just mentioned, all data from two-choice experiments using the contingent procedure appear to be inconsistent with the equal-alpha condition. The identity operator assumption for non-reward, on the other hand, leads to a model which is consistent with the bulk of the data presently available. However, we have not seriously tested the identity-operator assumption with the data; rather, we have shown that several sets of data indicate that $1 - \alpha_2$ is small compared to $1 - \alpha_1$. The two notable exceptions cannot be ignored of course.

*13.9   THREE-CHOICE EXPERIMENTS

As noted in Section 13.3, Neimark ran several groups of subjects in a three-choice situation. She used three telegraph keys and three lights with both the non-contingent and contingent procedures.

For the non-contingent groups, the schedules of appearance and non-appearance of lights were independent of the subjects' choices. The reward probabilities were $100 : 0 : 0$, $66 : 0 : 0$, $66 : 17 : 17$, and $66 : 8.5 : 8.5$. A model for this part of Neimark's experiment can be obtained from the case of experimenter-controlled events. In Sections 4.4 and 5.3 we discussed this case for more than two responses and more than two events. The symmetry among the three events suggests the equal alpha condition considered in Section 5.3. Furthermore, it seems plausible to choose the limit vectors of the three event operators so that, if event E_i occurs on every trial, response A_i will occur with a probability which tends to unity. Hence we assume that

$$(13.40) \qquad \lambda_1 = \begin{bmatrix} 1 \\ 0 \\ 0 \end{bmatrix}, \quad \lambda_2 = \begin{bmatrix} 0 \\ 1 \\ 0 \end{bmatrix}, \quad \lambda_3 = \begin{bmatrix} 0 \\ 0 \\ 1 \end{bmatrix}.$$

Equation 5.20 then gives, for the asymptotic vector of marginal means,

$$(13.41) \qquad V_{1,\infty} = \begin{bmatrix} \pi_1 \\ \pi_2 \\ \pi_3 \end{bmatrix},$$

where π_i is the probability of occurrence of event E_i. From this result we conclude that the asymptotic proportion of A_1 responses is π_1, and this result agrees with Neimark's data from her $100 : 0 : 0$ and $66 : 17 : 17$ groups.

In Neimark's $66 : 0 : 0$ and $66 : 8.5 : 8.5$ schedules, a number of "blank" trials appeared, that is, trials on which no light appeared. Thus we need to introduce a fourth event E_4 with probability π_4 which has the values 0.34 and 0.17, respectively, for the two groups. As in Section 13.4, we assume an identity operator for this event. This assumption, along with those previously made in this section, allow us to obtain from equation 4.39 the result

$$(13.42) \qquad V_{1,\infty} = \frac{1}{\pi_1 + \pi_2 + \pi_3} \begin{bmatrix} \pi_1 \\ \pi_2 \\ \pi_3 \end{bmatrix}.$$

Hence the asymptotic proportion of A_1 responses for Neimark's $66 : 0 : 0$ and $66 : 8.5 : 8.5$ groups should be 1.00 and 0.795, respectively. These conclusions are in close agreement with Neimark's data.

To describe the data from Neimark's three-key contingent groups, we need a model based upon experimenter-subject-controlled events. We have three responses corresponding to the three keys, and two outcomes, reward (light on) and non-reward (light not on). From the symmetry of the situation we assume

(13.43)
$$\alpha_{11} = \alpha_{21} = \alpha_{31} = \alpha_1,$$
$$\alpha_{12} = \alpha_{22} = \alpha_{32} = \alpha_2.$$

As before, we assume that reward can lead to perfect learning and so we take

(13.44) $$\lambda_{11} = \begin{bmatrix} 1 \\ 0 \\ 0 \end{bmatrix}, \quad \lambda_{21} = \begin{bmatrix} 0 \\ 1 \\ 0 \end{bmatrix}, \quad \lambda_{31} = \begin{bmatrix} 0 \\ 0 \\ 1 \end{bmatrix}.$$

When we assume the identity operator condition for non-reward ($\alpha_2 = 1$), it seems likely that the asymptotic vector of marginal means is

(13.45) $$V_{1,\infty} \cong \begin{bmatrix} 1 \\ 0 \\ 0 \end{bmatrix}$$

provided $\pi_1 > \pi_2 + \pi_3$ by a fair amount. This conclusion seems inconsistent with Neimark's data on her contingent 66 : 0 : 0, 66 : 34 : 34, and 66 : 17 : 17 groups; those groups did not appear to approach an asymptote of 1.00 for response A_1.

Except when we take $\alpha_2 = 1$, we need to make an assumption about the non-reward limit vectors λ_{j2}. When we assume that if A_j occurs and is not rewarded, its probability will tend towards zero, a special assumption needs to be made about how the probability divides up among the other responses. In Section 5.10 we introduced one possible assumption by the second of equations 5.108. For the model under discussion, that assumption becomes

(13.46) $$\lambda_{12} = \begin{bmatrix} 0 \\ 0.5 \\ 0.5 \end{bmatrix}, \quad \lambda_{22} = \begin{bmatrix} 0.5 \\ 0 \\ 0.5 \end{bmatrix}, \quad \lambda_{32} = \begin{bmatrix} 0.5 \\ 0.5 \\ 0 \end{bmatrix}.$$

When these limit vectors are assumed and we further let $\alpha_1 = \alpha_2$, we can compute the components of the asymptotic vector of marginal means from equation 5.113 and obtain for that vector

(13.47) $$V_{1,\infty} = \frac{1}{(1 - \pi_2)(1 - \pi_3) + (1 - \pi_1)(1 - \pi_3) + (1 - \pi_1)(1 - \pi_2)}$$
$$\times \begin{bmatrix} (1 - \pi_2)(1 - \pi_3) \\ (1 - \pi_1)(1 - \pi_3) \\ (1 - \pi_1)(1 - \pi_2) \end{bmatrix}.$$

For the reward probabilities of 66 : 0 : 0, 66 : 34 : 34, and 66 : 17 : 17, this equation gives asymptotes for A_1 of 0.60, 0.50, and 0.56, respectively. These computed asymptotes seem clearly less than the corresponding approximate asymptotes obtained from Neimark's data.

Neither set of assumptions just considered for the three-key contingent case is consistent with Neimark's data. The identity operator assumption for non-reward leads to asymptotes for A_1 that look high, whereas the equal alpha condition together with the special assumption about the non-reward limit vectors (equation 13.46) leads to asymptotes for A_1 that are too low.

13.10   EXTINCTION DATA

In both sets of experiments on the two-armed bandit described in Section 13.8, the acquisition trials were followed by a number of extinction trials during which neither side led to reward. We now wish to look at these extinction data and make some inferences about the rate of extinction following the various reinforcement schedules. But before analyzing the data, we examine the model to be used.

The most general operators we have considered applying in this chapter are given by equations 13.22. For extinction, $\pi_1 = \pi_2 = 0$, and so we have left only the two operators, Q_{12} and Q_{22}, specified by

$$(13.48) \qquad \begin{aligned} Q_{12}p &= \alpha_2 p, \\ Q_{22}p &= \alpha_2 p + (1 - \alpha_2). \end{aligned}$$

The first equation is appropriate when the originally favorable side is chosen and the second when the other choice is made. In Section 13.8 we took the non-reward parameter α_2 to be unity and thereby succeeded in fitting the acquisition data adequately. However, if we took $\alpha_2 = 1$ for extinction, we would conclude that extinction would never occur! Both operators Q_{12} and Q_{22} would be identity operators, and p would never change. The data cannot stand such a drastic assumption, and so we are forced to assume that extinction and acquisition are just different. We say more about this point in the next section, but first we estimate the extinction parameters α_2 from the data.

A procedure for estimating α_2 can be obtained directly from the total number of choices of the favorable side above chance. We need an expression for the means $V_{1,n}$. The operators defined by equations 13.48 are those discussed in Section 5.7. Hence we use equations 5.31, 5.57, and 5.58 to obtain the expression

$$(13.49) \qquad V_{1,n} = 1/2 + (V_{1,0} - 1/2)(2\alpha_2 - 1)^n.$$

The asymptotic mean, $V_{1,\infty}$, is 1/2. The expected number of choices of the previously favorable side, above chance, is $(V_{1,n} - 1/2)$ on trial n for each subject. We then sum over all trials from 0 to $K - 1$ and get

(13.50)
$$\sum_{n=0}^{K-1}(V_{1,n} - 1/2) = (V_{1,0} - 1/2) \sum_{n=0}^{K-1}(2\alpha_2 - 1)^n$$
$$= \frac{(V_{1,0} - 1/2)}{2(1 - \alpha_2)}[1 - (2\alpha_2 - 1)^K].$$

This quantity can be estimated by the mean number D of extinction responses above chance on trials 0 to $K - 1$, that is,

(13.51)
$$D \triangleq \frac{(V_{1,0} - 1/2)}{2(1 - \alpha_2)}[1 - (2\alpha_2 - 1)^K].$$

This equation must be solved numerically for the estimate of α_2 except when K is large enough to allow us to neglect $(2\alpha_2 - 1)^K$, in which case

(13.52)
$$\hat{\alpha}_2 \cong 1 - \frac{(V_{1,0} - 1/2)}{2D}.$$

For extinction after 100 percent reinforcement, the initial mean $V_{1,0}$ corresponds to the mean at the end of acquisition and so can be estimated from the acquisition data. For extinction after partial reinforcement, we do not know just when extinction begins from the point of view of the subject, but we accept the experimenter's definition of the beginning of extinction. In that case, $V_{1,0}$ for the extinction portion of the data can be estimated from the acquisition data also. We now apply the foregoing estimation formulas to the two-armed bandit data.

Table 13.7 summarizes the extinction data obtained from the Goodnow experiments, and Table 13.8 gives the corresponding data from the Robillard study. For each group we give the value of D obtained from the data, the value of $V_{1,0}$ obtained from the acquisition data, and the computed estimates $\hat{\alpha}_2$.

The estimates from the Goodnow data show that extinction occurs most rapidly after those training schedules in which one side was rewarded 100 percent of the trials. For the 100 : 0 play-free group and the 100 : 50 group, extinction was essentially "complete" at the end of the first block of 10 trials. As a result, we were not able to obtain useful estimates of α_2 from those data, but it seems clear that the values are less than say 0.9. The group most resistant to extinction was the 50 : 0 pay-to-play group; it extinguished more slowly than the 50 : 0 play-free group, and both 50 : 0 groups extinguished more slowly than the 75 : 25 group. When only the pay-to-play data are considered, the order of increasing resistance to extinction is 100 : 0, 75 : 25, and 50 : 0—this is the order of decreasing frequency of reward just prior to extinction.

The Robillard extinction data given in Table 13.8 lead to similar inferences regarding resistance to extinction after partial reinforcement schedules. For the moment consider only those groups for which reward occurred on one side only and for which the chips had no money value, namely, the 80 : 0, 50 : 0, and 30 : 0 zero-cent groups. The order of increasing resistance to extinction, as measured by the estimates $\hat{\alpha}_2$, is 80 : 0, 50 : 0, 30 : 0, although the difference between the first two groups

TABLE 13.7

Proportions of choices of the previously favorable side during extinction in the Goodnow "two-armed bandit" experiment. The quantity D is the mean number of extinction responses above chance on all 50 trials. The estimates of $V_{1,0}$ were obtained from the computed value of $V_{1,n}$ for the end of acquisition, and the estimates of α_2 were obtained from equation 13.51.

Trial Block	100 : 0 Pay to Play	100 : 0 Play Free	50 : 0 Pay to Play	50 : 0 Play Free	100 : 50 Pay to Play	75 : 25 Pay to Play
0–9	0.44	0.64	0.84	0.62	0.44	0.64
10–19	0.54	0.42	0.74	0.66	0.50	0.50
20–29	0.54	0.48	0.76	0.64	0.42	0.66
30–39	0.44	0.44	0.62	0.54	0.52	0.46
40–49	0.62	0.50	0.62	0.54	0.58	0.54
D	0.8	−0.2	10.8	5.0	−0.4	3.0
$\hat{V}_{1,0}$	1.000	0.972	0.945	0.828	0.995	0.922
$\hat{\alpha}_2$	0.688	—	0.979	0.967	—	0.930

is very slight. The 80 : 40 and 60 : 30 groups extinguished more rapidly than any of the other groups.

The general conclusion about resistance to extinction after partial reinforcement, inferred from the Goodnow and Robillard data, is in agreement with the results of a number of previous studies reviewed by Jenkins and Stanley [10]. Most of those studies employed experimental arrangements quite different from the symmetric two-choice situation, and the conclusion seems to be a rather general one: resistance to extinction increases monotonically with decreasing frequency of reward in the preceding acquisition training.

One other inference may be made from the estimates of α_2 for the Robillard extinction data. It may be noted from Table 13.8 that of the three 50 : 0 groups, the zero-cent group extinguished most rapidly and the five-cent group most slowly, with the one-cent group being intermediate. In Section 13.8 we noted that the zero-cent group learned most rapidly and the five-cent group most slowly. Therefore, our analysis of the data seems to show that increasing the amount of reward slows down both acquisition and extinction.

TABLE 13.8

Proportions of choices of the previously favorable side during extinction in the Robillard experiment. The quantity D is the mean number of extinction responses above chance on all 50 trials. The estimates of $V_{1,0}$ were obtained from the computed value of $V_{1,n}$ for the end of acquisition, and the estimates of α_2 were obtained from equation 13.51.

Trial Block	50 : 0 (0¢)	50 : 0 (1¢)	50 : 0 (5¢)	30 : 0 (0¢)	80 : 0 (0¢)	80 : 40 (0¢)	60 : 30 (0¢)
0–9	0.71	0.64	0.78	0.66	0.69	0.57	0.52
10–19	0.70	0.63	0.61	0.62	0.65	0.61	0.53
20–29	0.57	0.63	0.75	0.68	0.68	0.54	0.55
30–34	0.62	0.80	0.74	0.74	0.76	0.48	0.64
D	5.4	5.5	7.6	5.8	6.5	2.1	1.7
$\hat{V}_{1,0}$	0.891	0.825	0.794	0.729	0.990	0.731	0.681
$\hat{\alpha}_2$	0.967	0.976	0.991	0.990	0.965	0.946	0.948

13.11 COMPARISONS AND EVALUATIONS

In the preceding sections, data on behavior in symmetric choice situations have been analyzed. As a result, we are now in a position to make a number of comparisons and re-examine the usefulness of the general model.

The data from several experiments that used the non-contingent procedure were found to agree with a prediction obtained from a model based upon experimenter-controlled events and the equal alpha condition. This model was found satisfactory for data obtained in several different laboratories and with various physical arrangements for studying two-choice behavior. In addition, the generalized model was equally satisfactory when compared to Neimark's three-choice non-contingent data. In some of the two-choice and three-choice non-contingent experiments of Neimark, groups of subjects were run on schedules which included "blank" trials (trials on which no light appeared). By assuming that these trials did not alter the response probabilities, we again found agreement between the data and the equal-alpha experimenter-controlled event model. The large variety of data which agrees with this model is impressive.

Data from numerous experiments which used the contingent procedure of reinforcement present a somewhat different picture. We found no single model adequate for handling all these data. Most of the contingent two-choice data appears consistent with the identity operator assumption for non-reward, but two exceptions exist: the Neimark 66 : 34 data and the Detambel 50 : 0 data. In those two cases, an equal alpha assumption seems more tenable. The three-choice contingent data obtained by Neimark presents an even more serious problem; no assumptions considered led to a satisfactory model. Procedural and instructional differences among the various contingent studies may be responsible for the different results obtained, but such differences apparently had no appreciable effect in the non-contingent studies. In spite of the difficulties just mentioned we can make a number of inferences.

The studies of rats and human beings in similar situations give us an opportunity for comparing rats and people, if anyone is interested in such comparisons. In terms of the estimates of the success parameters α_1 we can say that on the whole Stanley's rats learned faster than or about as fast as high-school and college students in the 50 : 0 and 75 : 25 situations, but that the students learned much faster than the rats in the 100 : 0 situation. It is difficult to equate rats and people on such variables as amount of motivation or incentive, but we mention these comparisons for what they are worth. Differences between experimental procedures should also be emphasized.

The Goodnow data, given in Table 13.4, show that Harvard students learn faster in the 100 : 0 and 50 : 0 situations when they are required to pay to play rather than play free. (We hesitate to propose that these results have implications for educational policies.) On the other hand, the Robillard data in Table 13.5 show that increasing the amount of reward per trial tends to *decrease* the rate of learning. This effect may not be borne out by further experimental work (the differences are clearly insignificant), but the Robillard data do demonstrate that the amount of reward is not a major factor in this kind of learning experiment. All the comparisons just mentioned were made in terms of the relative estimated values of the parameter α_1. The comparisons are trivial to make once we have such a summary statistic of the data.

In Section 13.10 we analyzed some data on extinction and found results consistent with other data on extinction after partial reinforcement, reviewed by Jenkins and Stanley [10]. We found that the values of $\hat{\alpha}_2$

were indeed not unity for extinction, as was approximately the case for acquisition in the same experiments, and that the extinction values of $\hat{\alpha}_2$ increased with decreasing frequency of reward during the preceding training.

For a long time we tried to hold the position that an empirical event such as "turning right and finding food" or "turning right and not finding food" had associated with it a unique operator whose parameters were independent of the reinforcement schedule being used at the moment. We liked this principle of "event invariance" not because we had good psychological evidence or intuition to support it but rather because it provided a rather nice parsimony in the model—the identification between model event and empirical event could be made once and for all in a given series of experiments. For example, the non-reward parameter α_2 would have the same value in 100 percent reinforcement, 30 percent reinforcement, and in extinction for a particular apparatus, type of organism, strength of drive, etc. But we have been forced to abandon this view in light of the experimental evidence. The two-armed bandit extinction data clearly show that α_2 gets nearer to unity the smaller the reward probability in the *preceding* acquisition training when reward occurred on only one side.

Furthermore, the acquisition data described here require that α_2 *be* unity (or very near unity) during acquisition. If we were to cling to our earlier view of "event invariance" we would deduce that the asymptote of acquisition increased monotonically with the reward probability on the favorable side, and that extinction occurred at a rate independent of the preceding reinforcement schedule. These deductions would then imply that the total number of extinction responses would *increase* with the preceding reward probability because the probability at the beginning of extinction would be close to the acquisition asymptote. These deductions are completely inconsistent with the experimental findings and so we have had to take a broader and less parsimonious view of event identification. We conclude that the appropriate event identification does not simply involve "turning right and finding no food" but rather is "turning right and finding no food, given a particular psychological set or expectancy." This conclusion will be no surprise to those who are aware of the large body of experimental work on the effects of set in influencing behavior. Nevertheless, it seems worthwhile to record our unsuccessful attempt to avoid this complication.

13.12 SUMMARY

Data from experiments on rats and human beings in symmetric choice situations are analyzed in this chapter. The reference experiments are those by Humphreys [3], in which subjects predicted whether or not a light would come on during each trial, and by Brunswik [1], in which rats were run in a simple T-maze. A distinction is made between two experimental procedures, called contingent and non-contingent. With the contingent technique, the probabilities of certain kinds of environmental changes depend upon the subject's choice, whereas in the non-contingent experiments, those changes occur independent of the subject's choice.

A model with experimenter-controlled events and equal alphas is presented for analyzing data from non-contingent experiments. All the data examined agree with the model prediction that the asymptotic proportion of A_1 responses is equal to the probability of occurrence of event E_1. For the contingent experiments, a model using experimenter-subject-controlled events is developed. Most of the data is consistent with the assumption that the operators associated with non-reward are identity operators.

From estimates of the alphas obtained from various sets of data, comparisons of several kinds are made: rats with people, paying to play with playing free, small reward with larger reward, and college students with high-school students. Some data on extinction are also analyzed.

REFERENCES

1. Brunswik, E. Probability as a determiner of rat behavior. *J. exp. Psychol.*, 1939, **25**, 175–197.
2. Stanley, J. C., Jr. *The differential effects of partial and continuous reward upon the acquisition and elimination of a runway response in a two-choice situation.* Ed. D. Thesis, Harvard University, 1950.
3. Humphreys, L. G. Acquisition and extinction of verbal expectations in a situation analogous to conditioning. *J. exp. Psychol.*, 1939, **25**, 294–301.
4. Grant, D. A., Hake, H. W., and Hornseth, J. P. Acquisition and extinction of a verbal conditioned response with differing percentages of reinforcement. *J. exp. Psychol.*, 1951, **42**, 1–5.
5. Grant, D. A., Hornseth, J. P., and Hake, H. W. The influence of the inter-trial interval on the Humphreys' "random reinforcement" effect during the extinction of a verbal response. *J. exp. Psychol.*, 1950, **40**, 609–612.
6. Jarvik, M. E. Probability learning and a negative recency effect in the serial anticipation of alternative symbols. *J. exp. Psychol.*, 1951, **41**, 291–297.
7. Hake, H. W., and Hyman, R. Perception of the statistical structure of a random series of binary symbols. *J. exp. Psychol.*, 1953, **45**, 64–74.
8. Detambel, M. H. *A re-analysis of Humphreys' "acquisition and extinction of verbal expectations."* M.A. Thesis, Indiana University, 1950.
9. Neimark, E. D. *Effects of type of non-reinforcement and number of alternative responses in two verbal conditioning situations.* Ph.D. Thesis, Indiana University, 1953.
10. Jenkins, W. O., and Stanley, J. C., Jr. Partial reinforcement: a review and critique. *Psychol. Bull.*, 1950, **47**, 193–234.

CHAPTER 14

Runway Experiments

14.1 THE EXPERIMENTS

One of the simplest demonstrations of animal learning is found in the runway experiment reported by Graham and Gagné [1]. A hungry rat is placed at one end of a straight alley and is allowed to run to the other end to obtain food. Times spent by the rat in various portions of the alley are recorded on each of a series of trials, and it is found that the times decrease very rapidly at first and then tend to stabilize at some minimum value. In this chapter we develop a model for analyzing data from such experiments.

In the Graham-Gagné experiments, the apparatus consisted of a starting box and a food box connected by a straight alley 3 feet long. The length of time the rat remained in the starting box was called the latency, and the time consumed in traversing the alley was called the running time. A group of 21 albino rats was run through 15 trials of acquisition followed by 5 trials of extinction.

A more extensive runway experiment has been carried out by Weinstock [2]. The basic design of the apparatus was the same as that used by Graham and Gagné, but 3 times were recorded; the latency was defined as the time required to pass a point 6 inches from the starting end, and the running times to points 18 inches and 30 inches from the starting end were also recorded by photo-cells and timers. Acquisition consisted of 108 trials, with only one trial per rat per day. In addition to the 100 percent group of 23 animals, Weinstock ran 5 partial reinforcement groups. Extinction was studied in all groups.

We analyze in some detail the latency data obtained by Weinstock on the 100 percent group. We are indebted to him for making the original data available to us. The model attempts to handle not only the mean latencies from trial to trial but also the distributions of latencies. We consider fitting a theoretical curve to the mean latencies a minimal program which uses only a small part of the information in the data.

14.2 IDENTIFICATION PROBLEMS

From the general model described in Part I, we wish to construct a specific model for the runway experiment. As in preceding chapters we need to make identifications between elements in the general model and features of the experiment being described, but the runway presents some special identification problems. What shall we identify with alternatives A_1 and A_2? The runway does not appear to be a choice situation at all; the rat does not choose right or left alleys. On each experimental trial a rat runs from the starting position to the food box, for a trial is so defined by the experimenter. Our task is to construct a model for the runway by creating a choice situation in our mind even though there may not be one in the rat's.

There are several ways to view the runway as a choice situation. For example, we could postulate that the rat "chooses" a reaction time or latency from a distribution of times. We would need to allow many possible values of latency (perhaps an infinite number) and so we would have a very large number of alternatives A_1, A_2, $\cdots$, A_r. Another possibility leading to a simpler model is to choose an arbitrary time τ and to define A_1 as the occurrence of a latency less than τ and A_2 as the occurrence of a latency greater than or equal to τ. These definitions would reduce the data to 0's and 1's, similar to data from other experiments, but we would lose all the fine detail of the time measurements. Still another possibility is to postulate that two distributions of reaction times are available to the rat and that on each trial a random sample is drawn from one distribution or the other; A_1 would be identified with drawing a sample from the distribution with small mean latency and A_2 with drawing a sample from the other. If this were the model, the rat would gradually reduce his latencies by choosing his latency more and more often from the distribution with the smaller reaction times. This double-distribution model has its attractions because it would help describe some of the extremely large observations that occur in latency data. The disadvantage stems from the failure to describe the gradual decrease in latencies after many trials. This disadvantage could be overcome by introducing two sequences of distributions, members of the sequences corresponding to trials.

The models just mentioned have one feature in common: one of a set of alternatives occurs on an *experimental trial*. Operationally, we would observe a latency and assert that A_1 or A_2 occurred (or A_3, A_4, $\cdots$), depending on the value of the latency and the criterion being used. The latency data from trial to trial would thereby be translated into a sequence of A_j's for each rat. The important point is that an experimental trial corresponds to a trial as defined in Chapter 1—an opportunity for choosing among r alternatives—and those alternatives correspond to values of latencies. In spite of this advantage we do not choose such models.

A rather different set of models can be generated by assuming that a *sequence* of A_1's and A_2's occurs on *each experimental trial*. If we rather loosely identified A_1 with a "goal-directed movement" and A_2 with all other behavior, we would conceive that an experimental trial consists of a series of movements, some goal directed and some not. The model originally proposed by Estes [3] and discussed by Bush and Mosteller [4] was essentially of this type. It was assumed that an experimental trial was terminated by the first occurrence of A_1 and that it was preceded by a certain number of A_2's. If it takes more than one goal-directed response to complete a trial, it is no trouble to generalize the model to require some fixed number k of A_1's for trial termination. The basic difference between such models and the ones previously mentioned is that an experimental trial now consists of a series of *steps*, and a choice between A_1 and A_2 occurs at each such step. As a result, A_1 and A_2 correspond to overt acts or movements rather than the selection of a latency value. In this chapter we use this class of models.

Two main models of the type just described are developed in the following two sections. In both we assume that an experimental trial consists of a sequence of A_1's and A_2's and that the trial is terminated when the kth A_1 occurs. The number of A_2 occurrences varies during the course of learning. The fundamental difference between these two models lies in the assumptions about the times required for the A_1 and A_2 occurrences. The early model described by Estes and by Bush and Mosteller assumed that each act required a constant time h; the model developed in Section 14.3 employs this assumption. Clearly, the notion of a constant time h is an analytic device which at best leads to an adequate approximation. Latencies can have only values which are integral multiples of h, but if h is sufficiently small we would not be seriously worried about this apparent absurdity. The model described in Section 14.4 involves a different assumption, namely, that the time required for an A_1 or A_2 act is a random variable which has an assumed distribution. This model leads to a continuous distribution of latencies and so is slightly more realistic. A portion of the data obtained by Weinstock [2] is analyzed, using this continuous model.

14.3 A MODEL WITH DISCRETE TIMES

We first discuss a model in which it is assumed that time is quantized into intervals of length h (where h is measured in seconds) and that during each such interval either A_1 or A_2 occurs. Response A_1 is identified with a goal-directed movement, and k such acts are required to bring the rat from the starting position to the point at which the latency is clocked. Interspersed among the k occurrences of A_1 are a number of A_2 occurrences. If a total of N acts of type A_1 or A_2 occur, the latency L is

(14.1) $$L = Nh.$$

Now on experimental trial n, suppose that there is a constant probability p_n that A_1 occurs during an interval of length h. In terms of this probability p_n we wish to compute the probability $P(N_n)$ that precisely N_n responses occur before k goal-directed responses have taken place. This probability is given by the well-known negative binomial distribution [5, p. 61]:

(14.2) $$P(N_n) = \binom{N_n - 1}{k - 1} p_n^{\,k}(1 - p_n)^{N_n - k} ,$$

where

(14.3) $$\binom{N_n - 1}{k - 1} = \frac{(N_n - 1)!}{(N_n - k)!(k - 1)!}$$

is a binomial coefficient. We are concerned with the mean latency, and so we want an expression for the expected value of N_n in terms of p_n and k. It can be shown that

(14.4) $$E(N_n) = k/p_n.$$

It follows that the expected value of the latency L_n is

(14.5) $$E(L_n) = hE(N_n) = hk/p_n.$$

This expression for the mean latency involves the time interval h and the necessary number k of A_1 occurrences only as a product hk. We denote this product by

(14.6) $$L^0 = hk,$$

and the expected latency can be written as

(14.7) $$E(L_n) = L^0/p_n.$$

When analyzing latency data we could consider L^0 a parameter to be estimated from the data.

Any good model for the runway should be able to reproduce the mean latencies reasonably well, but we consider this a minimal requirement because curve-fitting methods are adequate for this purpose. Estes, in his early paper [3], fitted the means of the Graham-Gagné data as did Graham and Gagné [1]; we have done the same with Weinstock's data, though we do not present the results here. But to show its worth in analyzing such data, a model should reproduce the general character of the whole distributions of latencies. In Fig. 14.1 we show an example of the distribution function defined by equation 14.2. An important property of this distribution is its variance given by the formula

$$(14.8) \qquad \sigma^2(N_n) = k \, \frac{1 - p_n}{p_n^{\,2}} \, .$$

The variance of the latencies L_n is h^2 times this variance:

$$(14.9) \qquad \sigma^2(L_n) = k \, \frac{1 - p_n}{p_n^{\,2}} \, h^2 = \frac{1 - p_n}{k} \left(\frac{hk}{p_n} \right)^2 .$$

From this expression we find that the variance of the latencies is zero when $p_n = 1$. This result is intuitively obvious since when $p_n = 1$ an

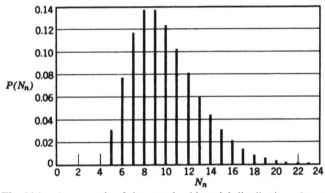

Fig. 14.1. An example of the negative binomial distribution of equation 14.2. The values $k = 5$ and $p_n = 0.5$ were used in constructing the figure.

A_1 is certain to occur during each interval of length h, and so the latency has the constant value hk. This conclusion creates a rather serious objection to the model described in this section: the model predicts that there will be no variability in latency when learning is complete provided that we assume that p_n approaches an asymptote of unity. If we find experimentally that the latency distributions stabilize after many trials but that the variance is appreciable, we must either assume that λ, the asymptote of p_n, is not unity or admit that the model has failed to describe this detail of the data. There is no *a priori* reason for insisting that $\lambda = 1$, but we feel that this assumption has some intuitive appeal. Observed reaction times are notoriously variable, and we object to ascribing deviations from split-second precision to imperfect learning ($\lambda \neq 1$). This objection, weak as it may appear to some readers, in addition to the more obvious objection that this model permits only discrete values of latencies, led us to develop the continuous model presented in the following section.

14.4 A MODEL WITH CONTINUOUS TIMES

The runway model given in the preceding section assumed that each A_1 and A_2 response requires a fixed time of length h. A generalization of this model may be made as follows. Let the time required for an A_1 act be t_1 and assume that t_1 is some constant h_1 plus a random variable τ_1:

$$(14.10) \qquad t_1 = h_1 + \tau_1.$$

Similarly, let the time required for an A_2 act be

$$(14.11) \qquad t_2 = h_2 + \tau_2,$$

where h_2 is a constant, not necessarily equal to h_1, and where τ_2 is a random variable. When $p_n = 1$, an experimental trial consists of precisely k acts of type A_1. The latency in this case is kh_1 plus the sum of k random variables τ_1, and thus the variance of the asymptotic latencies is unrestricted—it depends upon the assumed distribution of the random variable τ_1.

We wish to provide a model which specifies the precise form of the distributions of latencies, and so we must make some further assumptions. We could merely fit the data with a sensible distribution function, but we prefer to derive the distribution of latencies from a simple assumption about the distributions of the random variables τ_1 and τ_2. We assume that both τ_1 and τ_2 are distributed according to the function

$$(14.12) \qquad f(\tau) = se^{-s\tau}.$$

Several plausibility arguments can be made in favor of such a distribution of τ; an analogous assumption is made in many time problems such as in the theory of Geiger-Müller counters. This distribution may be derived from a pseudo-neurological model as follows. Consider τ to be a reaction time which follows the "decision" to make an A_1 act or an A_2 act. A decision is made, and τ seconds later the act begins. Then suppose a neuron in the nervous system of the organism fires every δ seconds but that the act begins only when a neuron of an appropriate class fires. Let π be the probability that a neuron of the appropriate class fires. The probability that this will occur after precisely $j - 1$ inappropriate neurons have fired is

$$(14.13) \qquad P(j) = (1 - \pi)^{j-1}\pi.$$

This, then, is the probability that the reaction time will be $j\delta$ seconds. We now wish to go to the limit as δ becomes vanishingly small. But we will take the limit while maintaining constant π/δ, the probability per second of the firing of an appropriate neuron. We let

(14.14) $$\tau = j\delta$$

and

(14.15) $$s = \pi/\delta.$$

Thus, equation 14.13 may be written

(14.16) $$P(\tau) = \frac{s\delta}{1 - s\delta} (1 - s\delta)^{\tau/\delta}.$$

Before going to the limit as $\delta \to 0$, we replace $P(\tau)$ with $f(\tau)\Delta\tau$, where $\Delta\tau = \delta$. We do this because the discrete distribution is being replaced by a continuous one, so the density $P(\tau)$ is being approximated by the ordinate $f(\tau)$ of a continuous distribution times the width of the interval $\Delta\tau$. Thus we have

(14.17) $$f(\tau) = \frac{s}{1 - s\delta} (1 - s\delta)^{\tau/\delta}.$$

We next expand the right-hand factor:

$$(1 - s\delta)^{\tau/\delta} = 1 - (\tau/\delta)s\delta + \frac{(\tau/\delta)(\tau/\delta - 1)}{2!} (s\delta)^2$$

(14.18)
$$- \frac{(\tau/\delta)(\tau/\delta - 1)(\tau/\delta - 2)}{3!} (s\delta)^3 + \cdots$$

$$= 1 - s\tau + \frac{(s\tau)(s\tau - s\delta)}{2!} - \frac{(s\tau)(s\tau - s\delta)(s\tau - 2s\delta)}{3!} + \cdots$$

When we then take the limit as $\delta \to 0$ the terms involving $s\delta$ vanish in the expansion on the right, and we get

(14.19) $$(1 - s\delta)^{\tau/\delta} \to 1 - s\tau + \frac{(s\tau)^2}{2!} - \frac{(s\tau)^3}{3!} + \cdots = e^{-s\tau}.$$

Therefore, as $\delta \to 0$, we have from equation 14.17

(14.20) $$f(\tau) \to se^{-s\tau},$$

in agreement with equation 14.12. We do not consider this pseudo-neurological argument essential to the model being developed in this section. Rather, we include it for whatever plausibility value it may have.

We assume that both τ_1 and τ_2 of equations 14.10 and 14.11 are distributed according to $f(\tau)$ given by equation 14.12. Furthermore, we assume that the constant h_2 is zero, that is, that A_2 responses do not require time for their execution but do involve a reaction time τ_2. This suggests that we consider A_2 responses to be "doing nothing" or "standing still." The total latency on an experimental trial, then, is kh_1 plus the sum of N random variables τ from the distribution $f(\tau)$. The number N will also be a random variable distributed according to the negative binomial of equation 14.2. But first we need to compute the distribution of the sum of N independent random observations from $f(\tau)$ for a given value of N. If we denote the sum by t_N and the distribution function by $g_N(t_N)$, it is well known that

$$(14.21) \qquad g_N(t_N) = \frac{s}{(N-1)!} e^{-st_N}(st_N)^{N-1}.$$

This expression is the gamma distribution [5, p. 112] and is readily derived from $f(\tau)$ by the method of moment generating functions.

The probability of a given value of N is

$$(14.22) \qquad P(N) = \frac{(N-1)!}{(k-1)!(N-k)!} p^k(1-p)^{N-k},$$

where p is the probability of an A_1 response at each step in the process during an experimental trial. To obtain the unconditional distribution of t we multiply g_N by $P(N)$ to get the joint distribution of N and t, and then sum over values of N from k to ∞:

$$
\begin{aligned}
(14.23) \qquad g(t) &= \sum_{N=k}^{\infty} g_N(t)P(N) \\
&= \frac{se^{-st}p^k}{(k-1)!} \sum_{N=k}^{\infty} \frac{(st)^{N-1}(1-p)^{N-k}}{(N-k)!} \\
&= \frac{se^{-st}p^k(st)^{k-1}}{(k-1)!} \sum_{N=k}^{\infty} \frac{[(st)(1-p)]^{N-k}}{(N-k)!}.
\end{aligned}
$$

The summation has terms of the form $x^n/n!$ summed from 0 to ∞ with $x = st(1-p)$, that is, the series expansion of e^x. Therefore

$$(14.24) \qquad g(t) = \frac{se^{-st}p^k(st)^{k-1}}{(k-1)!} e^{st(1-p)}.$$

Re-arranging terms, we get finally

$$(14.25) \qquad g(t) = \frac{sp}{(k-1)!} (spt)^{k-1} e^{-spt}.$$

The latency on an experimental trial is t plus kh_1, the minimum possible time. We let

(14.26) $$c = kh_1$$

and

(14.27) $$L = t + c.$$

Thus the distribution of the latencies L is

(14.28)
$$\phi(L) = \frac{sp}{(k-1)!} [sp(L-c)]^{k-1} e^{-sp(L-c)} \qquad L \geq c$$
$$= 0 \qquad\qquad\qquad\qquad\qquad\qquad L < c.$$

An example of this distribution function is shown in Fig. 14.2. It is the

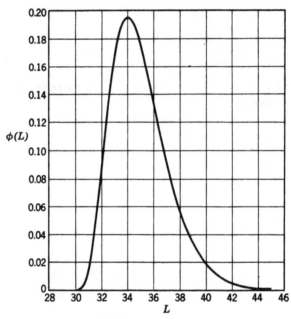

Fig. 14.2. An example of the gamma distribution of equation 14.28. The values $k = 5$, $sp = 1$, and $c = 30$ were used in constructing the figure.

generalized gamma distribution. It can be shown that the mean is

(14.29) $$E(L) = c + \frac{k}{sp},$$

and the variance is

(14.30) $$\sigma^2(L) = \frac{k}{(sp)^2}.$$

We use the distribution function $\phi(L)$ defined by equation 14.28 to analyze some of the data obtained by Weinstock [2]. Before doing this we consider the problem of estimating the parameters of $\phi(L)$ from data.

14.5 ESTIMATION OF PARAMETERS OF THE ASYMPTOTIC DISTRIBUTION

We attempt to use the gamma distribution of equation 14.28 to describe the observed asymptotic distribution of latencies. We assume that $p = 1$ during the latter trials in the experiment, and we wish to estimate s, c, and k from these data. However, we do not assume that the zero point c and the scale factor s are the same for all animals, but we do assume that the number k is the same for all animals in the experiment. The subscript $i = 1, 2, \cdots$ is used to distinguish the animals, and the subscript n will be used for trials as usual. The latency of the ith animal on trial n is then L_{in}. Thus, from equation 14.28 we write for $p = 1$,

$$(14.31) \qquad \phi_i(L_{in}) = \frac{s_i}{(k-1)!} [s_i(L_{in} - c_i)]^{k-1} e^{-s_i(L_{in}-c_i)} \qquad L_{in} \geq c_i$$

$$= 0 \qquad L_{in} < c_i.$$

We wish to estimate the parameter k for a group of animals and the parameters c_i and s_i for each animal.

Various properties of the distribution can be used to estimate the parameters; one easy procedure is to use the moments. Equations 14.29 and 14.30 give the mean and variance, which could be used along with a third property to carry out the estimation. For example, we could estimate the parameter c_i by the smallest latency observed for the ith animal, but this estimate clearly has a positive bias. The objection to using the distribution moments for the estimation is that latency data frequently contain a small percentage of exceedingly large observations which may result from uncontrolled or uncontrollable variations in experimental conditions; the moments are very sensitive to large observations. In the next section we present an example of such data.

The maximum likelihood procedure can be used to estimate the parameters of the distribution function given by equation 14.31. It can be shown that this procedure leads to expressions which involve three measures of location (cf.[7]): the arithmetic mean, the harmonic mean, and the geometric mean. The last two means are rather difficult to compute, as they involve the unknown parameters c_i, and so we do not propose this procedure, though we have carried it through.

An estimation procedure we have found relatively simple to carry out, and one which is not sensitive to very large observations, makes use of percentage points of the cumulative distribution. We first transform the distribution function of equation 14.31 by letting

$$(14.32) \qquad x_{in} = s_i(L_{in} - c_i).$$

We then obtain for the distribution of x_{in}, the gamma distribution:

$$(14.33) \qquad \psi(x_{in}) = \frac{x_{in}^{\,k-1}}{(k-1)!} e^{-x_{in}}, \qquad x_{in} \geq 0,$$

$$= 0, \qquad\qquad x_{in} < 0.$$

The percentage points are then defined by

$$(14.34) \qquad \int_0^u \psi(x_{in}) \, dx_{in} = F(u).$$

This integral, which is related to the incomplete gamma function, has been extensively tabulated [6]. The 25th percentile, u_{25}, for example, is the upper limit of the integral which corresponds to $F(u) = 0.25$. Percentiles to be used in later sections are shown in Table 14.1.

TABLE 14.1

Percentage points of the gamma distribution for
several values of k, and values of the ratio ρ.

k	u_{10}	u_{25}	u_{75}	$\rho = \dfrac{u_{75} - u_{25}}{u_{25} - u_{10}}$
3	1.10	1.73	3.92	3.48
4	1.75	2.53	5.11	3.31
5	2.44	3.37	6.27	3.12
6	3.15	4.22	7.42	2.99
7	3.90	5.08	8.56	2.95

The corresponding percentage points U of the distribution of L_{in} are obtained from equation 14.32:

$$(14.35) \qquad U = \frac{u}{s_i} + c_i.$$

As shown in Table 14.1, the u's are known functions of k, and so the U's are functions of k, s_i, and c_i. Hence three percentage points from the observed latency distribution for a single animal are sufficient to estimate the three parameters.

First consider estimating the parameter k. Let U', U'', and U''' be three observed percentage points such that $U' > U'' > U'''$. From equation 14.35 we see that

(14.36)
$$U' - U'' \triangleq \frac{u' - u''}{s_i},$$

$$U'' - U''' \triangleq \frac{u'' - u'''}{s_i},$$

and so

(14.37)
$$\frac{U' - U''}{U'' - U'''} \triangleq \frac{u' - u''}{u'' - u'''} = \rho.$$

Some values of the ratio ρ are shown in Table 14.1 for several values of k. The observed ratio can then be used to estimate k.

Once k has been estimated, the corresponding values of the u's can be obtained from Table 14.1. These, then, can be used in equation 14.35 to estimate s_i and c_i. In the next section we illustrate the procedure in detail.

14.6 ANALYSIS OF THE WEINSTOCK DATA

We now analyze the latency data from Weinstock's 100 percent reinforcement group. Twenty-three rats were run through 108 trials of acquisition; the latency on the first extinction trial reflects the previous reward training, and so the data on 109 trials for each animal are used. First we estimate the parameters of the asymptotic distributions by using the data from the last 40 trials; for the analysis, it is assumed that $p = 1$ for these trials.

As mentioned in the preceding section, latency data often contain a small percentage of extremely large observations. Preliminary analyses of the Weinstock data made it evident that the models described in this chapter could not account for these large observations while at the same time describing the major portion of the data. In the next section we discuss the problem of handling the large observations, but in this section we are not concerned with them. To estimate parameters of the model, we choose three percentiles—the 10th, 25th, and 75th—which are relatively insensitive to the extreme tail of the distribution. From the data on the last 40 trials we obtain these percentage points for each animal and then compute the ratio

(14.38)
$$\hat{\rho} = \frac{U_{75} - U_{25}}{U_{25} - U_{10}}.$$

In Table 14.2 we show the results. The mean value of the $\hat{\rho}$'s for the 23 animals is 3.148; in Table 14.1 we see that this is closest to the value of ρ for $k = 5$. Further computations are simplified if k is an integer.

Hence we take as our estimate of k, the number of goal-directed acts,

$$(14.39) \qquad\qquad \hat{k} = 5.$$

Using this estimate of k, we next estimate s_i and c_i for each animal from

TABLE 14.2

Data used for estimating k for the group of rats and s_i and c_i for each rat. The observed percentage points are U_{10}, U_{25}, and U_{75}. The ratio of equation 14.38 is $\hat{\rho}$, and the estimates from equations 14.41 are $\hat{s}_i$ and $\hat{c}_i$.

Rat	U_{10}	U_{25}	U_{75}	$\hat{\rho}$	$\hat{s}_i$	$\hat{c}_i$
1	29.7	30.3	33.1	4.67	1.04	27.0
2	31.9	32.9	37.0	4.10	0.71	28.1
3	38.4	40.1	42.9	1.65	1.04	36.8
4	32.9	34.8	41.6	3.58	0.43	26.9
5	36.2	37.9	41.1	1.88	0.91	34.2
6	33.1	35.7	45.5	3.77	0.30	24.3
7	32.8	33.8	38.1	4.30	0.67	28.8
8	34.8	35.8	38.8	3.00	0.97	32.3
9	37.0	38.3	42.8	3.46	0.64	33.1
10	36.0	37.5	44.7	4.80	0.40	29.1
11	30.2	31.7	33.8	1.40	1.38	29.3
12	39.3	40.2	43.1	3.22	1.00	36.8
13	34.5	36.0	38.8	1.87	1.04	32.7
14	34.0	35.3	39.2	3.00	0.74	30.8
15	33.6	35.8	40.1	1.95	0.67	30.8
16	36.0	36.9	39.9	3.33	0.97	33.4
17	38.9	40.8	45.5	2.47	0.62	35.3
18	34.0	35.9	38.3	1.26	1.21	33.1
19	31.8	33.0	37.3	3.58	0.67	28.0
20	31.9	32.8	37.5	5.22	0.62	27.3
21	44.6	47.2	55.1	3.04	0.37	38.0
22	34.1	35.5	40.0	3.21	0.64	30.3
23	31.5	34.1	43.6	3.65	0.31	23.1

the observed 25th and 75th percentiles. From equation 14.35 and Table 14.1 we obtain for $k = 5$,

$$U_{75} \triangleq \frac{6.27}{s_i} + c_i,$$

$$(14.40)$$

$$U_{25} \triangleq \frac{3.37}{s_i} + c_i.$$

Solving these for s_i and c_i, we have

$$\hat{s}_i = \frac{2.90}{U_{75} - U_{25}},$$

(14.41)

$$\hat{c}_i = U_{75} - 2.16(U_{75} - U_{25}).$$

These equations lead to the estimates shown in Table 14.2.

Once having obtained estimates of s_i and c_i for each rat, the observed latencies can be transformed by equation 14.32. These transformed

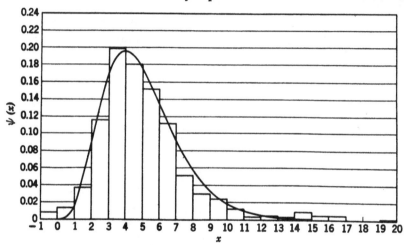

Fig. 14.3. Distribution of transformed latencies for the last forty trials of Weinstock's continuous reinforcement group. The theoretical distribution function was computed from equation 14.33, with $k = 5$.

latencies can then be combined to yield an empirical distribution for all animals on the last 40 trials. This distribution corresponds to the theoretical distribution given by equation 14.33. In Fig. 14.3 we compare the two distributions. The agreement appears satisfactory over most of the range, but two things can be noted. First, a small number of negative values of x appear in the empirical distribution; this presumably is a result of sampling errors in the estimates of the parameters. Second, the observed large values of x are not predicted by the theoretical distribution. About 5 percent of the observed values of x are 13.5 or more whereas the area under the computed curve is about 0.3 percent.

The preceding analysis has been concerned only with the asymptotic distribution for which it was assumed that $p = 1$. We now investigate the distributions for blocks of trials and estimate p for each block. For estimation purposes we proceed as if p were constant within a block.

The median (50th percentile) is used for this purpose. First, the median latency U_{50} of each rat is obtained for a particular block of trials. Then this latency is transformed by the equation

(14.42) $u_{50} \triangleq s_i(U_{50} - c_i)$.

In this way, we obtain 23 estimates of u_{50} for a given block of trials. (When $p = 1$ and $k = 5$, $u_{50} = 4.67$.) The median of these 23 estimates is then determined; denote this median by m. The estimate of p for that block of trials is then

(14.43) $\hat{p} = 4.67/m$.

In Table 14.3 we show the estimates obtained. The latencies were changing rapidly during the first five trials, and so we estimate p for each

TABLE 14.3

Estimates of p for blocks of trials obtained from equation 14.43 and values of p_n computed from equation 14.46, with $p_0 = 0.014$ and $\alpha = 0.974$.

Trials	$\hat{p}$	p_n
0	0.014	0.014
1	0.023	0.040
2	0.022	0.065
3	0.083	0.089
4	0.079	0.113
5–9	0.134	0.181
10–19	0.246	0.328
20–29	0.477	0.484
30–39	0.603	0.604
40–49	0.652	0.695
50–59	0.682	0.766
60–69	0.753	0.821
70–79	1.006	0.862
80–89	1.194	0.894
90–99	0.905	0.918
100–108	0.936	0.937

of these trials separately. These estimates are also given in Table 14.3. We need to transform equation 14.28 by equation 14.32; we obtain the result

(14.44) $\psi(x_{in}) = \dfrac{p_n}{(k-1)!} [p_n x_{in}]^{k-1} e^{-p_n x_{in}}$.

From this equation, we can compute the theoretical distribution for each block of trials by using the estimated value of p_n. However, we prefer to generate values of p_n from the model and thereby fit the entire sequence of distributions.

We assume that the reward that occurs on each trial of acquisition has an associated operator Q which is applied to p. Since we have already assumed that $p = 1$ asymptotically, we take

$$(14.45) \qquad Qp = \alpha p + (1 - \alpha).$$

We further assume that no other events alter p and so we have

$$(14.46) \qquad p_n = Q^n p_0 = \alpha^n p_0 + (1 - \alpha^n).$$

From the latency data we need to estimate the two parameters, p_0 and α. (It is assumed that all animals have the same values of p_0 and α.) We estimate p_0 from the data on trial 0. In Table 14.3 we see that

$$(14.47) \qquad \hat{p}_0 = 0.014.$$

The learning parameter α is estimated by the procedure used in Chapter 13. We sum p_n from trial 0 to $N - 1$ and get

$$(14.48) \qquad \sum_{n=0}^{N-1} p_n = N - (1 - p_0) \frac{1 - \alpha^N}{1 - \alpha}.$$

The sum on the left side of this equation is estimated by summing the estimates of p_n given in Table 14.3. (An estimate for a block of trials is taken as the estimate for each trial in the block.) This sum is 74.50. Taking $p_0 = 0.014$ and $N = 109$, we can solve equation 14.48 numerically to get

$$(14.49) \qquad \hat{\alpha} = 0.974.$$

Using the above estimates of p_0 and α we can compute the values of p_n from equation 14.46; these are shown in the last column of Table 14.3. (The middle value of n in each block was used for the computation of p_n.)

The theoretical distribution for each block of trials was computed from equation 14.44 by using $k = 5$ and the values of p_n just obtained. These distributions are compared with the distributions of transformed observations in Fig. 14.4. In several trial blocks there are noticeable discrepancies between the observed and computed distributions, but we wish to reiterate that we have fitted the whole sequence of distributions rather than just the means or even the distributions one at a time. Clearly, better fitting can be done by using the block estimates $\hat{p}$ rather than the computed values p_n.

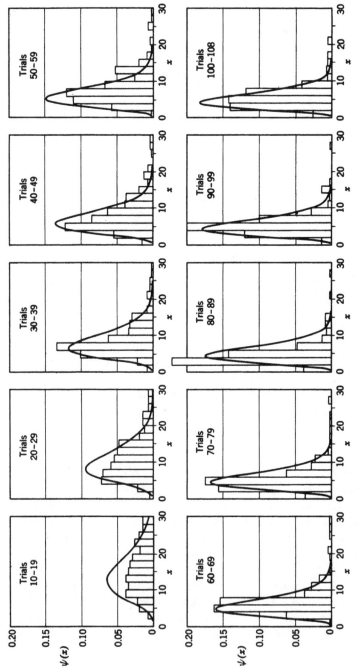

Fig. 14.4. Distributions of transformed latencies for ten blocks of trials. The histograms were obtained by transforming the raw data, using equation 14.32 and the estimates of s_i and c_i shown in Table 14.2. The smooth curves were obtained from equation 14.44, with $k = 5$ and the values of p_n given in Table 14.3.

A latency model should be concerned with the sequence of distributions and not primarily with obtaining close-fitting curves on single trials or separate blocks of trials. Nor is it enough for a stochastic model to fit just one parameter of the sequence of distributions. The model might do this very well, and still not do justice to the variability or the general shape of the distribution.

We do not proceed to make formal tests of significance for goodness-of-fit because it is obvious that there are consistently too many large observations. This matter is discussed further in the next section.

14.7   CONCLUDING REMARKS

In the analysis of the Weinstock latency data we were confronted with two serious problems. First, a casual inspection of the data shows marked differences among animals, and, second, a small percentage of extremely large observations was present. The between-animal differences were handled by estimating for each animal two parameters, a zero point and a scale factor, and then making appropriate linear transformations on the raw observations to make animals commensurate. The problem of the large observations was essentially circumvented by selecting estimation procedures that are insensitive to them. The result of this procedure is that the model gives a reasonably satisfactory description of the main portion of the data but does not properly predict the very large latencies.

We can take at least two views about the large observed latencies. We might argue that they arise from variations in experimental conditions which are of no interest, that is, that they result from environmental events which are irrelevant to the learning phenomena being studied. Another position, and probably a more tenable one, is that the large latencies are part of the data and cannot be dismissed. Accordingly, we would argue that the model applied in this chapter is only a first approximation. A more refined model would have a built-in mechanism for generating a small percentage of very large numbers. Some obvious sources of large latencies are animals turning around in the apparatus or other activities that tend to cancel previous goal-directed movements. The model we use in analyzing Weinstock's data does not include this possibility of cancellation.

Although it would be useful to develop a model that would fit the long tail as well as the main body of a latency distribution, we do not do so here. Such an analysis would be very specific to latency problems, rather than a direct application of the general model given in Part I. Our purpose in this chapter is to illustrate how the general model can be applied to time problems. We make no pretense to completeness.

14.8 SUMMARY

Two models for describing latency and running time data from simple runway experiments are described in this chapter. The first model allows only discrete values of times. The second model involves a continuous distribution and so overcomes some of the objections to the discrete model. Both models assume that an experimental trial consists of a sequence of acts or movements belonging to two classes: those which are "goal-directed" and those which are not. The probability of a goal-directed act (A_1) is transformed by an operator Q on each trial when reward is given in the goal box. The continuous time model is used to analyze the distributions of latencies obtained by Weinstock [2]. The main portions of those distributions in the sequence are adequately described, but the model does not satisfactorily handle the large observed values. The model, therefore, is regarded as a first approximation rather than an adequate description.

REFERENCES

1. Graham, C., and Gagné, R. M. The acquisition, extinction, and spontaneous recovery of a conditioned operant response. *J. exp. Psychol.*, 1940, **26**, 251–280.
2. Weinstock, S. Resistance to extinction of a running response following partial reinforcement under widely spaced trials. *J. comp. physiol. Psychol.*, 1954, **47**, 318–322.
3. Estes, W. K. Toward a statistical theory of learning. *Psychol. Rev.*, 1950, **57**, 94–107.
4. Bush, R. R., and Mosteller, F. A mathematical model for simple learning. *Psychol. Rev.*, 1951, **58**, 313–323.
5. Mood, A. M. *Introduction to the theory of statistics.* New York: McGraw-Hill, 1950.
6. Pearson, K. (ed.) *Tables of the incomplete Γ-function.* London: His Majesty's Stationery Office, 1922.
7. Fisher, R. A. *Contributions to mathematical statistics.* New York: Wiley, 1950, pp. 10.332–10.337.

CHAPTER 15

Evaluations

15.1 PURPOSE OF THIS CHAPTER

In Part I of this book we discussed some mathematical properties of a general model, and in Part II we describe several applications to experiments on learning. As a result we are now in a position to make some critical evaluations. In the last five chapters we have repeatedly pointed out the usefulness of the model for various purposes, but in this chapter we emphasize weaknesses and shortcomings of the model and indicate some unsolved problems. In the final section, for an ideal situation, we compare the use of the model with that of classical curve-fitting.

15.2 MEASURES OF BEHAVIOR

The fundamental measure of behavior used throughout this book is the probability of occurrence of a given class of responses. In the general model we introduced a set of mutually exclusive and exhaustive alternatives and assigned a probability measure to each. Such a starting point leads to certain special difficulties in handling some kinds of data; indeed, some readers may have felt in reading Chapter 1 that the general model was a very restricted one.

We are not greatly concerned about the requirement that the alternatives be mutually exclusive, even though we can easily find observable classes of behavior that occur simultaneously, for example, running and breathing. We have already argued that the set of all possible behavior elements can always be partitioned into classes which are disjunct by choosing appropriate classes. Moreover, a residual class can always be introduced in such a way that the set is exhaustive. Hence the set of alternatives leads to little trouble until we introduce the probability measure on this set. For the probability measure to make sense in applying the model to a particular experiment, we must define a sequence of trials such that one and only one alternative occurs on each trial. In many experiments, such as those discussed in Chapters 10 through 13, the trials were built into the experiment, but in Chapter 14, where we were interested in latency and running time, we were forced to make somewhat different identifications. A trial in the model was no longer an experimental trial but instead was identified with a small time increment. In this way we were able to relate latency and running time to our probability variables. (In some experiments we may study two or more measures of learning. For example, in the T-maze we could record latencies as well as choices. The model we describe offers no direct relation between those two measures. Cf. Section 13.2.)

Rate of responding can also be related to probabilities of appropriate response classes, as has been demonstrated in the literature [1, 2]. Furthermore, there is little difficulty in handling behavioral measures such as number of errors and resistance to extinction. On the other hand, an important experimental measure of behavior we have not successfully handled with our general model is *response intensity*. In this category we include amount of saliva flow in Pavlovian conditioning, galvanic skin reaction, kick deflection, and grams pull as used in conflict experiments. These measures are concerned with the strength or intensity of an operationally defined class of behavior, whereas our general model considers only the occurrence or non-occurrence of a particular behavior class. Although it is possible to conceive of identifications which would force intensity measures into our general framework, we have found no really satisfactory way of doing this. The principal suggestion that comes to mind is to set up a distribution function for the intensity and to make the operators change a parameter of the distribution. As in the latency problem, this method entails considerable bother.

15.3 BASIC ASSUMPTIONS

One of the basic axioms of the general model is the assumption of *path independence*: that the set of probabilities after an event has occurred depends only upon the set of probabilities just prior to the event and upon an operator associated with that event. For example, if some event occurs on each trial (and not at other times), and if we have only two alternatives with probabilities p_n and $1 - p_n$ on trial n, then p_{n+1} depends only on p_n and on what event occurred on trial n; the values of p_{n-1}, p_{n-2}, etc., are irrelevant. Heuristic objections can be raised against this assumption of path independence; the model does not seem to provide for "memory," for a "practice effect," or for long-range effects of "trauma."

Whether or not such objections are appropriate depends, we believe, upon the specific event identifications which are made. In principle, we can always make identifications that impose as long a "memory" as we want. For example, suppose we have a sequence of successes and failures,

SFFSSSFSS. The obvious identifications are to call *S* and *F* the two events, but we could just as well consider the sequence to contain four kinds of events, *SS, SF, FS, FF*; or we could consider triplets to be the events, etc. Therefore, we take the position that the general model is not seriously restricted by the path independence axiom, though our specific applications may be misleading here because they make the obvious single-trial identifications. The problem is to make the most appropriate event identifications for each application.

The other basic axiom of the general model is the assumption that the event operators are *linear*. We introduced this axiom very early in our development because we have never used non-linear operators in handling any specific set of data. Nevertheless we recognize a danger in being specific too soon: if the linearity axiom proved to be unsatisfactory, nearly everything we have said in this book would have to be modified. We could have established some rather general results without the linearity assumption and then *illustrated* the implications for the special case of linear operators. Or we could have introduced a less severe restriction, such as monotonicity. We chose to do otherwise, however; we introduced the linearity axiom early so that we could get to actual computations and data analysis as soon as possible. We might defend the linearity assumption on theoretical grounds by reminding the reader of the stimulus model described in Chapter 2 which yields linear operators and of the Savage theorem mentioned in Section 1.8 which shows that only linear operators can satisfy the combining-of-classes restriction. The defense on empirical grounds rests more with the reader's evaluation of goodness-of-fit in the examples given, and with similar evaluations in future applications.

If it should turn out that the linearity assumption is untenable for a certain class of problems, a possible remedy is the following. An intervening variable could change according to linear operators, and its relation to response probabilities and other behavioral measures could be determined (cf. Section 1.4, last paragraph). In effect, Estes and Burke [3] use such a strategy in their general model of stimulus variability.

15.4 MATHEMATICAL AND STATISTICAL PROBLEMS

In the Introduction we argued that we should investigate the mathematical properties of a model before applying it to actual data. In Part I we attempted to do just that, but many problems remain unsolved. For example, we know relatively little about the asymptotic distribution except in some special cases. Karlin [4] and Thompson [5] have obtained results which indicate that no simple exact computational scheme is possible. In Chapter 6, therefore, we developed some approximate methods for studying the asymptotic distribution. The computation of moments also raises some mathematical problems which we have not solved except in some very special cases. Another of the many problems incompletely solved in this book is the distribution of runs, discussed in Section 4.8.

Problems in the statistical estimation of parameters and in measuring goodness-of-fit have not been handled satisfactorily in this book. We have relied on minor modifications of standard techniques and improvised procedures for each specific problem. Furthermore, we usually imposed restrictions on the values of the limit points λ_i so as to obtain workable estimation procedures. Though we obtained estimators for the parameters in each specific model, we often knew little about their properties. To compare two estimates, we need better information about their distributions than we usually provide. More general methods are needed.

15.5 EXPERIMENTAL PROBLEMS

Undoubtedly the greatest disappointment to the psychologist in reading this book is our apparent lack of attention to a variety of variables which have been studied in learning experiments. Our model makes no reference to drive strength, amount of reward, amount of work, delay in reinforcement, etc. We would like to make quite clear that (1) we consider such variables to be important in the psychology of learning, and (2) we make no pretense of having anything to say about them in this book. We have focused attention on the stochastic aspects of learning and performance— the detailed properties of sequences of responses observed in experiments. In some problems we have illustrated how the parameters in the models change when experimental conditions are altered, but we make no attempt to explain such changes with the models (see Sections 11.8 and 13.11). We could readily introduce *ad hoc* assumptions to describe how certain parameters depend upon variables like drive strength. In effect this is what Hull has done in his system [6]. We have chosen to leave these questions, which are parametric in our model, both to experimental investigation and to psychological theories in which they are not parametric. On the other hand, we hope that our general stochastic model will provide a framework within which such questions can be discussed.

We have not attempted to handle experimental phenomena such as stimulus generalization, discrimination, spontaneous recovery, patterns of reinforcement, secondary reward, and response chaining.

15.6 THEORETICAL INTERPRETATIONS

Throughout this book we have attempted to divorce our model from particular psychological theories and have constantly talked about "describing" data rather than "explaining" it. We believe that many interpretations of our results are possible, and in a few places we have suggested some psychological interpretations. Nevertheless, we recognize that we have been strongly influenced by the Hullian and Guthrian schools of thought, and the general model undoubtedly exhibits this bias. We have not tried to relate the model to cognitive theory, for example. In some ways our approach is similar to Skinner's: we have attempted to stay close to data and observable acts and events. We have not presented a model of the organism but have developed models for experiments.

15.7   CONCLUDING REMARKS

We have often been asked to explain the sense in which our general model, or one of our specific models, represents anything more than curve-fitting. Furthermore, we have been asked to explain the sense in which the parameters of the model are psychologically meaningful. Unfortunately such a discussion could come only after applications had been provided. An explanation can best go forward if we think of an idealized situation in which the model is correct in every particular. People seriously asking about these issues understand perfectly well that, in the end, the question of goodness-of-fit will arise, but they are puzzled about the status of the model even if it consistently fits real data very closely. Therefore we set aside the goodness-of-fit question for the moment. Part of the trouble arises from the fact that if a sequence of rather similar experiments, say rote-learning studies, is planned, our model does not predict in advance the various parameters that will be found. Once this is understood, the original questions arise with still more force.

Suppose a learning experiment, involving 1000 subjects for 300 trials each, has recorded either a success or a failure for each subject on each trial. These data are turned over to a clerk for summary analysis. There are now literally thousands of independent questions that can be asked about these data without ever getting down to the level of the single cell (outcome for a particular subject on a particular trial). To make the point clear we shall list a few of the more usual questions and a few of the more esoteric ones. For our purpose it does not matter whether or not a psychologist would be interested in the entire range of such questions. The point is that each can be asked of the data and an answer could be obtained. Examples are:

1. What is the overall percentage of successes?
2. What are the trial-by-trial success percentages?
3. What is the best-fitting cubic curve that can be drawn through the trial-by-trial percentages?

4. What is the mean trial number on which the last failure occurred? (How about the median instead of the mean?)

5. What is the mean trial number on which the first success occurred? (The second, the third, $\cdots$?)

6. What is the average number of runs of successes? failures?

7. What percentage of subjects had at least twenty successes in a row?

8. What percentage of the subjects had failures on the third, fourth, ninth trials and successes on the eighteenth and twentieth? (This question can be varied thousands of ways.)

9. What is the correlation between numbers of successes in the first twenty and last twenty trials?

10. What is the variance of the subjects' total scores?

Even this small group of questions makes clear the rich variety of possible summary questions that can be asked of this simple data sheet. A complete summary could clearly take hundreds of pages. The clerk working over these data would just have to compute a new number in response to each question. There would be no necessary relation between the answers to the different questions. Knowing answers to one hundred questions would not help appreciably in answering the next hundred.

The clerk may find it convenient and useful to employ classical methods of curve-fitting in preparing to answer these questions. For example, in question 5 above, he might compute the mean trial of the nth success for each n in the data, and then fit one of the common curves (linear, quadratic, logarithmic, or exponential function) to those means. When asked about the mean trial number for the third success, he might report the fitted answer rather than the one computed from the raw data. This answer might be preferable because local fluctuations are smoothed out by the fitted function. The curve itself may be of interest because it summarizes the trend of the observed means. However, when the clerk is asked question 9, for example, he must return to the data, and if he chooses can do some additional curve fitting for correlation coefficients. Hence, for each class of questions a new curve could be fitted. Many classes of questions are possible, however, and therefore many numerical functions would be required to summarize all the information in the data.

A different level of analysis can be accomplished with a model. Suppose our model fitted closely the results of experiments like the one described. Then after a few questions had been asked and answered (say questions 1, 5, 6), we would obtain a few numbers, say three (we will regard these as parameters). With these three numbers we would be able to retire from the data and be prepared to answer question for question with the clerk. Our answers would not agree perfectly with the clerk's responses, but they would be generally close. Thus on the basis of three numbers we are prepared, in principle, to answer all the questions the original data sheet can answer provided that the questions do not get down to the level of a single cell. ("In principle," means it might take us a long time, but with computing machines we could do it.) This is saying a great deal. Furthermore, if we are given the three numbers we can work out in advance the answers to the questions, just in case someone ever comes across an experiment that has these three parameters. In addition, we would be glad to turn the rules for finding the three numbers over to someone else, and tell him how to generate the various answers so that when an experiment is done he could see for himself just how the generated answers agree with the results of the experiment. Thus for three numbers derived from the data we plan to be able to answer a wealth of questions. If these three numbers plus the model can answer every possible question above the level of the single cell, it is hard to see what more succinct way there is to summarize the data (unless another model can do it with fewer parameters).

Now what about the psychological meaningfulness of the three parameters? In almost any situation like the one we describe, if there are three numbers that will do all this heavy labor, there will be many other sets of three numbers derivable from the first three that can also be used to achieve the same end. Some of these sets of three numbers can be given familiar names and descriptions more easily than others. Furthermore some sets of parameters that appear strange initially seem, after maturer reflection and experience, more suitable than others. And, again, which set is more suitable sometimes depends on the specific nature of the question. In elementary mathematics we are used to having both rectangular and polar coordinates available (two of the many possible kinds), and we seldom, if ever, see a discussion as to which is more meaningful, though there is plenty of evidence that some problems that are easy in one set become vicious when discussed in the other. In the course of our own work we have gradually turned from the view that quantities like p_0, a_i, and b_i are the natural parameters to the view that p_0, λ_i, and α_i are easier to work with and lead more smoothly to generalizations. Yet each set has its own simple meaning, and one easy to explain and use. Thus preference for equivalent basic sets of parameters may be partly determined by individual taste, but more strongly determined by the services rendered by the particular sets.

The description of our model in an idealized situation is not unlike the situation with Kepler's laws for heavenly bodies. After working with a large set of numbers, he stated a few simple laws that explain (describe) a great deal of what had been measured. On the other hand, if a new heavenly body were to appear, the laws tell us very little that is absolutely quantitative about the motion. But given in addition to the laws a few accurate measurements for the particular body, its course both before and after the observations could be described in specific detail. There is another similarity too; the chemical composition of the body, its temperature, color, and other properties that would interest an observer were not included in the laws. Again, for our model there are many protocol features and quantitative measurements of deep importance to a psychologist that not only are not handled but that also we see no hope of handling.

Reluctantly, we rouse ourselves a little from this delightful dream world in which our model reproduces psychological data perfectly. Suppose that when we compare the clerk's detailed computations with the answers generated by the basic three numbers that there are indeed hundreds of pages of close agreement, but that there seems to be a class of questions which the most motherly comparison finds wanting. The usual procedure is to try to find what special conditions have given rise to such error. Indeed, reasoning from such a discrepancy in Kepler's laws, a new planet was found. The model suggested both where not to look further, and also where investigation was needed. Sometimes such discrepancies lead to reformulations of the model, sometimes to interesting discoveries. But in any case a model that predicts a great many things nearly correctly will almost certainly serve as a temporary baseline until a more comprehensive model appears. Nor is it always true that an outmoded model is always discarded; its use may merely be restricted to a narrower sphere.

The feature of a baseline provided by a model may be viewed in another important way. It changes considerably our practical view of data. In much of psychology the search is for statistically significant differences, and the aim is to show, or at least find out, whether different conditions lead to different results. Big differences are usually a source of great satisfaction. With a quantitative model the emphasis is usually in the reverse direction. We look for close agreement with the model, and regard large differences with dissatisfaction as owing to a lack of understanding. The lack may, of course, stem from the inadequacy or inappropriateness of the model; it might also stem from inadequacies of the design and execution of the experiment, but more hopefully it may stem from a principle soon to be discovered.

Lest some reader has entered this discussion *in medias res*, we remind him that we have not had a fit of arrogance, nor do we have delusions of grandeur; we are merely explaining the value of our type of model under the most favorable conditions imaginable.

REFERENCES

1. Estes, W. K. Toward a statistical theory of learning. *Psychol. Rev.*, 1950, **57**, 94–107.
2. Bush, R. R., and Mosteller, F. A mathematical model for simple learning. *Psychol. Rev.*, 1951, **58**, 313–323.
3. Estes, W. K., and Burke, C. J. A theory of stimulus variability in learning. *Psychol. Rev.*, 1953, **60**, 276–286.
4. Karlin, S. Some random walks arising in learning models I. *Pacific J. of Math.*, 1953, **3**. 725–756.
5. Thompson, G. L., unpublished work.
6. Hull, C. L. *Principles of behavior.* New York: Appleton-Century-Crofts, 1943.

Tables

TABLE A

The two functions,

$$\Phi(\alpha, \beta) = \sum_{\nu=0}^{\infty} \alpha^{\nu(\nu+1)/2}\beta^{\nu},$$

$$\Psi(\alpha, \beta) = \sum_{\nu=0}^{\infty} \nu\alpha^{\nu(\nu+1)/2}\beta^{\nu},$$

which occurred in Sections 4.8, 8.3, and 8.4 in the computation of run lengths. This table is used in Sections 9.6 and 11.4 for estimating a parameter and the variance of that estimate. The first entry in each cell is $\Phi(\alpha,\beta)$ and the second entry is $\Psi(\alpha,\beta)$. (The table was prepared by transforming functions computed by the Harvard Computation Laboratory.)

TABLE A

α \ β	.50	.52	.54	.56	.58	.60	.62
.50	1.2833	1.2961	1.3090	1.3220	1.3352	1.3485	1.3619
	.3186	.3345	.3506	.3670	.3837	.4006	.4179
.52	1.2977	1.3113	1.3250	1.3389	1.3529	1.3671	1.3814
	.3381	.3552	.3726	.3904	.4084	.4268	.4455
.54	1.3126	1.3270	1.3416	1.3563	1.3713	1.3863	1.4016
	.3586	.3771	.3959	.4151	.4346	.4546	.4749
.56	1.3280	1.3433	1.3587	1.3744	1.3902	1.4063	1.4225
	.3802	.4001	.4205	.4412	.4624	.4841	.5061
.58	1.3438	1.3600	1.3765	1.3931	1.4100	1.4271	1.4443
	.4030	.4245	.4465	.4690	.4920	.5155	.5395
.60	1.3602	1.3774	1.3949	1.4126	1.4305	1.4487	1.4671
	.4271	.4504	.4742	.4986	.5235	.5491	.5752
.62	1.3772	1.3955	1.4140	1.4328	1.4519	1.4712	1.4909
	.4527	.4779	.5037	.5301	.5573	.5851	.6136
.64	1.3949	1.4142	1.4339	1.4539	1.4742	1.4948	1.5158
	.4799	.5072	.5352	.5639	.5934	.6238	.6549
.66	1.4132	1.4338	1.4547	1.4759	1.4976	1.5195	1.5419
	.5090	.5385	.5689	.6002	.6324	.6655	.6996
.68	1.4324	1.4542	1.4764	1.4990	1.5221	1.5455	1.5695
	.5401	.5722	.6052	.6393	.6745	.7107	.7480
.70	1.4524	1.4755	1.4992	1.5233	1.5479	1.5730	1.5985
	.5735	.6084	.6444	.6816	.7201	.7598	.8009
.72	1.4733	1.4980	1.5231	1.5488	1.5751	1.6019	1.6293
	.6095	.6475	.6868	.7276	.7698	.8135	.8588
.74	1.4953	1.5215	1.5484	1.5758	1.6039	1.6327	1.6621
	.6484	.6899	.7330	.7778	.8243	.8725	.9225
.76	1.5184	1.5464	1.5750	1.6044	1.6345	1.6654	1.6971
	.6907	.7362	.7836	.8329	.8842	.9376	.9931
.78	1.5428	1.5727	1.6034	1.6349	1.6672	1.7004	1.7346
	.7369	.7869	.8391	.8936	.9505	1.0099	1.0718
.80	1.5687	1.6007	1.6336	1.6674	1.7022	1.7381	1.7750
	.7876	.8428	.9006	.9611	1.0245	1.0909	1.1603
.82	1.5963	1.6306	1.6659	1.7024	1.7400	1.7788	1.8188
	.8437	.9049	.9691	1.0367	1.1077	1.1823	1.2606
.84	1.6258	1.6626	1.7007	1.7401	1.7809	1.8231	1.8667
	.9061	.9742	1.0461	1.1220	1.2020	1.2865	1.3755
.86	1.6575	1.6972	1.7384	1.7812	1 8256	1.8716	1.9194
	.9763	1.0526	1.1336	1.2193	1.3103	1.4067	1.5088
.88	1.6918	1.7348	1.7796	1.8262	1.8748	1.9253	1.9780
	1.0558	1.1421	1.2340	1.3319	1.4362	1.5474	1.6658
.90	1.7291	1.7760	1.8250	1.8761	1.9296	1.9855	2.0440
	1.1472	1.2455	1.3509	1.4640	1.5851	1.7150	1.8543
.92	1.7703	1.8216	1.8755	1.9320	1.9914	2.0538	2.1194
	1.2538	1.3673	1.4897	1.6220	1.7648	1.9193	2.0862
.94	1.8161	1.8728	1.9326	1.9958	2.0624	2.1330	2.2076
	1.3806	1.5136	1.6585	1.8162	1.9881	2.1759	2.3810
.96	1.8680	1.9313	1.9985	2.0699	2.1460	2.2271	2.3137
	1.5358	1.6950	1.8703	2.0638	2.2771	2.5129	2.7744
.98	1.9280	1.9998	2.0766	2.1592	2.2480	2.3438	2.4474
	1.7320	1.9302	2.1508	2.3958	2.6744	2.9862	3.3400

TABLE A (continued)

α \ β	.64	.66	.68	.70	.72	.74	.76
.50	1.3755	1.3891	1.4029	1.4168	1.4309	1.4451	1.4594
	.4354	.4531	.4712	.4895	.5082	.5271	.5463
.52	1.3958	1.4104	1.4252	1.4400	1.4551	1.4703	1.4856
	.4645	.4839	.5036	.5236	.5439	.5646	.5857
.54	1.4170	1.4325	1.4483	1.4642	1.4803	1.4965	1.5130
	.4956	.5166	.5381	.5599	.5822	.6048	.6279
.56	1.4389	1.4556	1.4724	1.4894	1.5066	1.5240	1.5416
	.5286	.5516	.5750	.5989	.6232	.6481	.6733
.58	1.4619	1.4796	1.4976	1.5157	1.5342	1.5528	1.5717
	.5640	.5891	.6146	.6408	.6674	.6947	.7224
.60	1.4858	1.5047	1.5239	1.5434	1.5631	1.5831	1.6034
	.6020	.6293	.6573	.6859	.7152	.7451	.7757
.62	1.5108	1.5310	1.5516	1.5724	1.5936	1.6150	1.6368
	.6428	.6727	.7034	.7347	.7669	.7998	.8335
.64	1.5371	1.5587	1.5807	1.6030	1.6257	1.6487	1.6722
	.6868	.7196	.7533	.7878	.8232	.8595	.8968
.66	1.5647	1.5878	1.6114	1.6353	1.6597	1.6845	1.7097
	.7346	.7706	.8076	.8456	.8847	.9249	.9662
.68	1.5938	1.6186	1.6439	1.6696	1.6958	1.7225	1.7497
	.7865	.8261	.8670	.9090	.9523	.9968	1.0427
.70	1.6246	1.6512	1.6784	1.7061	1.7343	1.7631	1.7925
	.8433	.8870	.9322	.9787	1.0268	1.0764	1.1276
.72	1.6573	1.6859	1.7152	1.7450	1.7755	1.8067	1.8385
	.9056	.9540	1.0041	1.0560	1.1095	1.1650	1.2223
.74	1.6922	1.7230	1.7545	1.7868	1.8198	1.8536	1.8881
	.9744	1.0282	1.0841	1.1420	1.2020	1.2642	1.3286
.76	1.7295	1.7627	1.7968	1.8318	1.8676	1.9043	1.9420
	1.0509	1.1110	1.1735	1.2384	1.3060	1.3761	1.4491
.78	1.7696	1.8056	1.8426	1.8806	1.9196	1.9597	2.0009
	1.1365	1.2039	1.2742	1.3475	1.4240	1.5036	1.5867
.80	1.8130	1.8521	1.8923	1.9338	1.9764	2.0204	2.0656
	1.2330	1.3091	1.3887	1.4720	1.5591	1.6502	1.7456
.82	1.8601	1.9028	1.9468	1.9922	2.0391	2.0875	2.1376
	1.3429	1.4294	1.5202	1.6156	1.7157	1.8208	1.9312
.84	1.9118	1.9585	2.0069	2.0570	2.1088	2.1625	2.2182
	1.4695	1.5686	1.6731	1.7833	1.8996	2.0221	2.1514
.86	1.9690	2.0205	2.0740	2.1295	2.1873	2.2473	2.3097
	1.6171	1.7319	1.8535	1.9824	2.1190	2.2637	2.4171
.88	2.0328	2.0900	2.1497	2.2118	2.2767	2.3444	2.4152
	1.7921	1.9267	2.0701	2.2230	2.3860	2.5599	2.7452
.90	2.1052	2.1693	2.2364	2.3068	2.3805	2.4579	2.5391
	2.0038	2.1642	2.3364	2.5212	2.7198	2.9329	3.1619
.92	2.1885	2.2612	2.3378	2.4186	2.5038	2.5937	2.6887
	2.2668	2.4621	2.6736	2.9027	3.1509	3.4199	3.7117
.94	2.2867	2.3706	2.4596	2.5542	2.6548	2.7618	2.8760
	2.6053	2.8506	3.1188	3.4131	3.7355	4.0893	4.4779
.96	2.4063	2.5054	2.6117	2.7259	2.8487	2.9810	3.1238
	3.0644	3.3858	3.7439	4.1428	4.5869	5.0840	5.6393
.98	2.5598	2.6818	2.8146	2.9598	3.1188	3.2936	3.4862
	3.7428	4.1982	4.7179	5.3128	5.9962	6.7814	7.6913

TABLE A (continued)

α\β	.78	.80	.82	.84	.86	.88
.50	1.4738	1.4884	1.5031	1.5180	1.5329	1.5480
	.5658	.5856	.6058	.6262	.6469	.6680
.52	1.5011	1.5167	1.5325	1.5485	1.5646	1.5809
	.6071	.6288	.6509	.6734	.6962	.7194
.54	1.5296	1.5464	1.5633	1.5805	1.5978	1.6154
	.6514	.6753	.6996	.7243	.7495	.7751
.56	1.5595	1.5775	1.5957	1.6142	1.6328	1.6517
	.6991	.7254	.7522	.7795	.8073	.8356
.58	1.5908	1.6102	1.6298	1.6497	1.6698	1.6902
	.7508	.7798	.8093	.8394	.8702	.9015
.60	1.6239	1.6448	1.6659	1.6873	1.7090	1.7310
	.8069	.8389	.8715	.9049	.9389	.9737
.62	1.6589	1.6813	1.7041	1.7271	1.7506	1.7743
	.8681	.9034	.9395	.9765	1.0144	1.0531
.64	1.6959	1.7201	1.7447	1.7696	1.7949	1.8206
	.9350	.9742	1.0143	1.0555	1.0976	1.1409
.66	1.7353	1.7614	1.7880	1.8149	1.8424	1.8703
	1.0086	1.0521	1.0969	1.1428	1.1899	1.2383
.68	1.7774	1.8056	1.8343	1.8636	1.8934	1.9237
	1.0899	1.1385	1.1885	1.2399	1.2928	1.3472
.70	1.8225	1.8530	1.8842	1.9160	1.9484	1.9815
	1.1803	1.2348	1.2909	1.3487	1.4083	1.4698
.72	1.8710	1.9042	1.9381	1.9728	2.0082	2.0444
	1.2815	1.3427	1.4060	1.4713	1.5388	1.6086
.74	1.9235	1.9597	1.9967	2.0346	2.0734	2.1131
	1.3954	1.4646	1.5364	1.6106	1.6875	1.7672
.76	1.9806	2.0202	2.0608	2.1024	2.1451	2.1889
	1.5249	1.6036	1.6854	1.7703	1.8585	1.9501
.78	2.0432	2.0867	2.1314	2.1773	2.2245	2.2730
	1.6733	1.7635	1.8574	1.9553	2.0573	2.1635
.80	2.1123	2.1603	2.2098	2.2607	2.3132	2.3673
	1.8452	1.9494	2.0584	2.1722	2.2913	2.4156
.82	2.1892	2.2425	2.2977	2.3546	2.4134	2.4742
	2.0471	2.1687	2.2963	2.4302	2.5707	2.7182
.84	2.2758	2.3355	2.3974	2.4615	2.5280	2.5969
	2.2876	2.4313	2.5827	2.7423	2.9106	3.0880
.86	2.3745	2.4420	2.5122	2.5852	2.6612	2.7403
	2.5797	2.7520	2.9347	3.1282	3.3333	3.5505
.88	2.4890	2.5661	2.6467	2.7310	2.8191	2.9112
	2.9428	3.1535	3.3783	3.6179	3.8736	4.1463
.90	2.6244	2.7140	2.8081	2.9071	3.0112	3.1207
	3.4078	3.6722	3.9563	4.2618	4.5902	4.9435
.92	2.7892	2.8954	3.0079	3.1270	3.2533	3.3872
	4.0282	4.3716	4.7446	5.1497	5.5898	6.0683
.94	2.9977	3.1278	3.2667	3.4154	3.5747	3.7456
	4.9050	5.3750	5.8922	6.4621	7.0905	7.7842
.96	3.2782	3.4454	3.6268	3.8239	4.0385	4.2725
	6.2624	6.9614	7.7470	8.6324	9.6303	10.7581
.98	3.6992	3.9358	4.1992	4.4938	4.8244	5.1971
	8.7458	9.9742	11.4083	13.0912	15.0756	17.4229

TABLE A (*continued*)

α \ β	.90	.92	.94	.96	.98	1.00
.50	1.5633	1.5787	1.5942	1.6099	1.6257	1.6416
	.6893	.7110	.7330	.7553	.7780	.8009
.52	1.5973	1.6139	1.6306	1.6476	1.6646	1.6819
	.7430	.7669	.7913	.8160	.8411	.8666
.54	1.6331	1.6510	1.6690	1.6873	1.7058	1.7245
	.8011	.8276	.8546	.8820	.9099	.9382
.56	1.6708	1.6901	1.7097	1.7294	1.7494	1.7697
	.8644	.8938	.9237	.9542	.9852	1.0168
.58	1.7108	1.7317	1.7528	1.7742	1.7959	1.8178
	.9335	.9661	.9994	1.0333	1.0679	1.1031
.60	1.7532	1.7758	1.7987	1.8219	1.8454	1.8692
	1.0093	1.0456	1.0827	1.1205	1.1592	1.1986
.62	1.7984	1.8229	1.8477	1.8729	1.8984	1.9243
	1.0928	1.1333	1.1747	1.2171	1.2604	1.3047
.64	1.8468	1.8733	1.9003	1.9277	1.9555	1.9837
	1.1852	1.2305	1.2770	1.3246	1.3734	1.4233
.66	1.8987	1.9275	1.9569	1.9867	2.0171	2.0480
	1.2880	1.3390	1.3913	1.4450	1.5001	1.5565
.68	1.9546	1.9861	2.0181	2.0508	2.0840	2.1178
	1.4032	1.4607	1.5199	1.5807	1.6432	1.7074
.70	2.0153	2.0497	2.0847	2.1205	2.1570	2.1942
	1.5330	1.5983	1.6655	1.7347	1.8060	1.8794
.72	2.0813	2.1191	2.1576	2.1970	2.2373	2.2784
	1.6806	1.7549	1.8317	1.9110	1.9928	2.0772
.74	2.1538	2.1953	2.2379	2.2814	2.3260	2.3716
	1.8496	1.9349	2.0233	2.1147	2.2093	2.3071
.76	2.2338	2.2798	2.3270	2.3754	2.4250	2.4759
	2.0452	2.1439	2.2463	2.3525	2.4628	2.5772
.78	2.3229	2.3741	2.4268	2.4809	2.5365	2.5937
	2.2741	2.3892	2.5091	2.6339	2.7637	2.8988
.80	2.4230	2.4805	2.5396	2.6006	2.6635	2.7282
	2.5456	2.6814	2.8232	2.9714	3.1262	3.2878
.82	2.5370	2.6019	2.6690	2.7383	2.8100	2.8841
	2.8728	3.0351	3.2052	3.3837	3.5708	3.7670
.84	2.6684	2.7425	2.8194	2.8991	2.9818	3.0677
	3.2749	3.4719	3.6794	3.8981	4.1285	4.3713
.86	2.8226	2.9084	2.9977	3.0907	3.1876	3.2886
	3.7808	4.0248	4.2834	4.5574	4.8476	5.1552
.88	3.0076	3.1085	3.2141	3.3247	3.4405	3.5619
	4.4371	4.7475	5.0785	5.4317	5.8085	6.2106
.90	3.2360	3.3574	3.4853	3.6202	3.7624	3.9124
	5.3234	5.7322	6.1721	6.6454	7.1549	7.7035
.92	3.5293	3.6802	3.8406	4.0112	4.1926	4.3858
	6.5886	7.1548	7.7709	8.4416	9.1722	9.9682
.94	3.9289	4.1260	4.3379	4.5661	4.8121	5.0776
	8.5500	9.3966	10.3327	11.3692	12.5176	13.7904
.96	4.5283	4.8083	5.1154	5.4530	5.8248	6.2349
	12.0342	13.4811	15.1233	16.9914	19.1190	21.5464
.98	5.6190	6.0986	6.6465	7.2751	8.0000	8.8400
	20.2135	23.5438	27.5385	32.3474	38.1675	45.2450

TABLE B

The function

$$T(\alpha, \beta) = \frac{-\log \dfrac{1-\alpha}{1-\beta}}{\alpha - \beta}$$

defined in Section 8.2 and used in Sections 9.6 and 11.5 for estimating model parameters. Since $T(\alpha, \beta) = T(\beta, \alpha)$, the parameters α and β may be read along either margin. (The table was computed by Cleo Youtz and Lotte Bailyn.)

$\alpha \backslash \beta$	0.70	0.71	0.72	0.73	0.74	0.75	0.76	0.77	0.78	0.79
0.70	3.333	3.390	3.450	3.512	3.578	3.646	3.719	3.796	3.877	3.963
0.71		3.448	3.510	3.573	3.640	3.711	3.785	3.864	3.947	4.035
0.72			3.571	3.636	3.705	3.777	3.854	3.934	4.019	4.110
0.73				3.704	3.774	3.848	3.926	4.009	4.096	4.189
0.74					3.846	3.922	4.003	4.087	4.177	4.272
0.75						4.000	4.083	4.170	4.261	4.359
0.76							4.167	4.256	4.351	4.451
0.77								4.348	4.445	4.549
0.78									4.545	4.652
0.79										4.762

$\alpha \backslash \beta$	0.80	0.81	0.82	0.83	0.84	0.85	0.86	0.87	0.88	0.89
0.70	4.055	4.152	4.257	4.369	4.490	4.621	4.763	4.919	5.091	5.281
0.71	4.129	4.229	4.336	4.451	4.575	4.709	4.855	5.015	5.191	5.386
0.72	4.206	4.308	4.418	4.536	4.663	4.801	4.951	5.115	5.296	5.496
0.73	4.287	4.393	4.505	4.626	4.757	4.898	5.052	5.221	5.406	5.612
0.74	4.373	4.481	4.597	4.721	4.855	5.000	5.159	5.332	5.523	5.735
0.75	4.463	4.574	4.693	4.821	4.959	5.108	5.271	5.449	5.646	5.864
0.76	4.558	4.672	4.795	4.926	5.068	5.222	5.390	5.574	5.776	6.001
0.77	4.659	4.776	4.902	5.038	5.184	5.343	5.516	5.705	5.914	6.147
0.78	4.766	4.887	5.017	5.157	5.308	5.471	5.650	5.845	6.061	6.301
0.79	4.879	5.004	5.138	5.283	5.439	5.608	5.792	5.995	6.218	6.466
0.80	5.000	5.129	5.268	5.417	5.579	5.754	5.945	6.154	6.385	6.643
0.81		5.263	5.407	5.562	5.728	5.910	6.108	6.325	6.565	6.832
0.82			5.556	5.716	5.889	6.077	6.283	6.508	6.758	7.035
0.83				5.882	6.062	6.258	6.472	6.707	6.966	7.255
0.84					6.250	6.454	6.677	6.921	7.192	7.494
0.85						6.667	6.899	7.155	7.438	7.754
0.86							7.143	7.411	7.708	8.039
0.87								7.692	8.004	8.353
0.88									8.333	8.701
0.89										9.091

TABLE B (*continued*)

α \ β	0.90	0.91	0.92	0.93	0.94	0.95	0.96	0.97	0.98	0.99
0.70	5.493	5.733	6.008	6.327	6.706	7.167	7.750	8.528	9.672	11.728
0.71	5.604	5.850	6.133	6.461	6.850	7.324	7.924	8.726	9.904	12.026
0.72	5.720	5.974	6.264	6.601	7.002	7.490	8.108	8.934	10.150	12.341
0.73	5.843	6.103	6.402	6.750	7.162	7.665	8.302	9.155	10.411	12.676
0.74	5.972	6.240	6.548	6.906	7.332	7.851	8.508	9.389	10.687	13.032
0.75	6.109	6.385	6.703	7.072	7.511	8.047	8.727	9.638	10.981	13.412
0.76	6.253	6.539	6.866	7.248	7.702	8.256	8.959	9.902	11.295	13.818
0.77	6.407	6.702	7.040	7.435	7.904	8.478	9.206	10.184	11.630	14.252
0.78	6.570	6.876	7.226	7.634	8.121	8.715	9.471	10.486	11.989	14.719
0.79	6.745	7.061	7.424	7.847	8.352	8.969	9.754	10.811	12.376	15.223
0.80	6.931	7.259	7.636	8.076	8.600	9.242	10.059	11.160	12.792	15.767
0.81	7.132	7.472	7.864	8.321	8.867	9.536	10.388	11.536	13.243	16.358
0.82	7.347	7.702	8.109	8.586	9.155	9.853	10.743	11.945	13.733	17.002
0.83	7.580	7.950	8.375	8.873	9.468	10.198	11.130	12.390	14.267	17.708
0.84	7.833	8.220	8.664	9.185	9.808	10.574	11.553	12.877	14.853	18.484
0.85	8.109	8.514	8.980	9.527	10.181	10.986	12.016	13.412	15.499	19.343
0.86	8.412	8.837	9.327	9.902	10.591	11.440	12.528	14.004	16.216	20.300
0.87	8.745	9.193	9.710	10.317	11.046	11.944	13.096	14.663	17.016	21.375
0.88	9.116	9.590	10.137	10.780	11.553	12.507	13.733	15.403	17.918	22.590
0.89	9.531	10.034	10.615	11.300	12.123	13.141	14.452	16.241	18.942	23.979
0.90	10.000	10.537	11.158	11.889	12.771	13.863	15.272	17.200	20.118	25.584
0.91		11.111	11.778	12.566	13.515	14.695	16.219	18.310	21.487	27.465
0.92			12.500	13.353	14.384	15.667	17.329	19.617	23.105	29.706
0.93				14.286	15.415	16.824	18.654	21.183	25.055	32.432
0.94					16.667	18.232	20.274	23.105	27.465	35.835
0.95						20.000	22.315	25.542	30.543	40.236
0.96							25.000	28.768	34.657	46.210
0.97								33.333	40.546	54.931
0.98									50.000	69.315
0.99										100.000

TABLE C

The function $g_\nu(\alpha)$ defined by

$$g_\nu(\alpha) = \frac{\nu\alpha^\nu}{1 - \alpha^\nu}.$$

This function is introduced in Section 9.10 and is used in Section 11.4 to facilitate computation of a maximum likelihood estimate. (The table was computed by D. G. Hays and T. R. Wilson.)

α	ν									
	1	2	3	4	5	6	7	8	9	10
.50	1.000	.667	.429	.267	.161	.095	.055	.031	.018	.010
.51	1.041	.703	.459	.290	.179	.107	.863	.037	.021	.012
.52	1.083	.741	.491	.316	.198	.121	.073	.043	.025	.014
.53	1.128	.781	.525	.343	.218	.136	.083	.050	.030	.018
.54	1.174	.823	.561	.372	.241	.153	.095	.058	.035	.021
.55	1.222	.867	.599	.403	.265	.171	.108	.068	.042	.025
.56	1.273	.914	.639	.436	.291	.191	.123	.078	.049	.030
.57	1.326	.963	.682	.472	.320	.213	.140	.090	.058	.036
.58	1.381	1.014	.727	.510	.351	.237	.158	.104	.067	.043
.59	1.439	1.068	.775	.552	.385	.264	.179	.119	.079	.051
.60	1.500	1.125	.827	.596	.422	.294	.202	.137	.092	.061
.61	1.564	1.185	.881	.643	.461	.326	.227	.156	.106	.072
.62	1.632	1.249	.939	.694	.504	.361	.256	.179	.124	.085
.63	1.703	1.316	1.000	.748	.551	.400	.287	.204	.143	.099
.64	1.778	1.388	1.066	.806	.601	.443	.322	.232	.165	.117
.65	1.857	1.463	1.136	.869	.656	.489	.361	.263	.190	.136
.66	1.941	1.544	1.211	.937	.716	.541	.404	.299	.219	.159
.67	2.030	1.629	1.290	1.009	.780	.597	.452	.339	.252	.186
.68	2.125	1.720	1.376	1.088	.851	.658	.505	.383	.289	.216
.69	2.226	1.818	1.468	1.172	.927	.726	.563	.433	.331	.251
.70	2.333	1.922	1.566	1.264	1.010	.800	.628	.489	.378	.291
.71	2.448	2.033	1.672	1.363	1.101	.882	.700	.552	.432	.336
.72	2.571	2.153	1.787	1.470	1.200	.971	.780	.623	.494	.389
.73	2.704	2.282	1.910	1.586	1.308	1.070	.869	.702	.563	.449
.74	2.846	2.421	2.044	1.713	1.426	1.179	.968	.790	.642	.518
.75	3.000	2.571	2.189	1.851	1.556	1.299	1.078	.890	.731	.597
.76	3.167	2.735	2.347	2.003	1.698	1.432	1.201	1.002	.832	.687
.77	3.348	2.913	2.520	2.168	1.856	1.580	1.338	1.128	.946	.791
.78	3.545	3.107	2.709	2.351	2.030	1.744	1.492	1.270	1.077	.909
.79	3.762	3.321	2.918	2.552	2.222	1.927	1.664	1.431	1.226	1.046
.80	4.000	3.556	3.148	2.775	2.437	2.132	1.858	1.613	1.395	1.203
.81	4.263	3.816	3.403	3.023	2.677	2.362	2.076	1.820	1.589	1.384
.82	4.556	4.105	3.687	3.301	2.946	2.621	2.324	2.055	1.812	1.594
.83	4.882	4.429	4.006	3.613	3.250	2.915	2.607	2.326	2.069	1.837
.84	5.250	4.793	4.366	3.966	3.594	3.249	2.930	2.637	2.367	2.120
.85	5.667	5.207	4.775	4.368	3.988	3.633	3.303	2.996	2.713	2.451
.86	6.143	5.680	5.243	4.830	4.442	4.077	3.735	3.416	3.118	2.842
.87	6.692	6.227	5.785	5.365	4.969	4.594	4.241	3.909	3.597	3.305
.88	7.333	6.865	6.418	5.992	5.587	5.202	4.838	4.493	4.167	3.860
.89	8.091	7.620	7.168	6.736	6.323	5.928	5.552	5.194	4.854	4.531
.90	9.000	8.526	8.070	7.631	7.210	6.805	6.418	6.047	5.692	5.353
.91	10.111	9.635	9.174	8.729	8.299	7.885	7.486	7.102	6.732	6.378
.92	11.500	11.021	10.556	10.104	9.666	9.242	8.832	8.434	8.051	7.680
.93	13.286	12.804	12.334	11.876	11.431	10.997	10.575	10.165	9.766	9.379
.94	15.667	15.182	14.708	14.244	13.790	13.347	12.913	12.490	12.077	11.674
.95	19.000	18.513	18.034	17.564	17.102	16.649	16.205	15.769	15.341	14.921
.96	24.000	23.510	23.027	22.551	22.082	21.619	21.163	20.714	20.272	19.836
.97	32.333	31.841	31.354	30.871	30.394	29.922	29.455	28.993	28.536	28.084
.98	49.000	48.505	48.013	47.525	47.041	46.559	46.081	45.606	45.135	44.667
.99	99.000	98.503	98.007	97.513	97.020	96.529	96.040	95.553	95.067	94.583

TABLE D

The functions $F(\alpha, \beta, \Omega)$ and $G(\alpha, \beta, \Omega)$ defined by the equations

$$F(\alpha, \beta, \Omega) = \sum_{\nu=0}^{\Omega} \frac{\alpha^\nu \beta}{1 - \alpha^\nu \beta},$$

$$G(\alpha, \beta, \Omega) = \sum_{\nu=0}^{\Omega} \frac{\nu \alpha^\nu \beta}{1 - \alpha^\nu \beta}.$$

These functions are described in Sections 9.10 and 9.11, and are used for obtaining maximum likelihood estimates in Sections 10.5 and 10.6. The first entry in each cell is $F(\alpha, \beta, \Omega)$ and the second entry is $G(\alpha, \beta, \Omega)$. (The table was prepared by D. G. Hays.)

TABLE D

β	\multicolumn{10}{c}{α $\Omega = 4$}
	.50	.55	.60	.65	.70	.72	.74	.76	.78	.80
.50	1.57	1.70	1.84	2.01	2.21	2.30	2.39	2.50	2.61	2.74
	.95	1.20	1.51	1.89	2.35	2.57	2.81	3.07	3.35	3.67
.55	1.87	2.01	2.17	2.37	2.60	2.71	2.82	2.95	3.09	3.23
	1.06	1.35	1.70	2.13	2.67	2.92	3.20	3.50	3.84	4.21
.60	2.23	2.38	2.57	2.80	3.07	3.19	3.33	3.48	3.64	3.81
	1.18	1.50	1.90	2.39	3.01	3.30	3.62	3.97	4.37	4.80
.65	2.66	2.84	3.06	3.32	3.63	3.78	3.94	4.11	4.30	4.51
	1.30	1.66	2.11	2.67	3.37	3.70	4.07	4.48	4.94	5.45
.70	3.23	3.43	3.67	3.97	4.33	4.51	4.69	4.90	5.13	5.38
	1.43	1.83	2.33	2.96	3.76	4.14	4.57	5.05	5.58	6.18
.72	3.50	3.71	3.97	4.28	4.67	4.85	5.05	5.27	5.51	5.79
	1.49	1.90	2.43	3.08	3.93	4.33	4.78	5.28	5.85	6.49
.74	3.81	4.03	4.30	4.63	5.05	5.24	5.45	5.69	5.95	6.24
	1.54	1.97	2.52	3.21	4.10	4.52	5.00	5.53	6.14	6.82
.76	4.17	4.40	4.69	5.04	5.47	5.68	5.91	6.16	6.44	6.75
	1.60	2.05	2.62	3.34	4.27	4.72	5.23	5.79	6.43	7.16
.78	4.59	4.83	5.13	5.50	5.96	6.18	6.42	6.69	7.00	7.34
	1.65	2.12	2.72	3.48	4.46	4.93	5.46	6.07	6.75	7.53
.80	5.08	5.34	5.65	6.04	6.53	6.77	7.03	7.31	7.64	8.01
	1.71	2.20	2.82	3.61	4.65	5.15	5.71	6.35	7.08	7.91
.82	5.68	5.95	6.28	6.69	7.21	7.46	7.74	8.05	8.40	8.80
	1.77	2.28	2.93	3.76	4.85	5.37	5.97	6.65	7.42	8.31
.84	6.41	6.69	7.04	7.48	8.03	8.30	8.60	8.93	9.31	9.74
	1.83	2.36	3.04	3.91	5.05	5.61	6.24	6.96	7.78	8.74
.86	7.35	7.65	8.01	8.47	9.06	9.35	9.67	10.03	10.43	10.90
	1.89	2.44	3.15	4.06	5.27	5.85	6.53	7.29	8.17	9.19
.88	8.58	8.89	9.28	9.77	10.39	10.70	11.04	11.43	11.87	12.38
	1.95	2.53	3.27	4.22	5.49	6.11	6.82	7.64	8.58	9.67
.90	10.29	10.62	11.03	11.54	12.21	12.54	12.91	13.33	13.81	14.37
	2.02	2.61	3.39	4.39	5.72	6.38	7.14	8.01	9.01	10.19
.92	12.84	13.18	13.61	14.16	14.87	15.23	15.62	16.08	16.60	17.22
	2.08	2.71	3.51	4.56	5.97	6.67	7.47	8.39	9.47	10.74
.94	17.06	17.41	17.87	18.45	19.21	19.59	20.02	20.52	21.09	21.76
	2.15	2.80	3.64	4.74	6.23	6.97	7.82	8.81	9.97	11.34
.96	25.44	25.81	26.29	26.91	27.73	28.13	28.60	29.14	29.77	30.52
	2.22	2.89	3.77	4.93	6.50	7.29	8.20	9.26	10.50	11.99
.98	50.49	50.88	51.39	52.04	52.92	53.37	53.87	54.46	55.16	55.99
	2.29	2.99	3.91	5.12	6.78	7.62	8.60	9.73	11.08	12.70

TABLE D (*continued*)

β	.82	.84	.86	.88	.90	.92	.94	.96	.98
					α				Ω = 4
.50	2.87	3.02	3.18	3.36	3.56	3.78	4.03	4.31	4.63
	4.02	4.40	4.83	5.31	5.85	6.46	7.16	7.97	8.90
.55	3.40	3.58	3.77	3.99	4.24	4.52	4.83	5.19	5.61
	4.63	5.09	5.60	6.19	6.85	7.61	8.49	9.53	10.75
.60	4.01	4.23	4.47	4.74	5.05	5.39	5.80	6.27	6.82
	5.29	5.84	6.47	7.18	7.99	8.94	10.06	11.38	13.00
.65	4.75	5.01	5.31	5.64	6.02	6.46	6.98	7.60	8.35
	6.03	6.69	7.44	8.30	9.31	10.49	11.91	13.64	15.79
.70	5.66	5.98	6.34	6.76	7.24	7.80	8.47	9.29	10.32
	6.86	7.65	8.55	9.60	10.85	12.34	14.15	16.43	19.37
.72	6.09	6.44	6.83	7.28	7.81	8.43	9.19	10.11	11.29
	7.22	8.06	9.04	10.18	11.53	13.17	15.19	17.74	21.09
.74	6.57	6.95	7.37	7.87	8.45	9.14	9.98	11.03	12.39
	7.60	8.51	9.56	10.79	12.28	14.08	16.32	19.20	23.04
.76	7.11	7.52	7.99	8.53	9.17	9.94	10.88	12.08	13.64
	8.00	8.97	10.11	11.45	13.07	15.06	17.56	20.81	25.24
.78	7.72	8.16	8.67	9.27	9.98	10.84	11.91	13.27	15.10
	8.42	9.47	10.70	12.16	13.93	16.13	18.93	22.63	27.76
.80	8.43	8.91	9.47	10.13	10.92	11.88	13.09	14.66	16.81
	8.87	9.99	11.32	12.92	14.87	17.31	20.45	24.67	30.67
.82	9.25	9.78	10.39	11.12	12.00	13.09	14.46	16.29	18.84
	9.34	10.56	12.00	13.74	15.89	18.60	22.15	26.99	34.07
.84	10.24	10.81	11.49	12.30	13.29	14.51	16.10	18.23	21.30
	9.85	11.16	12.73	14.63	17.01	20.04	24.06	29.66	38.09
.86	11.44	12.08	12.83	13.73	14.84	16.24	18.07	20.59	24.33
	10.38	11.80	13.51	15.61	18.24	21.64	26.22	32.76	42.94
.88	12.98	13.67	14.51	15.52	16.78	18.39	20.53	23.54	28.19
	10.96	12.50	14.37	16.68	19.61	23.44	28.71	36.41	48.89
.90	15.02	15.79	16.73	17.87	19.30	21.17	23.70	27.37	33.27
	11.58	13.25	15.30	17.86	21.13	25.50	31.60	40.78	56.39
.92	17.93	18.79	19.84	21.14	22.80	24.99	28.03	32.59	40.31
	12.25	14.08	16.33	19.17	22.86	27.86	35.00	46.13	66.13
.94	22.56	23.53	24.71	26.20	28.14	30.75	34.48	40.29	50.84
	12.98	14.99	17.48	20.65	24.85	30.62	39.11	52.87	79.37
.96	31.41	32.50	33.86	35.59	37.88	41.06	45.76	53.49	68.81
	13.78	15.99	18.76	22.34	27.15	33.92	44.20	61.68	98.56
.98	57.00	58.24	59.81	61.85	64.63	68.60	74.78	85.71	110.55
	14.67	17.11	20.22	24.30	29.88	37.98	50.77	73.97	129.54

TABLE D (*continued*)

β	.50	.55	.60	.65	.70	.72	.74	.76	.78	Ω = 8 .80
.50	1.61	1.75	1.93	2.15	2.43	2.57	2.72	2.89	3.08	3.30
	1.12	1.50	2.02	2.74	3.73	4.23	4.80	5.46	6.22	7.11
.55	1.90	2.07	2.27	2.53	2.85	3.01	3.19	3.38	3.60	3.85
	1.25	1.68	2.26	3.07	4.19	4.76	5.41	6.16	7.03	8.05
.60	2.26	2.45	2.68	2.97	3.34	3.52	3.73	3.95	4.21	4.50
	1.39	1.86	2.52	3.42	4.68	5.31	6.05	6.90	7.89	9.05
.65	2.70	2.91	3.17	3.50	3.93	4.14	4.37	4.63	4.93	5.27
	1.53	2.06	2.78	3.78	5.19	5.90	6.73	7.69	8.81	10.12
.70	3.27	3.50	3.80	4.17	4.66	4.90	5.17	5.47	5.81	6.21
	1.67	2.26	3.06	4.17	5.73	6.53	7.46	8.54	9.80	11.29
.72	3.54	3.79	4.10	4.49	5.01	5.26	5.54	5.86	6.22	6.64
	1.73	2.34	3.17	4.33	5.96	6.79	7.76	8.89	10.21	11.78
.74	3.86	4.11	4.44	4.85	5.39	5.66	5.95	6.30	6.68	7.13
	1.79	2.42	3.29	4.49	6.19	7.06	8.07	9.25	10.64	12.29
.76	4.22	4.48	4.82	5.26	5.83	6.11	6.43	6.79	7.20	7.67
	1.86	2.51	3.41	4.66	6.43	7.34	8.40	9.63	11.09	12.81
.78	4.63	4.91	5.27	5.73	6.33	6.63	6.96	7.34	7.78	8.28
	1.92	2.60	3.53	4.83	6.68	7.62	8.73	10.02	11.55	13.36
.80	5.13	5.42	5.79	6.28	6.91	7.22	7.58	7.98	8.45	8.99
	1.98	2.69	3.66	5.01	6.93	7.92	9.07	10.43	12.03	13.93
.82	5.73	6.03	6.42	6.93	7.60	7.93	8.31	8.74	9.23	9.81
	2.05	2.78	3.79	5.19	7.19	8.22	9.43	10.84	12.52	14.52
.84	6.46	6.78	7.19	7.73	8.43	8.78	9.18	9.64	10.17	10.78
	2.12	2.87	3.91	5.38	7.46	8.54	9.80	11.28	13.04	15.14
.86	7.40	7.73	8.16	8.72	9.47	9.84	10.27	10.75	11.32	11.98
	2.18	2.97	4.05	5.57	7.74	8.86	10.18	11.73	13.58	15.79
.88	8.64	8.99	9.44	10.03	10.82	11.21	11.66	12.18	12.78	13.49
	2.26	3.07	4.19	5.76	8.03	9.20	10.58	12.20	14.14	16.47
.90	10.35	10.72	11.19	11.81	12.65	13.06	13.54	14.10	14.75	15.51
	2.33	3.16	4.33	5.97	8.32	9.55	10.99	12.70	14.73	17.18
.92	12.90	13.28	13.77	14.43	15.32	15.76	16.27	16.87	17.56	18.39
	2.40	3.27	4.47	6.18	8.63	9.92	11.42	13.21	15.35	17.94
.94	17.11	17.51	18.03	18.72	19.67	20.14	20.69	21.33	22.08	22.97
	2.47	3.37	4.62	6.40	8.96	10.30	11.88	13.76	16.01	18.74
.96	25.50	25.92	26.46	27.19	28.19	28.70	29.29	29.98	30.79	31.76
	2.55	3.48	4.78	6.62	9.29	10.69	12.35	14.33	16.71	19.60
.98	50.55	50.99	51.56	52.33	53.40	53.94	54.57	55.32	56.20	57.27
	2.63	3.59	4.94	6.85	9.65	11.11	12.86	14.94	17.45	20.52

TABLE D (*continued*)

β	.82	.84	.86	.88	.90	.92	.94	.96	.98
					α				Ω = 8
.50	3.54	3.81	4.13	4.50	4.93	5.44	6.05	6.81	7.76
	8.14	9.35	10.77	12.47	14.51	17.00	20.09	23.99	29.08
.55	4.14	4.47	4.84	5.28	5.80	6.42	7.18	8.13	9.35
	9.24	10.64	12.30	14.30	16.72	19.70	23.46	28.32	34.82
.60	4.84	5.22	5.67	6.19	6.81	7.57	8.50	9.69	11.27
	10.41	12.03	13.95	16.29	19.15	22.73	27.30	33.34	41.68
.65	5.66	6.11	6.64	7.26	8.02	8.93	10.09	11.60	13.66
	11.68	13.53	15.76	18.48	21.86	26.13	31.70	39.24	50.02
.70	6.66	7.19	7.82	8.56	9.47	10.60	12.04	13.97	16.70
	13.05	15.18	17.75	20.92	24.90	30.01	36.81	46.28	60.40
.72	7.13	7.69	8.36	9.16	10.15	11.37	12.95	15.08	18.16
	13.64	15.87	18.60	21.97	26.22	31.72	39.10	49.50	65.30
.74	7.64	8.25	8.97	9.83	10.89	12.22	13.96	16.33	19.81
	14.24	16.61	19.49	23.08	27.62	33.54	41.55	52.99	70.73
.76	8.22	8.87	9.64	10.57	11.72	13.17	15.08	17.72	21.67
	14.87	17.37	20.42	24.24	29.10	35.47	44.19	56.80	76.79
.78	8.87	9.56	10.39	11.40	12.65	14.24	16.34	19.29	23.80
	15.53	18.16	21.40	25.46	30.66	37.54	47.03	60.98	83.59
.80	9.62	10.36	11.25	12.34	13.70	15.44	17.77	21.08	26.26
	16.21	19.00	22.43	26.75	32.33	39.75	50.11	65.57	91.28
.82	10.48	11.28	12.24	13.42	14.90	16.82	19.41	23.15	29.14
	16.93	19.87	23.51	28.12	34.10	42.13	53.47	70.66	100.07
.84	11.50	12.37	13.41	14.69	16.31	18.43	21.33	25.57	32.56
	17.68	20.79	24.66	29.57	36.00	44.70	57.13	76.33	110.20
.86	12.75	13.68	14.81	16.22	18.00	20.35	23.60	28.46	36.70
	18.46	21.76	25.87	31.13	38.04	47.50	61.17	82.72	122.03
.88	14.33	15.33	16.57	18.10	20.08	22.70	26.38	31.99	41.83
	19.30	22.78	27.16	32.79	40.25	50.55	65.66	89.97	136.04
.90	16.41	17.51	18.86	20.55	22.74	25.69	29.90	36.45	48.41
	20.17	23.88	28.55	34.59	42.66	53.92	70.69	98.30	152.94
.92	19.37	20.57	22.05	23.93	26.39	29.74	34.60	42.37	57.23
	21.11	25.04	30.04	36.54	45.30	57.67	76.39	108.05	173.83
.94	24.04	25.36	27.00	29.10	31.88	35.74	41.45	50.87	69.90
	22.10	26.30	31.66	38.68	48.23	61.89	82.97	119.68	200.49
.96	32.94	34.39	36.22	38.59	41.79	46.30	53.17	64.95	90.52
	23.17	27.66	33.42	41.04	51.52	66.74	90.72	133.98	236.15
.98	58.57	60.18	62.25	64.97	68.70	74.11	82.66	98.18	135.61
	24.33	29.14	35.38	43.70	55.28	72.43	100.17	152.47	287.86

TABLE D (continued)

β	α									Ω = 12
	.50	.55	.60	.65	.70	.72	.74	.76	.78	.80
.50	1.61	1.75	1.94	2.17	2.49	2.64	2.81	3.01	3.24	3.51
	1.13	1.55	2.13	2.98	4.25	4.93	5.73	6.69	7.85	9.24
.55	1.90	2.07	2.28	2.55	2.91	3.09	3.29	3.52	3.78	4.09
	1.27	1.73	2.38	3.34	4.76	5.53	6.43	7.52	8.83	10.41
.60	2.26	2.45	2.69	3.00	3.41	3.61	3.84	4.10	4.41	4.76
	1.41	1.92	2.65	3.71	5.30	6.15	7.17	8.39	9.86	11.64
.65	2.70	2.92	3.19	3.54	4.00	4.23	4.49	4.79	5.14	5.55
	1.55	2.15	2.92	4.10	5.87	6.81	7.95	9.31	10.95	12.95
.70	3.27	3.51	3.81	4.21	4.73	5.00	5.30	5.64	6.04	6.51
	1.70	2.32	3.21	4.51	6.47	7.51	8.77	10.28	12.11	14.34
.72	3.55	3.79	4.11	4.52	5.08	5.36	5.67	6.03	6.46	6.95
	1.76	2.41	3.33	4.68	6.71	7.81	9.11	10.69	12.60	14.93
.74	3.86	4.12	4.45	4.89	5.47	5.76	6.09	6.48	6.92	7.45
	1.82	2.49	3.45	4.86	6.97	8.41	9.47	11.11	13.10	15.53
.76	4.22	4.49	4.84	5.30	5.91	6.21	6.57	6.97	7.44	8.00
	1.88	2.58	3.57	5.03	7.23	8.41	9.83	11.53	13.61	16.15
.78	4.63	4.92	5.29	5.77	6.41	6.73	7.10	7.54	8.03	8.62
	1.95	2.67	3.70	5.21	7.50	8.73	10.20	11.98	14.14	16.80
.80	5.13	5.43	5.81	6.32	6.99	7.33	7.73	8.18	8.71	9.33
	2.01	2.76	3.83	5.40	7.77	9.05	10.58	12.43	14.69	17.46
.82	5.73	6.04	6.44	6.97	7.69	8.05	8.46	8.94	9.50	10.16
	2.08	2.85	3.96	5.59	8.05	9.38	10.98	12.91	15.26	18.15
.84	6.47	6.79	7.21	7.77	8.52	8.90	9.34	9.85	10.44	11.15
	2.15	2.95	4.10	5.79	8.34	9.73	11.39	13.39	15.85	18.86
.86	7.40	7.74	8.18	8.77	9.56	9.96	10.43	10.97	11.60	12.35
	2.21	3.05	4.24	5.99	8.65	10.08	11.81	13.90	16.46	19.61
.88	8.64	8.99	9.45	10.07	10.91	11.33	11.82	12.40	13.07	13.87
	2.29	3.15	4.38	6.20	8.95	10.45	12.24	14.43	17.09	20.39
.90	10.35	10.72	11.21	11.85	12.74	13.19	13.71	14.32	15.04	15.90
	2.36	3.25	4.53	6.41	9.27	10.83	12.70	14.97	17.76	21.20
.92	12.90	13.29	13.79	14.47	15.41	15.89	16.45	17.10	17.87	18.79
	2.43	3.35	4.67	6.63	9.61	11.22	13.17	15.54	18.45	22.06
.94	17.11	17.52	18.05	18.77	19.76	20.27	20.87	21.56	22.39	23.38
	2.51	3.46	4.83	6.86	9.95	11.63	13.67	16.14	19.18	22.96
.96	25.50	25.92	26.48	27.24	28.29	28.83	29.47	30.21	31.11	32.18
	2.58	3.57	4.98	7.09	10.31	12.06	14.18	16.77	19.95	23.92
.98	50.55	51.00	51.58	52.38	53.50	54.08	54.76	55.56	56.53	57.70
	2.66	3.68	5.15	7.33	10.68	12.51	14.72	17.43	20.77	24.94

TABLE D (*continued*)

β	α .82	.84	.86	.88	.90	.92	.94	.96	Ω = 12 .98
.50	3.81	4.17	4.60	5.11	5.73	6.50	7.47	8.75	10.48
	10.94	13.01	15.57	18.77	22.81	28.01	34.86	44.19	57.52
.55	4.44	4.86	5.36	5.97	6.70	7.62	8.80	10.37	12.55
	12.34	14.71	17.64	21.33	26.03	32.14	40.31	51.67	68.37
.60	5.17	5.66	6.24	6.95	7.82	8.91	10.34	12.27	15.05
	13.82	16.51	19.85	24.08	29.52	36.68	46.39	60.17	81.15
.65	6.02	6.59	7.27	8.09	9.13	10.43	12.16	14.55	18.10
	15.40	18.43	22.23	27.06	33.33	41.68	53.20	69.96	96.43
.70	7.05	7.71	8.50	9.47	10.69	12.26	14.36	17.33	21.93
	17.09	20.50	24.80	30.31	37.53	47.26	60.93	81.35	115.03
.72	7.53	8.23	9.07	10.11	11.41	13.10	15.38	18.63	23.75
	17.80	21.37	25.89	31.70	39.33	49.67	64.32	86.46	123.67
.74	8.06	8.80	9.70	10.81	12.21	14.03	16.50	20.07	25.79
	18.53	22.28	27.02	33.14	41.21	52.21	67.92	91.94	133.13
.76	8.65	9.44	10.39	11.58	13.08	15.05	17.74	21.67	28.07
	19.29	23.22	28.20	34.64	43.18	54.89	71.74	97.84	143.57
.78	9.32	10.15	11.17	12.44	14.06	16.19	19.12	23.45	30.65
	20.08	24.19	29.42	36.21	45.25	57.72	75.81	104.21	155.13
.80	10.07	10.97	12.06	13.42	15.16	17.47	20.67	25.46	33.62
	20.89	25.20	30.70	37.85	47.43	60.71	80.17	111.13	168.03
.82	10.95	11.90	13.07	14.54	16.42	18.93	22.44	27.77	37.05
	21.74	26.25	32.03	39.58	49.72	63.89	84.84	118.67	182.51
.84	11.99	13.01	14.26	15.84	17.88	20.61	24.49	30.44	41.07
	22.62	27.35	33.43	41.39	52.15	67.28	89.88	126.94	198.92
.86	13.25	14.34	15.69	17.40	19.62	22.61	26.90	33.60	45.88
	23.54	28.51	34.90	43.31	54.74	70.92	95.34	136.07	217.70
.88	14.84	16.01	17.47	19.32	21.75	25.05	29.83	37.41	51.76
	24.50	29.72	36.45	45.35	57.50	74.85	101.31	146.24	239.45
.90	16.94	18.20	19.79	21.81	24.47	28.13	33.49	42.17	59.18
	25.52	30.99	38.10	47.52	60.47	79.10	107.88	157.69	265.01
.92	19.91	21.28	23.00	25.22	28.17	32.27	38.35	48.42	68.94
	26.58	32.35	39.85	49.85	63.68	83.77	115.18	170.75	295.67
.94	24.59	26.09	27.98	30.43	33.72	38.35	45.36	57.26	82.69
	27.72	33.79	41.73	52.38	67.20	88.93	123.43	185.92	333.43
.96	33.50	35.14	37.23	39.96	43.69	49.01	57.25	71.71	104.54
	28.92	35.34	43.77	55.13	71.09	94.74	132.93	204.02	381.82
.98	59.14	60.95	63.28	66.38	70.66	76.92	86.92	105.34	151.06
	30.22	37.02	45.99	58.19	75.46	101.43	144.20	226.58	448.29

TABLE D (*continued*)

β	α									Ω = 16
	.50	.55	.60	.65	.70	.72	.74	.76	.78	.80
.50	1.61	1.75	1.94	2.18	2.50	2.66	2.84	3.05	3.30	3.59
	1.14	1.55	2.15	3.04	4.42	5.19	6.11	7.25	8.66	10.42
.55	1.90	2.07	2.28	2.56	2.93	3.11	3.31	3.56	3.84	4.18
	1.27	1.73	2.40	3.41	4.95	5.81	6.85	8.14	9.73	11.71
.60	2.26	2.45	2.69	3.00	3.42	3.63	3.87	4.15	4.47	4.86
	1.41	1.93	2.67	3.78	5.51	6.46	7.63	9.06	10.84	13.06
.65	2.70	2.92	3.19	3.54	4.02	4.25	4.53	4.85	5.21	5.65
	1.55	2.16	2.95	4.18	6.09	7.15	8.45	10.04	12.01	14.49
.70	3.27	3.51	3.81	4.21	4.75	5.02	5.33	5.70	6.12	6.62
	1.70	2.33	3.24	4.60	6.71	7.88	9.31	11.07	13.26	16.01
.72	3.55	3.79	4.11	4.53	5.10	5.38	5.71	6.09	6.54	7.07
	1.76	2.41	3.36	4.77	6.96	8.18	9.67	11.50	13.78	16.64
.74	3.86	4.12	4.45	4.89	5.49	5.79	6.13	6.54	7.01	7.57
	1.82	2.50	3.48	4.95	7.22	8.49	10.03	11.94	14.31	17.29
.76	4.22	4.49	4.84	5.30	5.93	6.24	6.61	7.03	7.53	8.13
	1.88	2.59	3.60	5.13	7.49	8.81	10.41	12.39	14.86	17.96
.78	4.63	4.92	5.29	5.77	6.43	6.76	7.15	7.60	8.13	8.75
	1.95	2.68	3.73	5.31	7.77	9.13	10.80	12.86	15.43	18.66
.80	5.13	5.43	5.81	6.32	7.01	7.36	7.77	8.24	8.80	9.47
	2.01	2.77	3.86	5.50	8.05	9.46	11.20	13.34	16.01	19.37
.82	5.73	6.04	6.44	6.98	7.71	8.08	8.51	9.01	9.59	10.30
	2.08	2.86	4.00	5.69	8.34	9.81	11.61	13.83	16.61	20.11
.84	6.47	6.79	7.21	7.77	8.54	8.93	9.39	9.91	10.54	11.29
	2.15	2.96	4.13	5.89	8.64	10.16	12.03	14.34	17.23	20.87
.86	7.40	7.74	8.19	8.77	9.58	10.00	10.47	11.04	11.70	12.49
	2.22	3.06	4.27	6.09	8.94	10.53	12.47	14.87	17.88	21.67
.88	8.64	8.99	9.46	10.08	10.93	11.37	11.87	12.47	13.18	14.02
	2.29	3.16	4.41	6.30	9.26	10.90	12.92	15.42	18.55	22.50
.90	10.35	10.72	11.21	11.86	12.76	13.23	13.76	14.39	15.15	16.05
	2.36	3.26	4.56	6.52	9.59	11.29	13.39	15.99	19.24	23.37
.92	12.90	13.29	13.80	14.48	15.44	15.93	16.50	17.17	17.98	18.95
	2.43	3.36	4.71	6.74	9.93	11.70	13.88	16.58	19.97	24.27
.94	17.11	17.52	18.06	18.78	19.79	20.31	20.92	21.64	22.50	23.54
	2.51	3.47	4.87	6.97	10.27	12.12	14.39	17.21	20.74	25.22
.96	25.50	25.92	26.49	27.25	28.32	28.87	29.52	30.29	31.22	32.34
	2.59	3.58	5.02	7.21	10.64	12.56	14.92	17.85	21.54	26.23
.98	50.55	51.00	51.59	52.39	53.53	54.12	54.81	55.64	56.64	57.86
	2.66	3.69	5.19	7.45	11.02	13.02	15.48	18.54	22.39	27.30

TABLE D (*continued*)

β	α .82	.84	.86	.88	.90	.92	.94	.96	Ω = 16 .98
.50	3.93	4.34	4.84	5.45	6.22	7.21	8.50	10.28	12.86
	12.63	15.44	19.03	23.70	29.86	38.17	49.68	66.32	91.96
.55	4.57	5.05	5.63	6.35	7.25	8.41	9.96	12.12	15.34
	14.21	17.38	21.47	26.81	33.89	43.52	57.05	76.98	108.66
.60	5.31	5.86	6.53	7.37	8.42	9.79	11.64	14.26	18.30
	15.86	19.44	24.05	30.11	38.20	49.31	65.13	88.93	128.07
.65	6.18	6.81	7.59	8.55	9.79	11.41	13.61	16.80	21.87
	17.62	21.63	26.82	33.65	42.86	55.63	74.08	102.46	150.93
.70	7.22	7.95	8.85	9.97	11.41	13.33	15.97	19.87	26.31
	19.49	23.96	29.77	37.47	47.92	62.57	84.06	117.94	178.31
.72	7.71	8.48	9.43	10.62	12.16	14.21	17.05	21.29	28.40
	20.27	24.93	31.01	39.09	50.08	65.54	88.39	124.79	190.85
.74	8.24	9.06	10.07	11.34	12.98	15.17	18.24	22.86	30.73
	21.08	25.95	32.30	40.76	52.32	68.65	92.95	132.07	204.49
.76	8.84	9.70	10.77	12.13	13.88	16.23	19.55	24.58	33.32
	21.91	26.99	33.63	42.50	54.65	71.90	97.75	139.85	219.38
.78	9.51	10.42	11.56	13.01	14.88	17.41	20.99	26.50	36.24
	22.77	28.07	35.01	44.30	57.08	75.31	102.83	148.17	235.71
.80	10.27	11.24	12.46	14.00	16.01	18.73	22.62	28.65	39.55
	23.65	29.18	36.45	46.19	59.62	78.90	108.21	157.11	253.74
.82	11.15	12.19	13.48	15.13	17.29	20.23	24.46	31.11	43.37
	24.57	30.35	37.94	48.15	62.29	82.69	113.95	166.76	273.75
.84	12.19	13.30	14.68	16.45	18.78	21.97	26.59	33.93	47.80
	25.53	31.55	39.49	50.21	65.10	86.69	120.07	177.23	296.13
.86	13.46	14.64	16.13	18.03	20.55	24.01	29.08	37.25	53.07
	26.52	32.81	41.13	52.37	68.06	90.96	126.65	188.66	321.37
.88	15.05	16.32	17.92	19.97	22.71	26.49	32.08	41.23	59.43
	27.56	34.13	42.84	54.65	71.21	95.51	133.75	201.22	350.15
.90	17.15	18.52	20.25	22.47	25.45	29.62	35.83	46.17	67.39
	28.64	35.51	44.65	57.07	74.57	100.42	141.49	215.17	383.39
.92	20.13	21.60	23.47	25.91	29.18	33.80	40.77	52.59	77.73
	29.79	36.98	46.56	59.65	78.18	105.73	149.99	230.84	422.46
.94	24.82	26.42	28.46	31.13	34.76	39.93	47.87	61.63	92.13
	30.99	38.54	48.61	62.43	82.09	111.57	159.47	248.74	469.47
.96	33.73	35.48	37.72	40.68	44.75	50.63	59.85	76.27	114.69
	32.27	40.19	50.81	65.43	86.38	118.06	170.24	269.71	528.11
.98	59.38	61.30	63.79	67.11	71.75	78.59	89.61	110.11	162.01
	33.64	41.98	53.20	68.74	91.16	125.44	182.81	295.28	605.96

Glossary of Symbols Frequently Used

The numbers in parentheses are numbers of sections where the symbols are first introduced. English letters are listed first; Greek letters, next; and special symbols, last.

$A_1, A_2, \cdots, A_j, \cdots, A_r$	Alternatives or response classes (1.2).
a, a_i, a_{jk}	Gain parameter in the gain-loss form of the event operators (1.6).
b, b_i	Loss parameter in the gain-loss form of the event operators (1.6).
C	Projection matrix (1.8).
$E(x)$	Expected value of x (4.8).
$E_1, E_2, \cdots, E_i, \cdots, E_t$	Events which alter the response probabilities (1.3).
$\bar{F}_1, \bar{F}_2$	Mean number of trials before the first A_1 or A_2 occurrence, respectively (9.6).
$F(p)$	Cumulative distribution function (4.2).
$F(\alpha, \beta, \Omega)$	Function given in Table D (9.11).
$G(\alpha, \beta, \Omega)$	Function given in Table D (9.10).
$g_\nu(\alpha)$	Function given in Table C (9.10).
h	Time increment in runway model (14.2).
I	Identity (matrix) operator (1.5).
K	Trial index (4.8).
N	Trial index (9.4).
n	Trial index (3.3).
$O_1, O_2, \cdots, O_k, \cdots, O_s$	Outcomes of response occurrences (1.3).
$P_\nu, P_{\nu n}$	Probability of p-value sequence (3.9).
$p_1, p_2, \cdots, p_j, \cdots, p_r$	Response probabilities (1.2).
$\mathbf{p}$	Probability vector (1.5).
$p, p_n, p_{\nu n}$	Probability of alternative A_1 (1.6).
q, q_n	Probability of alternative A_2 (1.6).
Q_i, Q_{jk}	Row operator (for p) (1.6).
$\tilde{Q}_i, \tilde{Q}_{jk}$	Row operator (for $q = 1 - p$) (1.6).
$R_1, R_{1,n}, R_2, R_{2,n}$	Length of run of A_1's or A_2's (4.8).
T_1, T_2	Mean total number of A_1's or A_2's, respectively (9.6).
$T(\alpha, \beta)$	Function given in Table B (8.2).
$\mathbf{T}, \mathbf{T}_i, \mathbf{T}_{jk}$	Event operator (matrix) (1.5).
$V_m, V_{m,n}$	mth raw moment of p-value distribution (4.2).
$\mathbf{V}_{1,n}$	Vector of marginal means (4.4).
$x_{\mu\nu}, x_\nu, x_{in}$	Random variable representing response occurrence (9.4).
$\alpha, \alpha_i, \alpha_{jk}$	Operator parameter (measures "ineffectiveness" of event) (1.6).
Δ	Increment (3.4).
λ_i, λ_{jk}	Fixed point of row operator (1.6).

$\lambda_1, \lambda_2, \cdots, \lambda_r$	Elements of limit vector (1.8).
λ_i	Limit vector for event E_i (1.8).
Λ	Limit matrix (1.8).
π_i	Probability of event E_i (sometimes an abbreviation for π_{jk}) (3.9).
π_{jk}	Conditional probability of outcome O_k, given alternative A_j (3.13).
$\prod$	Product (4.8).
σ, σ^2	Standard deviation, variance (4.2).
$\Phi(\alpha, \beta)$	Function given in Table A (4.8).
$\Psi(\alpha, \beta)$	Function given in Table A (4.8).
χ^2	Chi-square statistic (9.13).
$\mathscr{M}(\ )$	Measure of the set named in the parentheses (2.1).
$\cup$	Set sum, join, or union (2.2).
$\cap$	Set product, meet, or intersection (2.2).
$\cong$	"Is approximately equal to" (6.4).
$\hat{\cong}$	"Estimates" or "is estimated by" (9.6).
∞	Infinity (3.3).
$Pr\{x|y\}$	Conditional probability of x, given y (3.11).
$|x|$	Absolute value or magnitude of x (6.3).
$\binom{N}{x}$	Binomial coefficient $(N!/x!(N-x)!)$ (4.2).
$\bar{p}, \bar{a}, \bar{x}$, etc.	Mean or average of p, a, x, etc. (3.9).
$\hat{p}, \hat{a}, \hat{x}$, etc.	Estimate of p, a, x, etc. (9.6).

Index

359